The Theory of Neutron Slowing Down in Nuclear Reactors

BY

JOEL H. FERZIGER

AND

P. F. ZWEIFEL

THE M.I.T. PRESS
MASSACHUSETTS INSTITUTE OF TECHNOLOGY
CAMBRIDGE, MASSACHUSETTS

First edition 1966

Library of Congress Catalog Card No. 66-13806

Printed in Great Britain

DEDICATION

To HARRIET AND CONNIE, without whose patience and understanding this book would not have been possible, this book is affectionately dedicated

CONTENTS

PREFACE

In this book, one small facet of nuclear reactor design is considered—the slowing down (or moderation) of neutrons from the high energies with which they are born in fission to the energies at which ultimately they are absorbed. In conjunction with the study of neutron moderation, however, some remarks must of necessity be made concerning the calculation of reactor *criticality*, for the *raison d'être* of slowing-down calculations (or experiments) is to provide information for criticality calculations. A criticality calculation is the computation of the amount of fuel and/or geometrical configuration required to bring a chain reacting system to steady state operation.

It is true that reactors can be built—and indeed have been built—with neither slowing down, criticality, nor almost any other type of calculation having been performed. But the present demands of nuclear power plant performance no longer permit this engineering *lèse majesté*. Today, reactors fall into three broad categories: experimental reactors, built to obtain information for future designs; high performance propulsion or auxiliary power systems (as for submarines, rocket engines, or satellites); and plants for the production of commercial power.

The first of these, the experimental reactor, while possibly amenable to slipshod design, must be analyzed rather carefully once it is operating, or else the information which the reactor was built to obtain will not be forthcoming. The second type clearly must be designed to high performance specifications including maximum power production and minimum weight and size, while the third type of plant, that designed to produce commercial power, must be optimized from the point of view of economics. In every case sophisticated mathematical techniques for the analysis of the reactor system are called into play, and the history of the world's atomic energy program since the end of the World War II includes a not insignificant portion devoted to the development of these techniques.

The names of the many scientists who contributed to the development of the methods described in subsequent chapters of this book could never be enumerated. However, the authors would like to recognize particularly the many contributions of Dr. Henry Hurwitz, Jr., and Dr. Alvin Weinberg. In addition, one of the authors (P.F.Z.) is indebted to Dr. T. J. Krieger and Dr. N. C. Francis for many fruitful discussions.

The authors would also like to thank the many people who made this book possible. This book was first written as lecture notes for an ASEE-AEC Summer Institute at the University of Michigan in the summer of 1959; the reactor physics portion of the course was taught by the authors and Dr. Sidney Yip, whose contribution to this work is quite significant. Thanks for aid at this phase of the work are due to the sponsors of the institute, the "students", who acted as guinea pigs for the draft, and to Mrs. Beverly Doane and Miss Linda Berns who did the typing. Later this book was much expanded as a doctoral dissertation for one of the authors (Joel H. Ferziger). At this stage, we were deeply indebted not only to Professor R. K. Osborn and Professor K. M. Case for their invaluable comments, but also to Professors J. S. King, and William Kerr who acted on the doctoral committee, to Dean Ralph A. Sawyer who made this work possible, and to Miss Harriett L. Gluckstein,† who, as typist, also made an excellent editor. Finally, the authors want to express their thanks to the publisher, Pergamon Press, whose patience has been nearly inexhaustible.

Ann Arbor, Michigan

† Now Mrs. Joel H. Ferziger.

CHAPTER I

SLOWING DOWN AND REACTOR CRITICALITY

A. The Role of Slowing Down

1. The Need for Neutron Moderation

In all reactors, the mean energy of neutrons which cause fission is lower than the mean energy with which neutrons are born as a result of fission. (The fission spectrum of U-235 is shown in Fig. I.1, and has a mean energy of 1.98 MeV. The spectra of other fissionable isotopes is not very different.)

In some reactors (called *fast* reactors) the mean fission energy is very high—sometimes above 100 keV. In such reactors, neutron slowing down is accomplished primarily by inelastic scattering in the fuel and structural materials of the reactor core; a smaller contribution comes from elastic and inelastic scattering by the coolant (usually a liquid metal, e.g., sodium). In other reactor types the mean energy of the fission-inducing neutrons may range anywhere from that of a fast reactor down to the mean energy associated with the thermal motion of the atoms of the materials comprising the reactor core (kT, at $20°C = 0.253$ eV). This last type of reactor is called a *thermal* reactor, while those in which the mean energy of fission-inducing neutrons is between those of thermal and fast reactors are called *intermediate* reactors. This distinction is, of course, arbitrary, but is nevertheless a useful guide.

In order to construct an intermediate reactor, one needs only to add quantities of light elements (e.g., hydrogen, deuterium, beryllium, and carbon or their compounds) called *moderators* to the core of a fast reactor. Elastic collisions between the neutrons and the moderator nuclei degrade the energy of the neutrons before the neutrons cause fission. The more moderator one adds, the lower the mean energy of fission-inducing neutrons will be (see Fig. I.2).

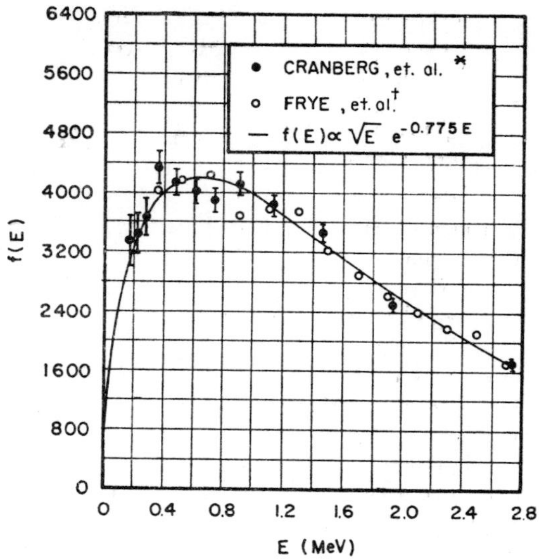

FIG. I.1. Fission spectrum of U-235. (After Weinberg and Wigner, p. 113.)

* L. Cranberg and N. G. Nereson, LA-1916, 1955.
† G. M. Frye, Jr., J. H. Gammel and L. Rosen, LA-1670, 1954.

Generally speaking, the energy spectrum of neutrons in a reactor has the following characteristics. In all reactor types the neutron population is high near the fission spectrum energies. In fast reactors, the neutron population falls off at lower energies, although there are always some neutrons at energies well below the fission spectrum. As moderator is added and elastic scattering becomes the predominant means of moderation, the neutron spectrum contains a larger proportion of intermediate (1 eV to 100 keV) neutrons. Any neutrons that survive long enough to reach thermal energies tend towards thermal equilibrium with the moderator and hence towards a Maxwellian spectrum, which is

$$M(E) = \frac{2\pi}{(\pi kT)^{3/2}} E^{1/2} \mathrm{e}^{-E/kT} \tag{I.1}$$

where $M(E)\,\mathrm{d}E$ is the number of neutrons with energies between E and $E + \mathrm{d}E$, k is Boltzmann's constant and T is the absolute temperature. Addition of sufficient moderator makes the reactor

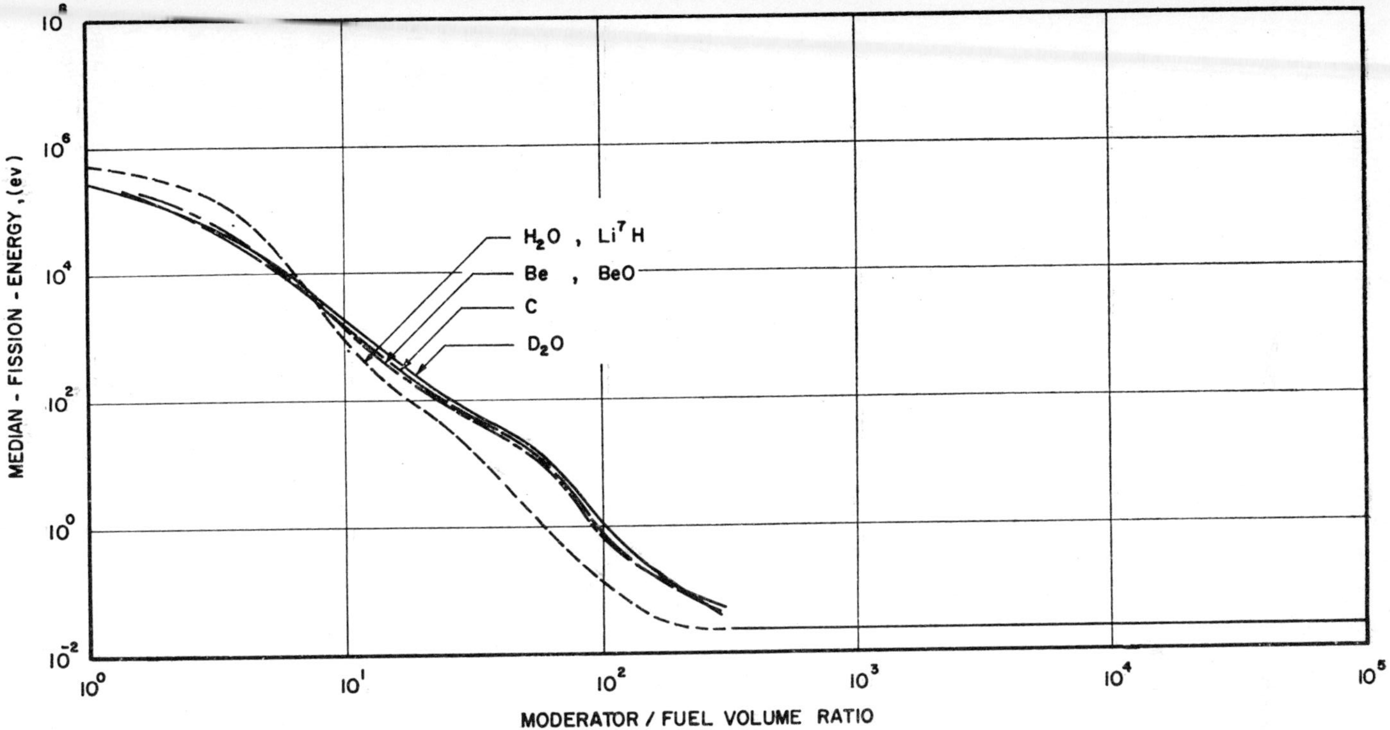

FIG. I.2. Variation of median fission energy with moderator to fuel ratio. Adapted from B. Pinkel and G. Young, Rand Corp. Report RM-2280.

thermal; the neutron spectrum still shows a peak near fission spectrum energies; it has approximately a "$1/E$" shape at intermediate energies and closely approximates a Maxwellian spectrum (Eqn. (I.1)) at thermal energies. Most of the fissions are induced by the neutrons in the Maxwellian component.

After a certain point, the introduction of more moderator only dilutes the fuel concentration without slowing the neutrons down any further since, at least in an infinite reactor, neutrons cannot be moderated to a spectrum lower than that of Eq. (I.1). This results in an increase rather than a decrease in fuel requirements (see Fig. I.3 and next paragraph). In practice, the thermal neutron spectrum is actually somewhat shifted toward higher energies than the Maxwellian because the presence of neutron sources and sinks prevents the neutrons from attaining a true equilibrium distribution [1, 2].

The advantage of decreasing the mean neutron fission energy is that the fission cross section of U-235 (and of other fissionable isotopes as well) is, in the large, a decreasing function of the neutron energy (see Fig. I.4). Thus, as the mean fission energy decreases, the fuel loading necessary to maintain the chain reaction decreases (compare Figs. I.2 and I.3). This saving in fuel is of more importance in reactors designed to produce economic power than in high performance propulsion and/or auxiliary power reactors. However, as was pointed out previously, even in fast reactors, which contain no moderator, some neutron moderation does occur, and must be taken into account if a proper design is to be obtained.

Because much of the cost of nuclear power is directly attributable to the fuel (including fuel fabrication costs, reprocessing costs, and inventory charges), thermal reactors, which have the lowest fuel loading, are most practical for commercial power purposes at present. Such reactors also have the tremendous advantage that they alone can be fueled with natural uranium (containing 0.7% U-235). At higher neutron energies, the fission cross section of U-235 is too low compared with the absorption cross section of the U-238 (99.3%) to maintain a chain reaction in natural uranium. Since natural uranium is far less expensive than enriched (> 0.7% U-235) fuel, the possibility of using natural uranium fuel should be

[1] E. P. Wigner and J. E. Wilkins, Jr., Report AECD-2275 (1944).

[2] N. Corngold, in *Proc. of Neutron Physics Conf.* at Boyne Mtn., Mich., 1962.

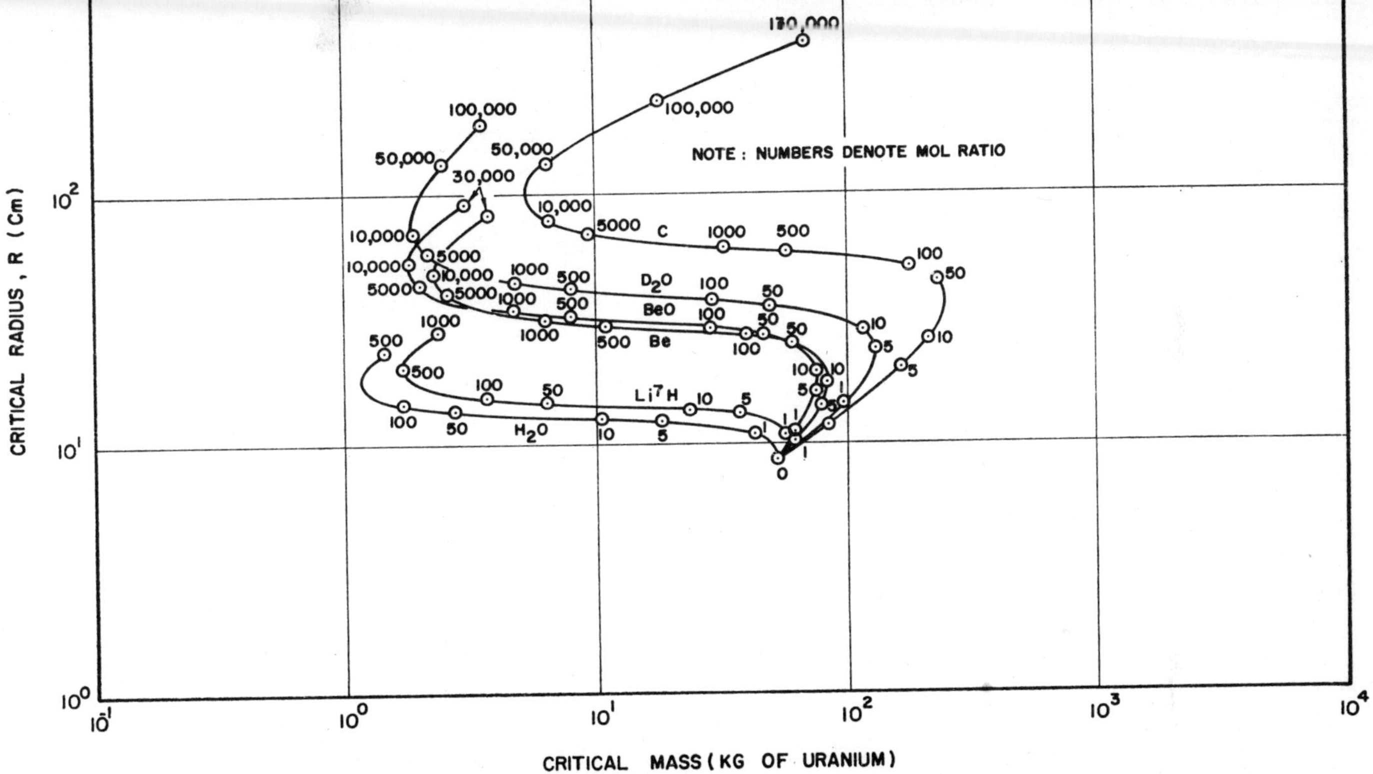

FIG. I.3. Relationship between size and fuel (U-235) investment required for criticality. Adapted from B. Pinkel and G. Young, Rand Corp. Report RM-2280.

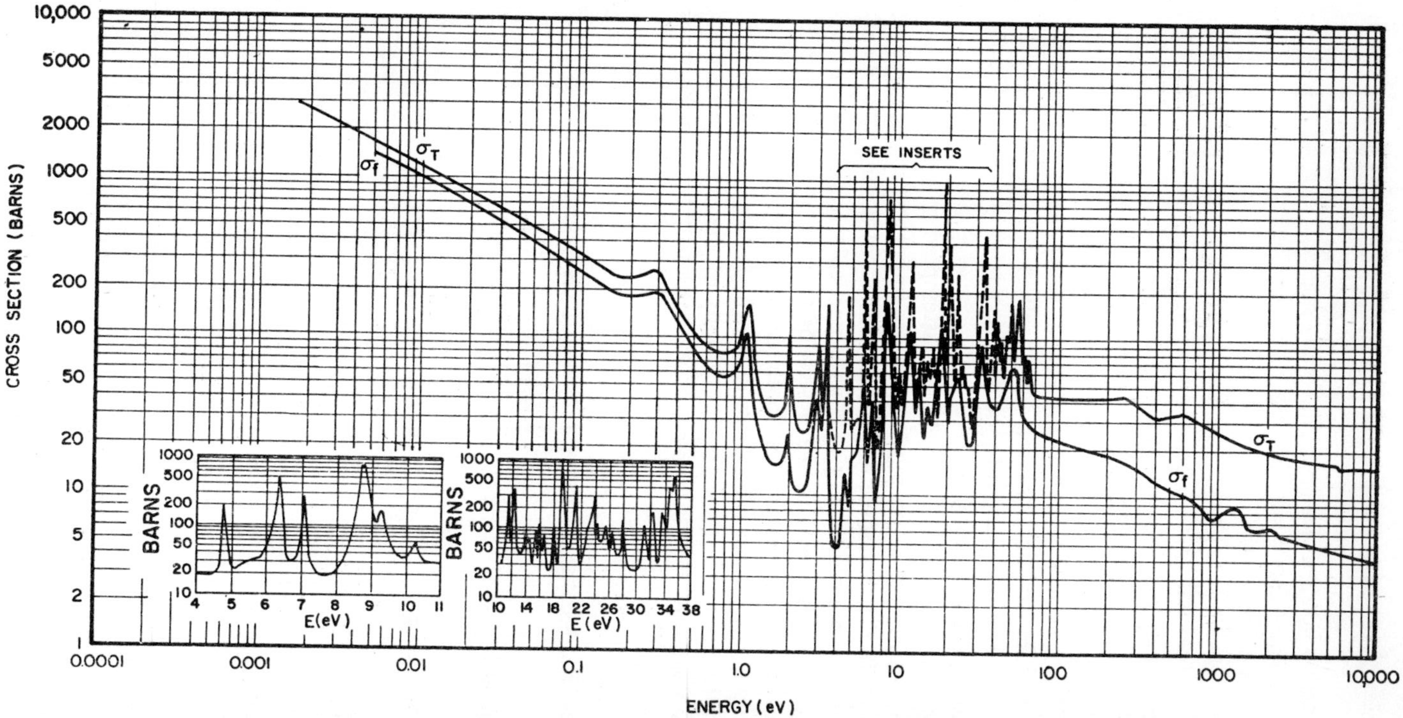

FIG. I.4. Total (σ_T) and fission (σ_f) cross sections of U-235 at low energies. From D. J. Hughes and J. Harvey, *Neutron Cross Sections*, BNL-325 (1958).

exploited whenever feasible. In many nations enriched fuel is unavailable, so this choice is reduced to a necessity.

Balanced against this is the fact that reprocessing and fuel fabrication costs are so high that high fuel burn-up is desirable. Clearly, enriched fuel will last longer than unenriched fuel, although the conversion of U-238 by neutron capture and two subsequent beta decays into Pu-239, a fissionable isotope, increases the lifetime of reactors containing U-238. In fact, in some designs, more fissionable isotope is produced than is destroyed, in which case one speaks of *breeding*.† In practical reactor design, all of these considerations must be taken into account, and the associated problems of analysis are quite formidable.

The discussion of this section has attempted to point out that the problems of neutron moderation are important in all types of reactors. In the rest of the book a mathematical description of the slowing-down process is given, with particular emphasis on the problems encountered in the design of thermal reactors. This is done because of the greater mathematical elegance possible in treating such reactors (i.e., those in which slowing down is predominantly due to elastic scattering). Many of the methods may be applied to any reactor type, however, as will be discussed in Chapter IV.

B. The Mathematical Description of the Neutron Density in a Reactor

2. The Boltzmann Equation

As has probably become clear by now, the primary problem of reactor physics is the prediction of the conditions under which a given reactor configuration will be critical. One method of solving this problem is to set up an equation which describes the time behavior of the neutron population in a reactor and to determine the conditions necessary for the existence of a steady state.‡ Such an equation is a simple statement of neutron conservation and is

† Only fast reactors can breed Pu-239 from U-235 because the ratio of fission to capture cross section in U-235 is too low at thermal and intermediate energies to permit breeding. However, U-233 can be bred (from itself) even in a thermal reactor, but here too the advantage is greater in a fast reactor.

‡ Some time-dependent problems are treated in Appendix A.

a linear *Boltzmann Equation*, sometimes called the *Transport Equation*. The derivation given here will be rather brief.

We define the *angular neutron flux* $\Phi(\mathbf{r}, E, \mathbf{\Omega}, t)$ relative to a fixed set of coordinates such that the number of neutrons in the volume element $d^3r (= dx\, dy\, dz)$ at $\mathbf{r}$ with energies between E and $E + dE$ and with velocity vectors in the solid angle element $d\mathbf{\Omega}$ $(\sin\theta\, d\theta\, d\varphi)$ about the unit vector $\mathbf{\Omega}(\theta, \varphi)$ at time t is given by

$$n(\mathbf{r}, E, \mathbf{\Omega}, t)\, d^3r\, dE\, d\mathbf{\Omega} = \frac{1}{v(E)} \Phi(\mathbf{r}, E, \mathbf{\Omega}, t)\, d^3r\, dE\, d\mathbf{\Omega} \qquad \text{(I.2a)}$$

Here $v(E)$ is the speed of a neutron of energy E:

$$v(E) = \left(\frac{2E}{m}\right)^{1/2} \qquad \text{(I.2b)}$$

where m is the neutron mass, so that the velocity of the neutron is given by

$$\mathbf{v}(E) = \mathbf{\Omega} v(E) \qquad \text{(I.2c)}$$

To obtain the Boltzmann equation, we consider the conservation of those neutrons which are in a small volume element d^3r, whose energies are between E and $E + dE$ and whose velocities are directed in $d\mathbf{\Omega}$ about $\mathbf{\Omega}$. The expression of neutron conservation obtained must apply for every $\mathbf{r}$, every E and every $\mathbf{\Omega}$ (see Fig. I.5).

If we consider the neutron balance in a small time interval Δt, we must have

Change of neutron density during Δt = Sources − Sinks
(in $d^3r\, dE\, d\mathbf{\Omega}$) (in $d^3r\, dE\, d\mathbf{\Omega}\, \Delta t$)

The change of neutron density during Δt is

$$\frac{\partial}{\partial t}\left[\frac{\Phi(\mathbf{r}, E, \mathbf{\Omega}, t)}{v(E)} d^3r\, dE\, d\mathbf{\Omega}\right] \Delta t$$

for sufficiently small Δt. Thus:

$$\frac{1}{v}\frac{\partial}{\partial t}[\Phi(\mathbf{r}, E, \mathbf{\Omega}, t)]\, d^3r\, dE\, d\Omega\, \Delta t = \text{Sources} - \text{Sinks}$$
$$(\text{in } d^3r\, dE\, d\mathbf{\Omega}\, \Delta t) \qquad \text{(I.3)}$$

Consider the right-hand side of the above equation. The sinks include two terms:

(1) Any neutron–nuclear interaction will, by definition, change the energy and/or direction of the neutron involved and hence

remove the neutron from dE or d$\mathbf{\Omega}$ or both and the neutron will thus be lost to the region under consideration.

The total number of neutron–nuclear events in d^3r per sec will be the total path length traveled by all neutrons in $d^3r \, d\mathbf{\Omega} \, dE$

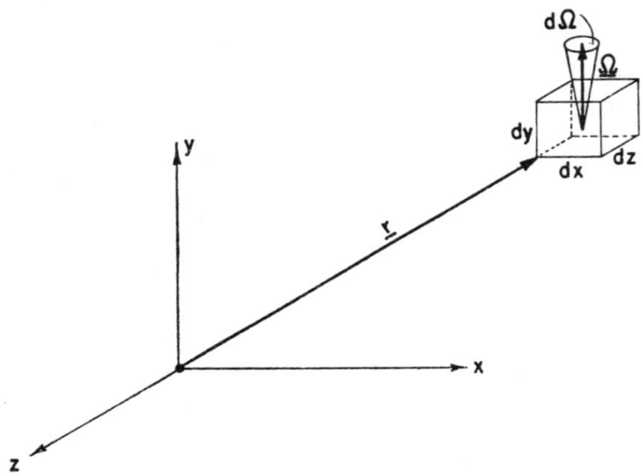

FIG. I.5. Coordinates for the Boltzmann equation.

per sec divided by the mean free path. Since a neutron of speed v travels v cm in one sec, this path length is just

$$vn(\mathbf{r}, E, \mathbf{\Omega}, t) \, d^3r \, dE \, d\mathbf{\Omega} = \Phi(\mathbf{r}, E, \mathbf{\Omega}, t) \, d^3r \, dE \, d\mathbf{\Omega} \quad \text{(I.4e)}$$

and the required number of events is:

$$\Sigma_t \, (\mathbf{r}, E) \, \Phi \, (\mathbf{r}, E, \mathbf{\Omega}, t) \, d^3r \, dE \, d\mathbf{\Omega} \, \Delta t \quad \text{(I.4f)}$$

where Σ_t, the total macroscopic cross section, is defined as the inverse mean free path.†

(2) Neutrons flow or *leak* from d^3r without collision in time Δt. Since the number of neutrons in d$\mathbf{\Omega}$ entering a small element **ds** of the surface in time Δt are those in the cylinder whose height is $v\Delta t$ and whose cross section is $\mathbf{\Omega} \cdot \mathbf{ds}$ (see Fig. I.6) the net leakage

† $\Sigma_t = \Sigma N^i \sigma_t^i$, where N^i is the number density of the ith material and σ_t^i is its total microscopic cross section. The microscopic cross section is in turn the sum of several partial cross sections, each referring to a different type of neutron-induced reaction.

can be written

$$\left[\int_S \boldsymbol{\Omega} \cdot \mathbf{ds}\,\Phi(\mathbf{r}, E, \boldsymbol{\Omega}, t)\right] \mathrm{d}\boldsymbol{\Omega}\, \mathrm{d}E\, \Delta t = \text{Leakage}$$

where S is the surface of d^3r. Applying the divergence theorem we have

$$\left[\int \boldsymbol{\nabla} \cdot \boldsymbol{\Omega}\Phi(\mathbf{r}', E, \boldsymbol{\Omega}, t)\, \mathrm{d}^3r'\right] \mathrm{d}\boldsymbol{\Omega}\, \mathrm{d}E\, \Delta t = \text{Leakage}$$

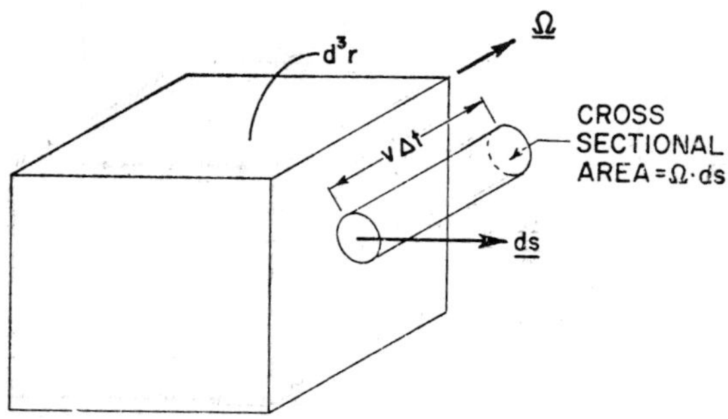

FIG. I.6. Geometry for the calculation of neutron leakage.

where now the integration is over the interior of d^3r. If d^3r is small enough, the integral may be approximated by

$$\boldsymbol{\nabla} \cdot \boldsymbol{\Omega}\Phi(\mathbf{r}, E, \boldsymbol{\Omega}, t)\, \mathrm{d}^3r\, \mathrm{d}\boldsymbol{\Omega}\, \mathrm{d}E\, \Delta t$$

or, applying a vector identity, this becomes

$$\boldsymbol{\Omega} \cdot \boldsymbol{\nabla}\Phi(r, E, \boldsymbol{\Omega}, t)\, \mathrm{d}^3r\, \mathrm{d}\boldsymbol{\Omega}\, \mathrm{d}E\, \Delta t \tag{I.5}$$

since $\boldsymbol{\Omega}$ is independent of $\mathbf{r}$. If we consider the balance we are setting up as one in a phase space with coordinates x, y, z, v_x, v_y, v_z and volume element

$$\mathrm{d}\tau = \mathrm{d}^3r\, \mathrm{d}E\, \mathrm{d}\boldsymbol{\Omega}$$

(I.4f) represents the loss of neutrons in the velocity space without change in the space coordinates; while (I.5) represents the loss of neutrons in configuration space without change in the velocity coordinates.

We now consider the sources required in Equation (I.3). These also consist of two terms:

(1) *Direct* ("*virgin*") *sources* of neutrons—those produced, for example, by fission or by the (γ, n) reaction on deuterium or

beryllium. We define a function $S(\mathbf{r}, E, \mathbf{\Omega}, t)$ such that

$$\text{Direct source} = S(\mathbf{r}, E, \mathbf{\Omega}, t)\, d^3r\, dE\, d\mathbf{\Omega}\, \Delta t \qquad \text{(I.6)}$$
$$(\text{in } d^3r\, dE\, d\mathbf{\Omega}\, \Delta t)$$

(2) *Scattering in* of neutrons from other energies and/or other directions into $dE\, d\mathbf{\Omega}$. Here, we define a *scattering frequency* $F(E' \to E, \mathbf{\Omega}' \to \mathbf{\Omega})$ such that this term is given by

$$\text{Scattering in} = \int dE' \int d\mathbf{\Omega}'\, \Sigma_s(\mathbf{r}, E')\, \Phi(\mathbf{r}, E', \mathbf{\Omega}', t)\, F(E' \to E, \mathbf{\Omega}' \to \mathbf{\Omega})\, d^3\mathbf{r}\, dE\, d\mathbf{\Omega}\, \Delta t \qquad (1.7)$$

from which it may be seen that $F(E' \to E, \mathbf{\Omega}' \to \mathbf{\Omega})\, d\mathbf{\Omega}\, dE$ represents the probability that a neutron which was scattered with energy E' and while traveling in direction $\mathbf{\Omega}'$ will have an energy between E and $E + dE$ and direction in $d\mathbf{\Omega}$ about $\mathbf{\Omega}$ after that collision. (Of course, F may be a function of $\mathbf{r}$ although this dependence is not displayed here.)

Substituting the source and sink terms (I.4f), (I.5), (I.6) and (I.7) into the balance relation (I.3) and eliminating the common differentials, we have the Boltzmann equation:

$$\frac{1}{v}\frac{\partial \Phi(\mathbf{r}, E, \mathbf{\Omega}, t)}{\partial t} = S(\mathbf{r}, E, \mathbf{\Omega}, t) + \int dE' \int d\mathbf{\Omega}'\, \Sigma_s(\mathbf{r}, E')\, \Phi(\mathbf{r}, E', \mathbf{\Omega}', t)\, F(E' \to E, \mathbf{\Omega}' \to \mathbf{\Omega}) - \mathbf{\Omega} \cdot \nabla \Phi(\mathbf{r}, E, \mathbf{\Omega}, t) - \Sigma_t(\mathbf{r}, E)\, \Phi(\mathbf{r}, E, \mathbf{\Omega}, t) \qquad \text{(I.8)}$$

In this derivation, the possibility of neutron–neutron interactions has been neglected. The justification for this lies in the fact that neutron densities are normally no greater than about 10^9 per cm^3 whereas atomic densities are of the order of 10^{22} per cm^3.

3. An Alternate Form of the Boltzmann Equation

For some purposes, the form (I.8) of the Boltzmann equation will be convenient, while for others it will be rather clumsy. In particular, the scattering-in term is often rather difficult to treat in general; a particular functional form must be obtained for the scattering frequency, $F(E' \to E, \mathbf{\Omega}' \to \mathbf{\Omega})$, before any calculations can be carried out.

Instead of writing the slowing-down contribution to the angular flux $\Phi(\mathbf{r}, E, \mathbf{\Omega}, t)$ in terms of the angular flux prior to the last collision, we can formulate it in terms of the sources at other energies and positions. To do this we must first define the *slowing-down kernel*, $P(\mathbf{r}' \to \mathbf{r}, E' \to E, \mathbf{\Omega}' \to \mathbf{\Omega}, t' \to t)$ such that $P(\mathbf{r}' \to \mathbf{r}, E' \to E, \mathbf{\Omega}' \to \mathbf{\Omega}, t' \to t)\, \mathrm{d}^3 r\, \mathrm{d}E\, \mathrm{d}\mathbf{\Omega}\, \Delta t$ is the probability that a neutron which was born at time t' at point $\mathbf{r}'$ in the reactor with energy E', with velocity in direction $\mathbf{\Omega}'$, will *appear in* (or in the case of fast neutrons, be slowed down into) the energy range $\mathrm{d}E$ about E and solid angle $\mathrm{d}\mathbf{\Omega}$ about $\mathbf{\Omega}$ in $\mathrm{d}^3 r$ at $\mathbf{r}$ in the time interval Δt at t. Then the slowing-down or scattering-in term (I.7) may be written as

$$\int_R \mathrm{d}^3 r' \int \mathrm{d}E' \int \mathrm{d}\mathbf{\Omega}' \int_{-\infty}^{\infty} \mathrm{d}t'\, S(\mathbf{r}', E', \mathbf{\Omega}', t') \times$$
$$\times P(\mathbf{r}' \to \mathbf{r}, E' \to E, \mathbf{\Omega}' \to \mathbf{\Omega}, t' \to t)$$

and the Boltzmann equation becomes

$$\frac{1}{v}\frac{\partial \Phi}{\partial t} + \mathbf{\Omega} \cdot \nabla \Phi + \Sigma_t \Phi$$
$$= S + \int_R \mathrm{d}^3 r' \int \mathrm{d}E' \int \mathrm{d}\mathbf{\Omega}' \int_{-\infty}^{t} \mathrm{d}t'\, S(\mathbf{r}', E', \mathbf{\Omega}', t') \times$$
$$\times P(\mathbf{r}' \to \mathbf{r}, E' \to E, \mathbf{\Omega}', \to \mathbf{\Omega}, t' \to t) \qquad \text{(I.9)}$$

Throughout this book we shall be interested primarily in critical reactors, i.e., those in which a time independent neutron population is possible. Furthermore, the only neutron source considered, except for a brief discussion of non-critical systems, will be fission, in which case the source is proportional to the neutron density. Thus, the basic problem of the book will be one of solving a homogeneous linear integro-differential equation for the time independent neutron flux $\Phi(\mathbf{r}, E, \mathbf{\Omega})$ with the boundary condition (for a finite convex assembly) that no neutrons enter the system from the surroundings

$$\Phi(\mathbf{r}_s, E, \mathbf{\Omega}) = 0 \qquad \text{(I.10)}$$

at all points $\mathbf{r}_s$ on the boundary if $\mathbf{\Omega}$ points into the body (i.e., for $\mathbf{\Omega} \cdot \mathbf{n} < 0$ where $\mathbf{n}$ is the unit outward normal to the surface). This leads to an eigenvalue problem which will yield a time-independent solution only for certain combinations of the physical parameters and geometry of the system. The determination of this relationship (the critical condition) in terms of the slowing-down kernel will be the major interest of the first part of this chapter. Thus, we shall deal with the time-independent version of Eq. (I.9), namely:

$$\mathbf{\Omega} \cdot \nabla\Phi(\mathbf{r}, E, \mathbf{\Omega}) + \Sigma_t(\mathbf{r}, E)\,\Phi(\mathbf{r}, E, \mathbf{\Omega}) = S(\mathbf{r}, E, \mathbf{\Omega}) + \int_R \mathrm{d}^3r' \int \mathrm{d}E' \int \mathrm{d}\mathbf{\Omega}'\, S(\mathbf{r}', E', \mathbf{\Omega}')\, P(\mathbf{r}' \to \mathbf{r}, E' \to E, \mathbf{\Omega}' \to \mathbf{\Omega}) \tag{I.11}$$

where the new kernel is defined by

$$P(\mathbf{r}' \to \mathbf{r}, E' \to E, \mathbf{\Omega}' \to \mathbf{\Omega}) = \int_{-\infty}^{t} \mathrm{d}t'\, P(\mathbf{r}' \to r, E' \to E, \mathbf{\Omega}' \to \mathbf{\Omega}; t' \to t)$$

This new kernel is time-independent since the time-dependent kernel may be expected to depend only on the time interval $t - t'$. Thus, the integration over t' may be converted into an integration over a new variable, $\tau = t - t'$, with limits zero and infinity, which plainly is independent of t. The new kernel represents the probability per unit volume, energy and solid angle, that a neutron born with energy E' and direction $\mathbf{\Omega}'$ at $\mathbf{r}'$ will appear with energy E and direction $\mathbf{\Omega}$ at $\mathbf{r}$.

The difference between the two forms of the Boltzmann equation (I.8) and (I.9) is, of course, only a formal one. Equation (I.9) will prove to be more amenable to the extraction of a general critical condition (in terms of the slowing down kernel P), but the calculation of the slowing-down kernel is extremely difficult and can generally be carried out in terms of the scattering frequency. On the other hand, the scattering frequency F is readily derived from existing cross section data. Hence our program will be first to develop a critical condition (to some approximation) from Equation (I.9) and then to calculate the slowing-down kernel in terms of the scattering frequency (again, to some approximation).

4. The Total Neutron Flux and Current

In the context of the reactor model used throughout most of this book, it is necessary to solve Equation (I.8) or (I.11) only for thermal energies (i.e., for E in the range below about 1 eV) in order to obtain all necessary criticality information. However, this equation is, in its full generality, not amenable to analytical solution; thus approximations must be sought. We are aided by the fact that thermal cross sections are large—thus a typical reactor may be expected to be many *thermal* mean free paths thick. We note that in an infinite, homogeneous medium with uniform, isotropic neutron sources (such as fission), the angular flux must be *isotropic* since there exists no preferred direction in space. On the other hand, at a reactor–vacuum interface as in a bare (unreflected) reactor the boundary condition (I.10) must be obeyed; thus the angular flux will be grossly anisotropic there. Physically, this anisotropy can be expected to be propagated into the system up to a few mean free paths from the surface. When the reactor becomes very large with respect to a mean free path (as is usually required for criticality) the flux far from the boundary is almost isotropic. Thus, as a starting point we may say that the angular flux is almost isotropic throughout the reactor;† hence we shall deal only with large reactors for the present. It would be incorrect to assume a completely isotropic flux, as it would predict no net neutron transport. Thus, in order to account for neutron flow out of the system, calculations in a finite assembly require an anisotropic component of the angular flux. The simplest representation of such an angular flux is an expansion in spherical harmonics up to the first order (using the first four harmonics), which may be written compactly as

$$\Phi(\mathbf{r}, E, \mathbf{\Omega}) = \frac{\varphi_0(\mathbf{r}, E)}{4\pi} + \frac{3}{4\pi}\boldsymbol{\varphi}_1(\mathbf{r}, E)\cdot\mathbf{\Omega} \qquad \text{(I.12a)}$$

The second term is equivalent to an expansion in the three spherical harmonics of order unity:

$$\boldsymbol{\varphi}_1\cdot\mathbf{\Omega} = \sum_{m=-1}^{+1} \varphi_{1m}(\mathbf{r}, E)\, Y_1^m(\mathbf{\Omega}) \qquad \text{(1.12b)}$$

where the $Y_l^m(\mathbf{\Omega})$ are the normalized spherical harmonics

$$Y_l^m(\mathbf{\Omega}) = \left(\frac{2l+1}{4\pi}\,\frac{(l-|m|)!}{(l+|m|!)}\right)^{\frac{1}{2}} P_l^{|m|}(\mu)\, e^{im\varphi} \qquad \text{(I.12c)}$$

and the $P_l^{|m|}(\mu)$ are the associated Legendre Polynomials[3]. An expression of the form (I.12b) is possible because the components

† We shall use this approximation only for thermal neutrons here, but it may just as easily be applied to fast neutrons as well. (See Chapter III.)

[3] For definitions and properties of these functions, see any of the standard mathematical texts, for example, E. T. Whittaker and G. N. Watson, *Modern Analysis*, Oxford, 1927.

of a unit vector can be written in terms of the $Y_l^m(\Omega)$:

$$\Omega_x = \sin\theta\cos\varphi = \left(\frac{2\pi}{3}\right)^{\frac{1}{2}}[Y_1^1(\mathbf{\Omega}) + Y_1^{-1}(\mathbf{\Omega})]$$

$$\Omega_y = \sin\theta\sin\varphi = -\left(\frac{2\pi}{3}\right)^{\frac{1}{2}} i\,[Y_1^1(\mathbf{\Omega}) - Y_1^{-1}(\mathbf{\Omega})] \quad \text{(I.12d)}$$

$$\Omega_z = \cos\theta = \left(\frac{4\pi}{3}\right)^{\frac{1}{2}} Y_1^0(\mathbf{\Omega})$$

Before applying (I.12a), it will be informative to inquire as to the physical meaning of the quantities $\varphi_0(\mathbf{r})$ and $\boldsymbol{\varphi}_1(\mathbf{r})$.

First we define the *total neutron flux* (usually simply called the flux), $\varphi(\mathbf{r}, E)$, such that $\varphi(\mathbf{r}, E)\,\mathrm{d}^3r\,\mathrm{d}E$ is the sum of the path lengths travelled per second by all neutrons with energies between E and $E + \mathrm{d}E$ contained within a volume element d^3r at $\mathbf{r}$.† Since $\Phi(\mathbf{r}, E, \mathbf{\Omega})$ is the path length traveled in one second by those neutrons traveling in directions $\mathrm{d}\mathbf{\Omega}$ about $\mathbf{\Omega}$, this quantity is

$$\varphi(\mathbf{r}, E) = \int \mathrm{d}\mathbf{\Omega}\,\Phi(\mathbf{r}, E, \mathbf{\Omega}) \equiv n(\mathbf{r}, E)\,v(E) \qquad \text{(I.13a)}$$

where $n(\mathbf{r}, E)$ is the density of neutrons per unit volume at $\mathbf{r}$ per unit energy at E, and $v(E)$ is the neutron speed defined by Equation (I.2b). By integration of Equation (I.12a) over $\mathbf{\Omega}$, we find

$$\varphi(\mathbf{r}, E) \equiv \varphi_0(\mathbf{r}, E) \qquad \text{(I.13b)}$$

since‡

$$\int \mathrm{d}\mathbf{\Omega} = 4\pi \qquad \text{(I.13c)}$$

$$\int \mathbf{A}\cdot\mathbf{\Omega}\,\mathrm{d}\mathbf{\Omega} = 0 \qquad \text{(I.13d)}$$

for any constant vector $\mathbf{A}$.

Secondly, we define the *angular neutron current* $\mathbf{j}(\mathbf{r}, E, \mathbf{\Omega})$ such that

$$\mathbf{j}(\mathbf{r}, E, \mathbf{\Omega})\cdot\mathbf{ds}\,\mathrm{d}E\,\mathrm{d}\mathbf{\Omega} \qquad \text{(I.14a)}$$

represents the number of neutrons per second which cross area element $\mathbf{ds}$ located at $\mathbf{r}$ with energies in $\mathrm{d}E$ about E and velocities in $\mathrm{d}\mathbf{\Omega}$ about $\mathbf{\Omega}$. By using the artifice used in the derivation of the Boltzmann equation, namely considering all neutrons in the

† Then $\Sigma_t(\mathbf{r})\varphi_0(\mathbf{r})$ is the total reaction rate per unit volume at $\mathbf{r}$ for neutrons with energies in $\mathrm{d}E$. Cf. Equation (I.4f).

‡ Proof of these and the vector identities to follow are given in Appendix B.

cylinder of volume ($v\mathbf{\Omega} \cdot \mathbf{ds}$) as crossing $\mathbf{ds}$ in one second, we find that this term is also:

$$\Phi(\mathbf{r}, E, \mathbf{\Omega})\,(\mathbf{\Omega} \cdot \mathbf{ds})\,\mathrm{d}E\,\mathrm{d}\mathbf{\Omega}$$

so that comparison with the definition of $\mathbf{j}(\mathbf{r}, E, \mathbf{\Omega})$ gives:

$$\mathbf{j}(\mathbf{r}, E, \mathbf{\Omega}) = \mathbf{\Omega}\Phi(\mathbf{r}, E, \mathbf{\Omega}) = \mathbf{v}n(\mathbf{r}, E, \mathbf{\Omega}) \tag{I.14b}$$

By analogy to the total neutron flux we define the *total neutron current* (or simply the current) $\mathbf{J}(\mathbf{r}, E)$ so that

$$\mathbf{J}(\mathbf{r}, E) \cdot \mathbf{ds}\,\mathrm{d}E \tag{I.14c}$$

represents the *net* number of neutrons which cross the area element $\mathbf{ds}$ at $\mathbf{r}$ per second with energies in $\mathrm{d}E$ about E. Clearly

$$\begin{aligned}\mathbf{J}(\mathbf{r}, E) &= \int \mathrm{d}\mathbf{\Omega}\,\mathbf{j}(\mathbf{r}, E, \mathbf{\Omega}) \\ &= \int \mathrm{d}\mathbf{\Omega}\,\mathbf{\Omega}\Phi(\mathbf{r}, E, \mathbf{\Omega})\end{aligned} \tag{I.14d}$$

Multiplying Equation (I.12) by $\mathbf{\Omega}$ and integrating over $\mathbf{\Omega}$, we find

$$\mathbf{J}(\mathbf{r}, E) \equiv \varphi_1(\mathbf{r}, E) \tag{I.14e}$$

where the results

$$\int \mathbf{\Omega}\,\mathrm{d}\mathbf{\Omega} = 0 \tag{I.14f}$$

and

$$\int \mathrm{d}\mathbf{\Omega}\,\mathbf{A} \cdot \mathbf{\Omega}\,\mathbf{\Omega} = \frac{4\pi}{3}\mathbf{A} \tag{I.14g}$$

have been employed. This argument also shows that an isotropic flux would yield $\mathbf{J}(\mathbf{r}, E) = 0$ and hence no net neutron transport.

5. Reduction of the Boltzmann Equation

With these interpretations of the expansion coefficients in Equation (I.12), we are ready to insert the expansion into the time independent version of the Boltzmann Equation (I.8). A scalar equation may be obtained by integrating over all $\mathbf{\Omega}$:

$$\begin{aligned}\int \mathrm{d}\mathbf{\Omega}\,\mathbf{\Omega} \cdot \nabla\Phi(\mathbf{r}, E, \mathbf{\Omega}) + \Sigma_t(\mathbf{r}, E)\int \mathrm{d}\mathbf{\Omega}\,\Phi(\mathbf{r}, E, \mathbf{\Omega}) = \int \mathrm{d}\mathbf{\Omega}\,S(\mathbf{r}, E, \mathbf{\Omega}) \\ + \int \mathrm{d}\mathbf{\Omega} \int \mathrm{d}E' \int \mathrm{d}\mathbf{\Omega}'\,\Sigma_s(\mathbf{r}, E')\,\Phi(\mathbf{r}, E', \mathbf{\Omega}')\,F(E' \rightarrow E, \mathbf{\Omega}' \rightarrow \mathbf{\Omega})\end{aligned} \tag{I.15a}$$

Inserting Equation (I.12) for Φ in the first term of Equation (I.15a) and applying (I.13d) and

$$\int d\mathbf{\Omega}\,(\mathbf{A}\cdot\mathbf{\Omega})\,(\mathbf{B}\cdot\mathbf{\Omega}) = \frac{4\pi}{3}\mathbf{A}\cdot\mathbf{B} \tag{I.15b}$$

(treating the gradient operator ∇ as a vector) we obtain just $\nabla \cdot \mathbf{J}(\mathbf{r}, E)$. The integral appearing in the second term on the left side of Equation (I.15a) is the definition of $\varphi(\mathbf{r}, E)$ so that this term becomes $\Sigma_t(\mathbf{r}, E)\,\varphi(r, E)$. Under the assumption of isotropic sources, the second term on the right-hand side is simply

$$S(\mathbf{r}, E) = \int S(\mathbf{r}, E, \mathbf{\Omega})\,d\mathbf{\Omega} = 4\pi\, S(\mathbf{r}, E, \mathbf{\Omega}) \tag{I.15c}$$

In the first term on the right-hand side of Equation (I.15a) the integral

$$\int d\mathbf{\Omega}\, F(E' \to E, \mathbf{\Omega}' \to \mathbf{\Omega}) \tag{I.15d}$$

occurs. In most cases, anisotropies due to crystalline effects are absent and the scattering frequency $F(E' \to E, \mathbf{\Omega}' \to \mathbf{\Omega})$ is rotationally invariant; that is, it depends only on the angle between $\mathbf{\Omega}'$ and $\mathbf{\Omega}$ (i.e., only on $\mathbf{\Omega}' \cdot \mathbf{\Omega} = \mu$) and not on $\mathbf{\Omega}'$ and $\mathbf{\Omega}$ individually:

$$F(E' \to E, \mathbf{\Omega}' \to \mathbf{\Omega}) = F(E' \to E, \mathbf{\Omega}' \cdot \mathbf{\Omega})$$

Then the integral (I.15d) will be independent of $\mathbf{\Omega}'$ and it is possible to write:

$$\int d\mathbf{\Omega}\, F(E' \to E, \mathbf{\Omega}' \cdot \mathbf{\Omega}) = F_0(E' \to E) \tag{I.15e}$$

which is the zeroth moment of the scattering frequency.† The remainder of the second term on the right-hand side of Equation (I.15a) is

$$\int dE'\, F_0(E' \to E)\,\Sigma_s(\mathbf{r}, E') \int d\mathbf{\Omega}'\, \Phi(\mathbf{r}, E', \mathbf{\Omega}')$$

which, in light of the definition of $\varphi(\mathbf{r}, E)$ becomes

$$\int dE'\, F_0(E' \to E)\,\varphi(\mathbf{r}, E)\,\Sigma_s(\mathbf{r}, E') \tag{I.15f}$$

† Physically, $F_0(E' \to E)\,dE$ is the probability that a neutron scattered at E' will have an energy in dE about E after the collision.

Combining all of these results, we see that Equation (I.15a) can be rewritten as:

$$\nabla \cdot \mathbf{J}(\mathbf{r}, E) + \Sigma_t(\mathbf{r}, E)\,\varphi(\mathbf{r}, E) = \int dE'\,\Sigma_s(\mathbf{r}, E')\,\varphi(\mathbf{r}, E')\,F_0(E' \to E) + S(\mathbf{r}, E) \qquad \text{(I.15g)}$$

Another equation between $\varphi(\mathbf{r}, E)$ and $\mathbf{J}(\mathbf{r}, E)$ is obtained by multiplying Equation (I.9) by $\mathbf{\Omega}$ before integrating:

$$\int d\mathbf{\Omega}\,\mathbf{\Omega}\,\mathbf{\Omega} \cdot \nabla \Phi(\mathbf{r}, E, \mathbf{\Omega}) + \Sigma_t(\mathbf{r}, E) \int d\mathbf{\Omega}\,\mathbf{\Omega}\,\Phi(\mathbf{r}, E, \mathbf{\Omega}) = \int d\mathbf{\Omega}\,\mathbf{\Omega}\,S(\mathbf{r}, E, \mathbf{\Omega}) + \int \mathbf{\Omega}\,d\mathbf{\Omega} \int dE' \int d\mathbf{\Omega}' \int \Sigma_s(\mathbf{r}, E') \times \times \Phi(\mathbf{r}, E', \mathbf{\Omega}')\,F(E' \to E, \mathbf{\Omega}' \cdot \mathbf{\Omega}) \qquad \text{(I.16a)}$$

The source term is seen to vanish in light of the assumed isotropy of $S(\mathbf{r}, E, \mathbf{\Omega})$ and Equation (I.14f), while the first term is readily evaluated by means of Equation (I.14g) and the identity

$$\int d\mathbf{\Omega}(\mathbf{A} \cdot \mathbf{\Omega})\,(\mathbf{B} \cdot \mathbf{\Omega})\,\mathbf{\Omega} = 0 \qquad \text{(I.16b)}$$

with ∇ again treated as a vector: the result is $(1/3)\,\nabla\varphi(r, E)$. The integral in the second term on the left is just the definition of $\mathbf{J}(\mathbf{r}, E)$. Now the last term on the right-hand side contains the integral

$$\int d\mathbf{\Omega}\,\mathbf{\Omega}\,F(E' \to E, \mathbf{\Omega}' \cdot \mathbf{\Omega}) \qquad \text{(I.16c)}$$

where the rotational invariance already assumed is explicitly exhibited. This integral must be a vector proportional to $\mathbf{\Omega}'$, since $\mathbf{\Omega}$ is the variable of integration. Call the result $A\mathbf{\Omega}'$. Then, taking the scalar product of (I.16c) with $\mathbf{\Omega}'$, we obtain (note that $\mathbf{\Omega}' \cdot \mathbf{\Omega}' = 1$):

$$A = \int d\mathbf{\Omega}\,\mathbf{\Omega} \cdot \mathbf{\Omega}'\,F(E' \to E, \mathbf{\Omega} \cdot \mathbf{\Omega}') \equiv F_1(E' \to E) \qquad \text{(I.16d)}$$

which cannot depend on $\mathbf{\Omega}'$ as a result of the assumed rotational invariance of $F(E' \to E, \mathbf{\Omega}' \to \mathbf{\Omega})$.† Thus,

$$\int d\mathbf{\Omega}\,\mathbf{\Omega}\,F(E' \to E, \mathbf{\Omega}' \cdot \mathbf{\Omega}) = \mathbf{\Omega}'\,F_1(E' \to E) \qquad \text{(I.16e)}$$

and the triple integral (I.16a) reduces to

$$\int dE'\,\Sigma_s(\mathbf{r}, E')\,F_1(E' \to E) \int d\mathbf{\Omega}'\,\mathbf{\Omega}'\,\Phi(\mathbf{r}, E, \mathbf{\Omega}') = \int dE'\,\Sigma_s(\mathbf{r}, E')\,\mathbf{J}(\mathbf{r}, E')\,F_1(E' \to E) \qquad \text{(I.16f)}$$

† Unfortunately, $F_1(E' \to E)$, the first moment of the scattering frequency, does not have the simple physical significance that the zeroth moment $F_0(E' \to E)$ has.

where the definition of $\mathbf{J}(\mathbf{r}, E)$, Equation (I.14d), has been used. Equation (I.16a) can now be written in the following form:

$$\tfrac{1}{3}\nabla\varphi(\mathbf{r}, E) + \Sigma_t(\mathbf{r}, E)\,\mathbf{J}(\mathbf{r}, E) = \int \Sigma_s(\mathbf{r}, E')\,\mathbf{J}(\mathbf{r}, E')\,F_1(E' \to E)\,\mathrm{d}E' \tag{I.16g}$$

Equations (I.15g) and (I.16g) are two equations in the two unknowns $\varphi(\mathbf{r}, E)$ and $\mathbf{J}(\mathbf{r}, E)$. The procedure used here is called the P_1 approximation as the angular flux was expanded in terms of spherical harmonics of order zero and one. A more complete discussion of these equations, which constitute one of the major calculational procedures of reactor physics, is found in Chapter III. It will be shown there that Equation (I.15g) is rigorous, i.e., the expansion (1.12a) need not have been used in deriving it, whereas Equation (I.16g) is not rigorous, i.e. the inclusion of higher order spherical harmonics in Equation (I.12a) would result in an additional term in Equation (I.16g).

6. Reduction to One-velocity Equations for Thermal Neutrons

At thermal neutron energies, i.e., energies of the order of the energies associated with the thermal motion of the atoms of the media in which the neutrons are diffusing, neutrons gain or lose energy in collisions with nuclei with roughly equal probability. (Certainly, the behavior of thermal energy neutrons is considerably different from that of high energy (fast) neutrons, which lose energy in every collision with a nucleus.) Of course, there is no really clear-cut energy at which a neutron can be said to be transformed from a fast neutron into a thermal neutron, but the distinction is an intuitively clear one and a "cutoff energy", E_T, above which a neutron is a fast neutron and below which it is a "thermal" neutron, can generally be chosen without introducing a great deal of error [4]. In order to simplify the treatment of the thermal neutron group, we reduce the P_1 equations already obtained to a one-velocity diffusion equation by "averaging" the equation over energies

[4] A convenient cutoff is that energy above which the neutron has no appreciable probability of gaining energy in a collision; this definition will be adopted here. The determination of this energy and many other problems concerning thermal neutrons, which are but briefly mentioned in this book, are discussed at great length by N. Corngold in the *Proceedings of the Neutron Physics Conference* at Boyne Mtn., Michigan, 1962.

in the thermal range. Before proceeding, we define the *thermal neutron flux*:

$$\varphi_T(\mathbf{r}) = \int_0^{E_T} \varphi(\mathbf{r}, E)\,\mathrm{d}E \tag{I.17a}$$

the *thermal neutron current*:

$$\mathbf{J}_T(\mathbf{r}) = \int_0^{E_T} \mathbf{J}(\mathbf{r}, E)\,\mathrm{d}E \tag{I.17b}$$

and the *thermal average total macroscopic cross section*:

$$\Sigma_{t_T}(\mathbf{r}) = \frac{\int_0^{E_T} \Sigma_t(\mathbf{r}, E)\,\varphi(\mathbf{r}, E)\,\mathrm{d}E}{\int_0^{E_T} \varphi(\mathbf{r}, E)\,\mathrm{d}E} \tag{I.17c}$$

with similar definitions for the thermal average absorption, fission, and scattering cross sections. Integrating Equation (I.15g) from $E = 0$ to $E = E_T$, and noting that fission produces essentially no neutrons at thermal energies, we obtain:

$$\int_0^{E_T} \nabla \cdot \mathbf{J}(\mathbf{r}, E)\,\mathrm{d}E + \int_0^{E_T} \Sigma_t(\mathbf{r}, E)\,\varphi(\mathbf{r}, E)\,\mathrm{d}E$$
$$= \int_0^{E_T} \mathrm{d}E \int_0^{\infty} \Sigma_s(\mathbf{r}, E')\,\varphi(\mathbf{r}, E')\,F_0(E' \to E)\,\mathrm{d}E' \tag{I.18a}$$

The first term is simply the divergence of the thermal current, $\nabla \cdot \mathbf{J}_T(\mathbf{r})$, since we expect $\mathbf{J}(\mathbf{r}, E)$ to be sufficiently well behaved to allow interchange of the integration and divergence operations. The second term is easily evaluated using definition (I.17c) and is $\Sigma_{t_T}(\mathbf{r})\,\varphi_T(\mathbf{r})$, the thermal *collision density* or the total number of thermal neutron–nuclear interactions per second per unit volume at $\mathbf{r}$. Only the term on the right-hand side causes any difficulty. To treat that term, we first decompose it as follows:

$$\int_0^{E_T} \mathrm{d}E \int_0^{\infty} \mathrm{d}E'\,\Sigma_s(\mathbf{r}, E')\,\varphi(\mathbf{r}, E')\,F_0(E' \to E)$$
$$= \int_0^{E_T} \mathrm{dE} \int_0^{E_T} \mathrm{d}E'\,\Sigma_s(\mathbf{r}, E')\,\varphi(\mathbf{r}, E')\,F_0(E' \to E)$$
$$+ \int_0^{E_T} \mathrm{d}E \int_{E_T}^{\infty} \mathrm{d}E'\,\Sigma_s\,(\mathbf{r}, E')\,\varphi(\mathbf{r}, E')\,F_0(E' \to E) \tag{I.18b}$$

In the first term on the right-hand side of Equation (I.18b), consider first the integral

$$\int_0^{E_T} F_0(E' \to E)\,\mathrm{d}E$$

which is the probability that a thermal neutron which has scattered (note the limits on the E' integral) will remain within the thermal group after that collision. Now E_T can be chosen such that the number of neutrons with energies less than E_T which scatter to energies above E_T is vanishingly small. Clearly, the optimum choice of E_T will be the lowest energy for which this statement holds to some desired approximation [5]. With this choice of E_T the above integral is unity by definition and the first term on the right-hand side of Equation (I.18b) becomes:

$$\int_0^{E_T} \Sigma_s(\mathbf{r}, E)\,\varphi(\mathbf{r}, E)\,\mathrm{d}E = \Sigma_{sT}(\mathbf{r})\,\varphi(\mathbf{r}) \qquad \text{(I.18c)}$$

as a consequence of the definition of Σ_{sT}.

Returning to the last term of Equation (I.18b), we note that it represents physically the number of neutrons entering the thermal group due to collisions at higher energies. In order to obtain a more convenient form of the equation for thermal neutrons, we formally introduce a thermalization kernel $P_T(\mathbf{r}' \to \mathbf{r}, E', \mathbf{\Omega})$ defined such that

$$P_T(\mathbf{r}' \to \mathbf{r}, E', \mathbf{\Omega})\,\mathrm{d}\mathbf{\Omega}\,\mathrm{d}^3r \qquad \text{(I.19a)}$$

is the expected number of neutrons traveling in directions in $\mathrm{d}\mathbf{\Omega}$ about $\mathbf{\Omega}$ thermalized (i.e., entering the thermal group or, equivalently, slowing down past E_T) in d^3r at $\mathbf{r}$ per neutron born *isotropically* with energy E' at $\mathbf{r}'$. Then

$$\int_R \mathrm{d}^3r' \int_{E_T}^{\infty} \mathrm{d}E'\, S(\mathbf{r}', E')\, P_{T_0}(\mathbf{r}' \to \mathbf{r}, E') \qquad \text{(I.19b)}$$

is the total number of neutrons thermalized per second per unit volume at $\mathbf{r}$ and is equal to the last term of Equation (I.18b) by physical identity. Here, $P_{T_0}(\mathbf{r}' \to \mathbf{r}, E')$ is the zeroth moment,

$$P_{T_0}(\mathbf{r}' \to \mathbf{r}, E') = \int P_T(\mathbf{r}' \to \mathbf{r}, E', \mathbf{\Omega})\,\mathrm{d}\mathbf{\Omega} \qquad \text{(I.19c)}$$

[5] See N. Corngold, *op. cit.*

Hence, we arrive at the first thermal equation:

$$\nabla \cdot \mathbf{J}_T(\mathbf{r}) + \Sigma_{aT}(\mathbf{r})\, \varphi_T(\mathbf{r}) = \int_R \mathrm{d}^3 r' \int \mathrm{d}E'\, S(\mathbf{r}', E')\, P_{T_0}(\mathbf{r}' \to \mathbf{r}, E') \tag{I.20}$$

since

$$\Sigma_{tT} - \Sigma_{sT} = \Sigma_{aT}$$

The adaptation of Equation (I.16 g) to the thermal group proceeds in a like manner. Integration from $E = 0$ to $E = E_T$ yields:

$$\tfrac{1}{3}\int_0^{E_T} \nabla \varphi(\mathbf{r}, E)\, \mathrm{d}E + \int_0^{E_T} \Sigma_t(\mathbf{r}, E)\, \mathbf{J}(\mathbf{r}, E)\, \mathrm{d}E$$

$$= \int_0^{E_T} \mathrm{d}E \int_0^{\infty} \mathrm{d}E'\, \Sigma_s(\mathbf{r}, E')\, \mathbf{J}(\mathbf{r}, E')\, F_1(E' \to E) \tag{I.21 a}$$

The first term on the left-hand side is simply $\tfrac{1}{3}\nabla\varphi_T(\mathbf{r})$, again provided the functions are sufficiently well behaved to permit the interchange of operations. Consideration of the second term is delayed for the present. Again, the scattering-in term is decomposed:

$$\int_0^{E_T} \mathrm{d}E \int_0^{\infty} \mathrm{d}E'\, \Sigma_s(\mathbf{r}, E')\, \mathbf{J}(\mathbf{r}, E')\, F_1(E' \to E)$$

$$= \int_0^{E_T} \mathrm{d}E \int_0^{E_T} \mathrm{d}E'\, \Sigma_s(\mathbf{r}, E')\, \mathbf{J}(\mathbf{r}, E')\, F_1(E' \to E)$$

$$+ \int_0^{E_T} \mathrm{d}E \int_{E_T}^{\infty} \mathrm{d}E'\, \Sigma_s(\mathbf{r}, E')\, \mathbf{J}(\mathbf{r}, E')\, F_1(E' \to E) \tag{I.21 b}$$

In the first term on the right-hand side, the integral

$$\int_0^{E_T} F_1(E' \to E)\, \mathrm{d}E = \int_0^{E_T} \mathrm{d}E \int \mathrm{d}\boldsymbol{\Omega}\, (\boldsymbol{\Omega} \cdot \boldsymbol{\Omega}')\, F(E' \to E, \boldsymbol{\Omega}' \cdot \boldsymbol{\Omega})$$

is encountered. From the definition of $F(E' \to E, \boldsymbol{\Omega}' \to \boldsymbol{\Omega})$, this term is seen to be the average value of $\mu = \boldsymbol{\Omega} \cdot \boldsymbol{\Omega}'$, the cosine of the angle through which neutrons scatter if they suffer a collision at energy E'. We denote this quantity by $\bar{\mu}(E')$. The remaining integral over E' may be combined with the second term on the left-hand side of Equation (I.21 a) and after defining the *transport cross section* by

$$\Sigma_{tr}(\mathbf{r}, E) = \Sigma_t(\mathbf{r}, E) - \Sigma_s(\mathbf{r}, E)\, \bar{\mu}(\mathbf{r}, E) \tag{I.21 c}$$

we find that these terms reduce to:

$$\int_0^{E_T} (\Sigma(\mathbf{r}, E) - \Sigma_s(\mathbf{r}, E)\,\bar{\mu}(\mathbf{r}, E))\,\mathbf{J}(\mathbf{r}, E)\,\mathrm{d}E = \Sigma_{tr_T}\,\mathbf{J}_T(\mathbf{r}, E) \qquad \text{(I.21 d)}$$

which defines the thermal average cross section $\Sigma_{tr_T}(\mathbf{r})$. Note that $\Sigma_{tr_T}(\mathbf{r})$ is obtained by averaging over the current as opposed to the flux averaging used for the other cross sections. The second term on the right-hand side of Equation (I.21 b) can be expressed in terms of the first Legendre moment $\mathbf{P}_{T_1}(\mathbf{r}' \to \mathbf{r}, E')$ of the thermalization kernel $P_T(\mathbf{r}' \to \mathbf{r}, E', \mathbf{\Omega})$ just as the zeroth moment was introduced into Equation (1.20). In this fashion Equation (I.21 b) reduces to

$$\tfrac{1}{3}\,\nabla \varphi_T(\mathbf{r}) + \Sigma_{tr_T}\,\mathbf{J}_T(\mathbf{r}, E) = \int \mathrm{d}^3 r' \int \mathrm{d}E'\, S(\mathbf{r}', E')\,\mathbf{P}_{T_1}(\mathbf{r}' \to \mathbf{r}, E) \qquad \text{(I.21 e)}$$

Equations (I.20) and (I.21 e) are then the two simultaneous (P_1) thermal group equations for the two unknowns—the thermal flux and the thermal current.†

7. The Thermal Diffusion Equation

The pair of Equations (I.20) and (I.21 e) (or the pair (I.15 g) and (I.16 g)) are sometimes called the (consistent) P_1 approximation (see Chapter III, Section 5) because the angular flux was assumed to contain only the first order terms in a spherical harmonics expansion (i.e., Legendre Polynomials up to P_1). To obtain the thermal *diffusion* approximation it is necessary to approximate further by ignoring the integral term on the right-hand side of Equation (I.21 e).‡ If this is done, we obtain Fick's Rule which is the basic equation of diffusion theory:

$$\mathbf{J}_T(\mathbf{r}, E) = -D_T(\mathbf{r})\,\nabla \varphi_T(\mathbf{r}) \qquad \text{(I.21 f)}$$

where D_T, the thermal diffusion coefficient, is defined by

$$D_T(r) = \frac{1}{3\,\Sigma_{tr_T}(\mathbf{r})} \qquad \text{(I.21 g)}$$

† These Equations [(I.20) and (I.21 e)] could also be obtained by making the P_1 approximation, i.e., expanding the flux as in Equation(I.12 a), in the alternate form of the Boltzmann Equation (I.11).

‡ Equivalently, this assumes that the neutrons slowing down into the thermal group are distributed isotropically. For a discussion of the error introduced by this approximation see Appendix D.

The term "diffusion" arises out of the similarity of Equation (I.21 f) to other diffusion equations, e.g., these encountered in the kinetic theory of gases. The more often quoted result for the thermal diffusion coefficient

$$D_T(\mathbf{r}) = \frac{\int_0^{E_T} D(\mathbf{r}, E)\,\varphi(\mathbf{r}, E)\,\mathrm{d}E}{\int_0^{E_T} \varphi(\mathbf{r}, E)\,\mathrm{d}E} \tag{I.21 h}$$

with

$$D(\mathbf{r}, E) = [3\,\Sigma_{tr}(\mathbf{r}, E)]^{-1} \tag{I.21 i}$$

is obtained only if the neutrons are assumed to obey Fick's Rule at every energy in the thermal region. The diffusion equation for thermal neutrons is obtained by combining Equation (I.20) and (I.21 f):

$$-\nabla \cdot D_T(\mathbf{r})\,\nabla\,\varphi_T(\mathbf{r}) + \Sigma_{aT}(\mathbf{r})\,\varphi_T(\mathbf{r}) = \int_R \mathrm{d}^3 r' \int \mathrm{d}E'\,S(\mathbf{r}, E')\,P_{T_0}(\mathbf{r}' \to \mathbf{r}, E') \tag{I.21 j}$$

For assemblies in which the application of diffusion theory to the thermal neutron group is not valid, a "Thermal Transport Equation" can be derived in the same manner as the thermal diffusion equation, i.e., by integrating the Transport Equation over the thermal group. This procedure again leads to the same two difficulties which appeared in the derivation of the thermal diffusion equation:

(a) Cross section averages over the angular-dependent flux appear. Unless the flux is separable in energy and angle the average cross section becomes a function of neutron direction, $\mathbf{\Omega}$. Alternatively, if one uses a spherical harmonics expansion of the angular flux, cross sections appear averaged over the various spherical harmonics components of the flux, just as in the case of the diffusion equation, the total cross section that appears is a flux average value while the transport cross section is current averaged. This problem has never been treated satisfactorily and, indeed, may be the reason that higher order approximations to the Boltzmann equation for thermal neutrons frequently provide less accurate answers than does diffusion theory.

(b) The other difficulty is concerned with the higher order anisotropies associated with neutron slowing down into the thermal group. However, since (see Appendix D) it may be argued that these anisotropy terms are smaller than those terms included in the diffusion approximation, their neglect in higher order calculations is probably not a severe error.

C. Asymptotic Reactor Theory

8. The Asymptotic Reactor Model

It would be desirable to obtain a simple mathematical expression which would tell one when any reactor is critical, but, unfortunately, this is impossible. However, one specific reactor class, for which the critical conditions can be formulated simply, does exist. This type of reactor, called an "asymptotic reactor", is not a practical reactor design, but it is a useful first approximation to many reactors. We shall consider it in detail because it is a useful first approximation and because it provides considerable insight into more complex and hence, more realistic, reactor designs. Criticality calculations for actual reactors are generally performed numerically by the methods discussed in Chapter IV, but even in these calculations, frequent recourse is made to the results of the analytical theory.

An asymptotic reactor is defined as follows:

(a) The reactor is composed of a single, convex region in which all materials are distributed homogeneously.

(b) The reactor is large compared to a neutron mean free path. This is generally a requirement for criticality.

It would be natural, at this point, to question the assumption of homogeneity made above in view of the fact that most reactors today consist of solid fuel elements, separate solid or liquid moderators, and, when the moderator is not also the coolant, a fluid coolant. However, if the dimensions of the inhomogeneities (fuel plates, coolant channels, etc.) are small compared to the mean free path of a neutron, the neutron "does not know" that the system is inhomogeneous and treatment of the system as homogeneous can be justified. If this is not true, but the inhomogeneities are periodic within the core, it is frequently possible to find a set of physical constants which yield the same overall reactor behavior (critical mass, outer dimensions, etc.) while destroying only the details of the neutron density variation within one periodic substructure (*cell*).† For reactors which fall in neither of these categories, or vary in other ways from the simple model given below, analytical techniques of general applicability have not been developed and it is necessary to fall back upon the numerical methods to be described in Chapter IV.

The physical model of the neutron processes in an ideal thermal asymptotic reactor is as follows:

A fast neutron, born as a result of fission, suffers scattering collisions with moderator nuclei until it reaches an energy comparable to that of the moderator nuclei themselves. In the process of slowing down, neutrons may leak out or be absorbed but cannot cause fission. As long as a neutron is moving much faster than the moderator nucleus with which it collides, it will always lose energy in the collision; when the energies of the neutron and nucleus are comparable, i.e., when the neutron is in the thermal group, both loss and gain of energy are possible and the neutron diffuses at thermal energies until it either leaks out or is absorbed (possibly to cause a fission).

It will be possible to treat fissions caused by neutrons at energies above E_T with only slight loss of generality. This case is discussed in Appendix D. Some further restrictions will be required, but we shall reserve their application until the need for them is apparent.

9. Fission Sources

Up to now, the nature of the neutron source has been left unspecified. While it has been mentioned that fission is the only neutron source in a critical reactor, a general source term has been retained for simplicity.

Restricting our attention to the consideration of fission sources (which are the only sources of importance to critical reactors), we

† The usual procedure for accomplishing this is to treat the cell by means of a separate neutron diffusion calculation with appropriate boundary conditions (generally, no net neutron flow across the cell boundary). Even if all of the cells are not identical, this procedure may be used to smooth out the variation of cross sections with position, thus simplifying the numerical calculations considerably.

note that the number of fissions at **r** per second is, by definition,

$$\int_0^\infty \Sigma_f(\mathbf{r}, E)\, \varphi(\mathbf{r}, E)\, \mathrm{d}E = \Sigma_{fT}\varphi_T(\mathbf{r}) + \int_{E_T}^\infty \Sigma_f(\mathbf{r}, E)\, \varphi(\mathbf{r}, E)\, \mathrm{d}E \qquad (I.22a)$$

where Σ_{fT} is the thermal fission cross section. Thus the fission source may be written

$$S(\mathbf{r}, E) = \nu f(E)\left[\Sigma_{fT}\, \varphi_T(\mathbf{r}) + \int_{E_T}^\infty \Sigma_f(\mathbf{r}, E)\, \varphi\, (\mathbf{r}, E)\, \mathrm{d}E\right] \qquad (I.22b)$$

where ν represents the mean number of neutrons produced per fission† and $f(E)$ is the spectrum of fission neutrons [6].

In the remainder of this chapter we treat only the case in which all fissions are induced by thermal neutrons; the general case is treated in Appendix D. For this case Equation (I.22b) reduces to

$$S(\mathbf{r}, E) = \nu f(E)\, \Sigma_{fT}\varphi_T(\mathbf{r}) \qquad (I.22c)$$

Not all neutrons are produced simultaneously with a fission event. However, the effects of delayed neutron are unimportant in steady state situations so we shall neglect them.

10. The Asymptotic Reactor Equations

It was argued earlier that in a large reactor, the thermal angular neutron flux is nearly isotropic throughout a large portion of the reactor core. In such reactors, the system "looks" infinite to neutrons near the center, so that it might be possible to replace the slowing-down and thermalization kernels by the infinite medium kernels (which we denote by subscript "∞") to a good approximation. Since these kernels, in general, decrease rapidly with increasing source to field point distance, the integrals may be extended to all space, again to a good approximation. We defer the question to the meaning of the integral outside the reactor until later.

† ν is only slightly sensitive to the energy of the fission-inducing neutron and will be assumed constant. If it is not constant, one can treat $\nu\Sigma_f(E)$ as a "neutron production" cross section and carry through the same analysis without loss of generality.

[6] It is frequently given by the semi-empirical formula $f(E) = 0.484\, \mathrm{e}^{-E} \sinh \sqrt{2\, E}$ (E in MeV), the "Watt spectrum". B. E. WATT, *Phys. Rev.* **87**, 1037 (1952). An alternate form is given in Fig. I.1; these forms agree near the peak of the spectrum but in the high or low energy ranges the Watt spectrum seems to fit the experimental data more closely.

The major point is that the slowing down and thermalization kernels for an infinite homogeneous medium have definite symmetry properties. In particular, they are functions of the vector $(\mathbf{r} - \mathbf{r}')$ only.† The proof of this is obtained by noting the invariance properties of the Boltzmann equation for an infinite homogeneous medium; e.g., it is invariant under the coordinate transformation $\mathbf{r} \rightarrow \mathbf{r} + \mathbf{R}$, where $\mathbf{R}$ is a constant vector.

Under the approximation considered here, i.e., that all fissions take place at thermal energies and that the diffusion equation can be used to describe the thermal flux, only the zeroth Legendre moment of $P_{\infty T}$, the infinite medium thermalization kernel averaged over a fission spectrum, is required. Since $P_{\infty T}$ depends on $|\mathbf{r} - \mathbf{r}'|$ and $(\mathbf{r} - \mathbf{r}') \cdot \mathbf{\Omega}$,

$$\begin{aligned} P_{\infty T_0}(\mathbf{r} - \mathbf{r}') &= \int \mathrm{d}\mathbf{\Omega}\, P_{\infty T}(|\mathbf{r} - \mathbf{r}'|, (\mathbf{r} - \mathbf{r}') \cdot \mathbf{\Omega}) \\ &= \int \mathrm{d}\mathbf{\Omega} \int \mathrm{d}E'\, f(E') \cdot P_{\infty T}(|\mathbf{r} - \mathbf{r}'|, (\mathbf{r} - \mathbf{r}') \cdot \mathbf{\Omega}\, , E') \end{aligned} \qquad \text{(I.23a)}$$

depends only on $|\mathbf{r} - \mathbf{r}'|$. In order to limit the number of subscripts, we will write $P_{\infty T_0} = P_{\infty}$. We can now substitute (I.22c) and (I.23a) into (I.21j) to obtain the thermal diffusion equation for an asymptotic reactor

$$\begin{aligned} &- \nabla \cdot D_T \nabla (\mathbf{r})\, \varphi_T(\mathbf{r}) + \Sigma_{aT}(\mathbf{r})\, \varphi_T(\mathbf{r}) \\ &\qquad = \nu \int_{\text{all space}} \Sigma_{fT}(\mathbf{r}')\, \varphi_T(\mathbf{r}')\, P_{\infty}(|\mathbf{r} - \mathbf{r}|')\, \mathrm{d}^3 r' \end{aligned} \qquad \text{(I.23b)}$$

or, if D_T, Σ_{aT}, and Σ_{f_T} are not functions of $\mathbf{r}$,

$$- D_T \nabla^2 \varphi_T(\mathbf{r}) + \Sigma_{aT} \varphi_T(\mathbf{r}) = \nu\, \Sigma_{f_T} \int_{\text{all space}} \varphi_T(\mathbf{r}')\, P_{\infty}(|\mathbf{r} - \mathbf{r}'|)\, \mathrm{d}^3 r' \qquad \text{(I.23c)}$$

Equation (I.23c) can be written only if the neutron flux at thermal energies is separable into a product of a function of energy and a spatial distribution function. If this condition did not obtain, the thermal average cross sections might be functions of position even though the energy dependent cross sections were not (see Equation (I.17c)). This assumption will later be seen to be justified for the idealized type of reactor considered here.

† In fact, because of rotational invariance, the kernels are functions of $|\mathbf{r} - \mathbf{r}'|$ and $(\mathbf{r} - \mathbf{r}') \cdot \mathbf{\Omega}$ only. This fact will be used when it is needed (see Fig. I.7).

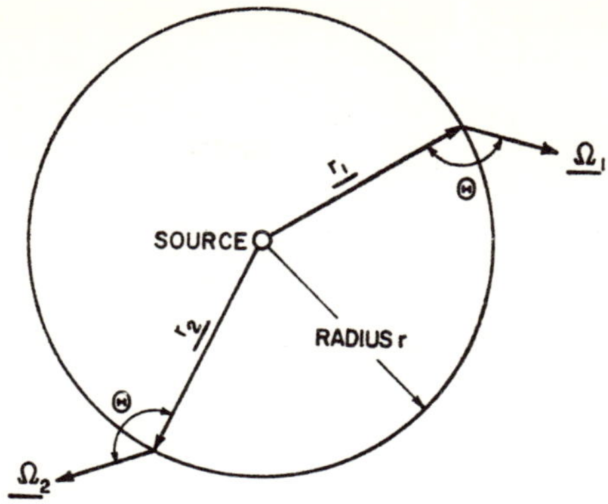

FIG. 1.7. For an isotropic source in a uniform isotropic medium the flux at any point on a given sphere will be the same in all directions which make the same angles with the radius vector.

11. Critical Conditions

While the assumptions made above are rather arbitrary, they do lead to significant and accurate results. On examination, the spatial integral in Equation (I.23c) is seen to be a convolution integral in the context of the Fourier transform [7]. Since this equation is linear and homogeneous in $\varphi_T(\mathbf{r})$ we can expect to arrive at a linear homogeneous algebraic equation in Fourier transform space. The condition that the Fourier transform of $\varphi_T(\mathbf{r})$ not vanish will then be a critical condition.

The Fourier transforms of the essential quantities are defined by

$$\bar{\varphi}_T(\mathbf{k}) = \int_{\text{all space}} e^{i\mathbf{k}\cdot\mathbf{r}}\,\varphi_T(\mathbf{r})\,d^3r \qquad \text{(I.24a)}$$

$$\bar{P}_\infty(k^2) = \int_{\text{all space}} e^{i\mathbf{k}\cdot\mathbf{r}} P(|\mathbf{r}|)\,d^3r \qquad \text{(I.24b)}$$

Note that $\bar{P}_\infty(\mathrm{k})$ is a function only of $|\mathrm{k}|$ (or k^2) since $P_\infty(r)$ depends on $|\mathbf{r}| = r$. Now, taking the Fourier transform of Equation (I.23c), that is, multiplying it by $e^{i\mathbf{k}\cdot\mathbf{r}}$ and integrating over all space, we have

$$(D_T k^2 + \Sigma_{aT})\,\bar{\varphi}_T(\mathbf{k}) = \nu\Sigma_{fT}\,\bar{P}_\infty(k^2)\,\bar{\varphi}(\mathbf{k}) \qquad \text{(I.25a)}$$

[7] For a discussion of the properties of Fourier transforms see, for example, I. N. SNEDDON, *Fourier Transforms*, New York, McGraw-Hill, 1957.

so that if $\overline{\varphi}_T(\mathbf{k})$ is to be non-vanishing, it follows that

$$\frac{(\nu\Sigma_{f_T}/\Sigma_{a_T})\,\overline{P}_\infty(k^2)}{1 + L^2k^2} \equiv k_{eff} = 1 \tag{I.25b}$$

Here $L^2 = D_T/\Sigma_{aT}$ is called the thermal diffusion area and can be shown to be equal to one-sixth the mean square distance traveled by a neutron from the time it is thermalized until the time it is absorbed in an infinite medium [8].

Equation (I.25b) is sometimes written in terms of the two quantities

$$\eta = \nu\frac{\Sigma_{f_T}^U}{\Sigma_{a_T}^U} \tag{I.25c}$$

and

$$f = \frac{\Sigma_{a_T}^U}{\Sigma_{a_T}} \tag{I.25d}$$

Here Σ_{aT}^U is the absorption (fission plus capture) cross section of the fuel. η is called the *fuel efficiency* and f the *thermal utilization.* The advantage of introducing these quantities is that η is a function only of the type of fuel and that any geometric effects that arise in the case of heterogeneous reactors [9], and which require the taking of cell averages before asymptotic reactor theory can be applied, are contained in f. With these definitions, Equation (I.25b) becomes

$$\frac{\eta f\overline{P}_\infty(k^2)}{1 + L^2k^2} = 1 \tag{I.25e}$$

This equation has a simple physical interpretation. η represents the number of fission neutrons emitted per neutron absorbed in the fuel. Since f is the fractional number of neutrons absorbed in the fuel, ηf is the number of fission neutrons emitted per thermal neutron absorbed. $\overline{P}_{\infty_0}(k^2)$ will be shown to be the number of neutrons thermalized in the reactor per fast neutron born and $(1 + L^2k^2)^{-1}$ will be shown to be the probability that a neutron thermalized will

[8] See, for example, S. Glasstone and M. Edlund, *The Elements of Nuclear Reactor Theory*, Van Nostrand (1952).

[9] In this case space averaged cross sections must be used in the critical equation. See, for example, A. M. Weinberg and E. P. Wigner, *The Physical Theory of Neutron Chain Reactors*, University of Chicago Press (1958).

be absorbed before it leaks from the system. Clearly then

$$\frac{\eta f \bar{P}_\infty(k^2)}{1 + L^2k^2} \equiv k_{eff}$$

represents the number of thermal neutrons absorbed per thermal neutron absorbed in the preceding generation. This quantity must equal unity if a steady state is to be maintained. If $k_{\text{eff}} < 1$, the reactor is subcritical while if $k_{\text{eff}} > 1$ the reactor is supercritical. In either case no steady state neutron flux can exist.

12. The Material Buckling

Equation (I.25e), the criticality equation, is an implicit equation for k^2 which has at most one positive root $k^2 = B_m^2$, the so-called material buckling of the reactor. This is easily seen when we note that $\bar{P}_\infty(k^2)$ is expected to be a monotonically decreasing function of k^2, a fact which will be proved in a later section. Since $(1 + L^2k^2)^{-1}$ is also monotonically decreasing, the product can equal unity at most once, and then only if $\eta f \bar{P}_\infty(0) > 1$.

13. The Helmholtz Equation and Boundary Conditions

Another important result is that the thermal flux obeys the Helmholtz equation.

$$\nabla^2 \varphi_T(\mathbf{r}) = -B_m^2 \varphi_T(\mathbf{r}) \qquad \text{(I.26)}$$

To prove this, we write

$$\varphi_T(\mathbf{r}) = (2\pi)^{-3} \int e^{-i\mathbf{k}\cdot\mathbf{r}}\, \bar{\varphi}_T(\mathbf{k})\, d^3k \qquad \text{(I.27a)}$$

Then operating with ∇^2:

$$\nabla^2 \varphi_T(\mathbf{r}) = -(2\pi)^{-3} \int k^2\, e^{-i\mathbf{k}\cdot\mathbf{r}}\, \bar{\varphi}_T(\mathbf{k})\, d^3k \qquad \text{(I.27b)}$$

But we have shown in Sections 11 and 12 that $\bar{\varphi}_T(k) = 0$ unless $k^2 = B_m^2$. Therefore we may remove k^2 from the integral, replacing it by B_m^2 and find

$$\nabla^2 \varphi_T(\mathbf{r}) = -B_m^2\, (2\pi)^{-3} \int e^{-i\mathbf{k}\cdot\mathbf{r}} \bar{\varphi}_T(\mathbf{k})\, d^3k \qquad \text{(I.27c)}$$

or, comparing with Equation (I.27a):

$$\nabla^2 \varphi_T(\mathbf{r}) = -B_m^2\, \varphi_T(\mathbf{r}) \qquad \text{(I.27d)}$$

This is a general property of a function whose Fourier transform vanishes everywhere except on a spherical surface of radius B_m in transform space.

Thus, in order to calculate the thermal flux shape in a reactor, we need only solve the criticality Equation (I.25e) for B_m^2 and then solve the Helmholtz equation (I.26). The flux shape thus obtained must, however, satisfy a condition at the outer boundary, as well as other physically necessary conditions.

Clearly, two conditions on $\varphi(\mathbf{r}, E)$ are that it be finite and non-negative everywhere within the reactor. Being a second-order differential equation, the Helmholtz equation will possess two independent solutions for every B_m. Generally, these conditions will suffice to eliminate one of these solutions. The outer boundary condition then serves to relate reactor geometry to material composition.

The exact condition to be satisfied at the outer boundary has already been discussed and is represented by Equation (I.10). Clearly, this condition cannot be satisfied exactly by any function of the form of Equation (I.12a); that is, the P_1 approximation is incapable of treating the boundary conditions exactly. However, we may demand that the *net inward* current vanish, i.e.

$$\int_0^{E_T} \mathrm{d}E \int \Phi(\mathbf{r}_s, E, \mathbf{\Omega})\, \mathbf{\Omega} \cdot \mathbf{n}\, \mathrm{d}\mathbf{\Omega} = 0 \qquad \text{(I.28a)}$$

$$\mathbf{\Omega} \cdot \mathbf{n} < 0$$

On substituting Equation (I.12a) into Equation (I.28a) and noting the interpretations of φ_0 and φ_1, we have:

$$\varphi_T(\mathbf{r}_s) - 2 \cdot \mathbf{n} \cdot \mathbf{J}_T(\mathbf{r}_s) = 0 \qquad \text{(I.28b)}$$

If Fick's rule, Equation (I.21f) is applied, we then have a condition on $\varphi_T(\mathbf{r}_s)$:

$$\frac{1}{\varphi_T} \mathbf{n} \cdot \nabla \varphi_T = -\frac{1}{2D} = -\frac{3\,\Sigma_{tr}}{2} \qquad \text{(I.28c)}$$

The difficulty here is that Fick's rule was obtained by neglecting anisotropic slowing down. But earlier discussion has pointed out that this approximation is least likely to hold near the boundary so that Equation (I.28c) is not likely to be very accurate.

A means of obtaining another boundary condition is provided by the Milne problem. In solving the problem of a semi-infinite medium bounded by a plane, with a source at infinity, it is found that the flux obeys the condition [10]

$$\frac{1}{\varphi}\mathbf{n}\cdot\nabla\varphi = -\frac{\Sigma_{tr}}{0.7104} = -\frac{\Sigma_{tr}}{z_o} \tag{I.28d}$$

provided that

$$c = \frac{\nu\Sigma_f + \Sigma_s}{\Sigma_t} = 1 \tag{I.28e}$$

a condition which is nearly valid in most reactors. Equations (I.28c) and (I.28d) are quite similar; the difference is just 0.7104 vs. 0.6667 on the right-hand sides (see footnote).

Since we may expect the difference between a linear extrapolation of the flux and the analytic continuation of its shape inside the core to be small, we may simply allow the flux to vanish at the point at which the linear extrapolation (rather than the analytic continuation) vanishes. From Fig. I.8 we have:†

$$\tan\theta = -\mathbf{n}\cdot\nabla\varphi = \partial\varphi/\partial n$$

$$d = \frac{\varphi(\mathbf{r}_s)}{\tan\theta} = \frac{\varphi(\mathbf{r}_s)}{\partial\varphi(\mathbf{r}_s)/\partial n} = \frac{0.7104}{\Sigma_{tr}} \tag{I.29}$$

Thus, to a reasonable approximation, we may use the boundary condition that $\varphi(\mathbf{r}, E)$ and $\varphi_T(\mathbf{r})$ vanish at a distance $0.7104/\Sigma_{tr}$ from the boundary. As the extrapolation distance is usually small compared to reactor dimensions the difference between the points at which the linear extrapolation and the analytic continuation of the flux go to zero will be exceedingly small. If this is not true, then the reactor is probably not large enough to permit use of asymptotic reactor theory at all. As written, the extrapolation distance, d, is a function of energy, since Σ_{tr} is a function of energy. If is of extreme importance, however, to have the flux vanish on the same

† Had condition (I.28c) been applied, the value 0.7104 in Equation (I.29) would be replaced by 0.6667. Thus the Milne problem boundary condition gives an extrapolation distance which is 0.05 mean free paths larger than the P_1 boundary condition. This difference is negligible for all but very small systems, to which the present development is inapplicable anyway.

[10] Strictly speaking, it is the flux far from the boundary which obeys this condition. This *asymptotic* flux is shown in Appendix C to be the component with which asymptotic reactor theory deals. See B. DAVISON, *op. cit.* Davison also gives z_0 (here 0.7104) as a function of c.

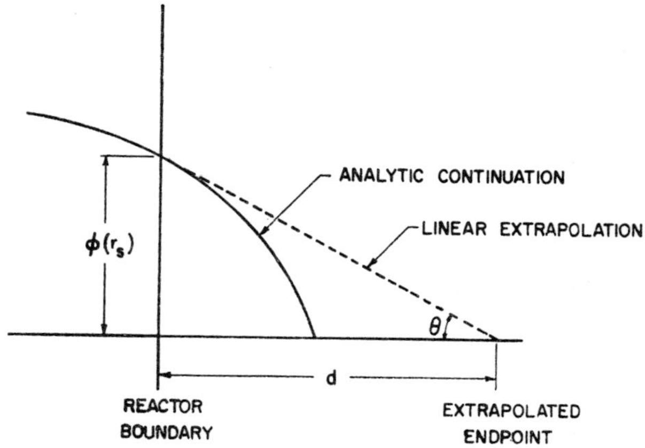

FIG. 1.8.

surface at all energies. Otherwise, the shape of the flux will vary with energy and will not be separable in space and energy. Again, d is usually much smaller than the reactor dimensions so that this effect is rather small unless d varies strongly with energy. For that case İnönü [11] has used multigroup methods (see Chapter IV) to define an energy independent average extrapolation length. As his analysis is rather lengthy only his result is quoted:

$$d_{av} = \frac{\int d(E)\, \varphi(E)\, dE}{\int \varphi(E)\, dE} \tag{I.30}$$

14. The Spatial Flux Shape

By virtue of the discussion of the previous section, the problem of finding flux shape has been reduced to the problem of solving the Helmholtz equation with the boundary conditions of finiteness, continuity, and non-negativity everywhere within a convex surface on which the flux must vanish. This is a well known problem in mathematical physics and the properties of its solutions have been studied in rather great detail. It is known that the solutions of this problem form an infinite set of eigenfunctions, which may be made orthonormal over the reactor volume (including the small extra-

[11] E. İNÖNÜ, *Proc. Int. Conf., Geneva*, P/2344 (1958). The result quoted here is not precisely his result, but is easily derived from it.

polation distance); all of these eigenvalues are positive and only the first eigenfunction, which corresponds to the smallest eigenvalue, does not vanish anywhere within the bounding surface [12]. This lowest eigenvalue is called the geometric buckling, B_g^2, since it depends only on the geometry of the system.

For a bare reactor, we have seen that the only value of the wave number in the Helmholtz equation which permits the establishment of a steady state flux is B_m. Obviously, only one of the eigenvalues of the problem discussed above can be equal to B_m^2; the flux must, therefore, have the spatial behavior of a single eigenfunction of the Helmholtz equation which must furthermore be the first eigenfunction as a consequence of the physical restriction that $\varphi(\mathbf{r}, E)$ be non-negative. Hence, a non-trivial solution for the neutron flux is possible if, and only if:

$$B_m^2 = B_g^2 \tag{I.31}$$

Since B_m^2 is a function of material properties alone and B_g^2 is a function solely of geometry, this is a relation between the material composition and geometry required to achieve criticality.

The eigenfunctions and eigenvalues for infinite slab, spherical, and cylindrical reactors are given in Table 1. Note that in each case the dimensions include the extrapolation distance.

All of the other eigenvalues of the Helmholtz equation are larger than B_g^2. Since k_{eff} is a monotonically decreasing function of k^2 (or B_m^2), any of these eigenvalues leads to values of k_{eff} less than unity.

The Boltzmann equation is linear and homogeneous in $\Phi(\mathbf{r}, E, \mathbf{\Omega})$ and the superposition principle may be applied to it. That is to say, the sum of independent solutions (which obey the proper boundary conditions) is also a solution. Hence a sum of solutions of the Helmholtz equation is a mathematically acceptable solution for the flux, again provided all of the physically imposed restrictions are met. However, only one of these solutions can meet the critical condition; all others have $k_{\mathrm{eff}} < 1$, which a study of the time

[12] A. E. Heins, private communication. These results can be proved by converting the problem into a Fredholm integral equation on the interior of the volume. The quoted theorems are known to hold for this type of equation. See R. Courant and D. Hilbert, *Methods of Mathematical Physics*, New York, Interscience, 1950.

TABLE 1

Reactor geometry	Lowest eigenvalue	First eigenfunction
Semi-infinite slab, width l	$B_g^2 = (\pi/l)^2$	$\cos(\pi x/l)$
Sphere of radius R	$B_g^2 = (\pi/R)^2$	$1/r \sin(\pi r/R)$
Cylinder of radius R, height H	$B_g^2 = (\pi/H)^2 + (2.405/R)^2$	$\cos(\pi x/H) J_0(2.405\varrho/R)$
Rectangular parallelepiped of dimensions l_1, l_2, l_3	$B_g^2 = (\pi/l_1)^2 + (\pi/l_2)^2 + (\pi/l_3)^2$	$\cos(\pi x/l_1)\cos(\pi y/l_2)\cos(\pi z/l_3)$

dependent equation will reveal to lead to solutions which decay exponentially in time [13]. Again, it is emphasized that the time independent flux has the shape of the first eigenfunction of the Helmholtz equation.

It should be pointed out that all of these results depend upon the replacement of the finite medium slowing-down kernel by an infinite medium kernel, which depends only on $|\mathbf{r} - \mathbf{r}'|$, and extension of the integrals over the reactor volume to all space. Intuitively, this replacement is valid only for a large reactor; that this replacement is also valid in more general cases is shown in Appendix E.

15. The Slowing-down Density

A quantity of extreme importance in reactor calculations is the *escape probability* defined as the probability that a neutron will not leak out of the reactor or be absorbed before reaching energy E. However, as high energy absorptions in U-238 may lead to fission, a correction must be made for this effect. We define the *fast fission factor*, ε, to be the number of neutrons that would slow down past the U-238 fission threshold (which is about 1 MeV) per neutron produced by U-235 fission in an infinite medium of the reactor material.

Returning to the consideration of the escape probability, we note that it is the number of neutrons *slowing down past energy E* in the core per fission neutron produced at energies above E. To facilitate

[13] See for example S. GLASSTONE and M. C. EDLUND, *op. cit.*, or Appendix A.

computation of this quantity, we now introduce the *slowing-down density*, $q(\mathbf{r}, E)$, defined such that $q(\mathbf{r}, E)\,\mathrm{d}^3\mathbf{r}$ is the number of neutrons slowing down *past* energy $E(>E_T)$ per second in volume element d^3r at $\mathbf{r}$. To derive an expression for $q(\mathbf{r}, E)$ we remember that $\Sigma_{s_0}(E' \to E)$ (where $E' > E$) is the macroscopic cross section for neutron scattering from energy E' to E. Then the number of neutrons slowing down past energy E per cm^3 which had their last collision in the energy range $\mathrm{d}E'$ at E' $(>E)$ is

$$\mathrm{d}q(\mathbf{r}, E, E') = \left[\int_0^E \Sigma_{s_0}(E' \to E'')\,\varphi(\mathbf{r}, E')\,\mathrm{d}E''\right]\mathrm{d}E'$$

where
$$\Sigma_{s_0}(E' \to E) = \Sigma_s(E')\,F_0(E' \to E) \tag{I.32a}$$

and $F_0(E' \to E)$ is defined by Equation (1.15e).

The total number of neutrons slowing down past E is therefore

$$q(\mathbf{r}, E) = \int_E^\infty \mathrm{d}E' \int_0^E \mathrm{d}E''\,\Sigma_{s_0}(E' \to E'')\,\varphi(\mathbf{r}, E') \tag{I.32b}$$

16. The Slowing-down Density Kernel

By its very nature the slowing-down density is a much more slowly varying function of energy than the flux and for this reason we will have occasion to employ it frequently in the ensuing chapters. Hence we introduce a *slowing-down density kernel*, $P^*(\mathbf{r}' \to \mathbf{r}, E)$, which is the slowing-down density per unit volume at $\mathbf{r}$ at energy E due to a point source of fission neutrons at point $\mathbf{r}'$. To be consistent with the conventions employed earlier with respect to the slowing-down kernel, the slowing-down density kernel must include the effects of fast fission, that is, the slowing-down density described by the kernel P^* is somewhat enhanced by U-238 fission.

It follows from the definition of the kernel that the slowing-down density is given by

$$q(\mathbf{r}, E) = \int_R S(\mathbf{r}')\,P^*(\mathbf{r}' \to \mathbf{r}, E)\,d^3r' \tag{I.33a}$$

Following the arguments used earlier, the asymptotic reactor model may be invoked. To do this, one need only replace the kernel P^* in Equation (I.33a) by the infinite medium kernel $P^*_\infty(|\mathbf{r} - \mathbf{r}'|, E)$

and extend the integral to all space:

$$q(\mathbf{r}, E) = \int_{\text{all space}} S(\mathbf{r}')\, P_{\infty}^{*}(|\mathbf{r} - \mathbf{r}',|\, E)\, d^3r' \tag{I.33b}$$

In a critical reactor, $S(\mathbf{r}')$ is, of course, the fission source and has the shape of the first eigenfunction of the Helmholtz equation $\chi(\mathbf{r})$.

At this point, we must make use of the following lemma: If $\chi(\mathbf{r})$ is a solution of the Helmholtz equation with wave number B_n, then

$$\int_{\text{all space}} f(|\mathbf{r} - \mathbf{r}'|)\, \chi(\mathbf{r}')\, d^3r' = \chi(\mathbf{r})\bar{f}(B_n^2) \tag{I.34}$$

The proof is short. Taking the Fourier transform of the left-hand side of Equation (I.34), we obtain, by virtue of the convolution theorem, $\bar{f}(k^2)\, \chi(\mathbf{k})$. But, because $\chi(\mathbf{r})$ obeys the Helmholtz equation, $\chi(\mathbf{k}) = 0$ unless $k^2 = B_n^2$. Hence the Fourier transform becomes $f(B_n^2)\, \chi(k)$ and inverting, we obtain $f(B_n^2)\, \chi(\mathbf{r})$, Q.E.D. An application of the lemma to Equation (I.33b) quickly yields:

$$q(\mathbf{r}, E) = S(\mathbf{r})\, \bar{P}_{\infty}^{*}(B_g^2, E) \tag{I.35a}$$

which shows that, within the bounds of asymptotic reactor theory, $q(\mathbf{r}, E)$ also has the spatial distribution of the first eigenfunction of the Helmholtz equation and that it is separable in energy and space. The first result is to be expected from the close relationship between the slowing-down density and the flux. Further, it can be argued that since Equation (I.32b) relates $q(\mathbf{r}, E)$ and $\varphi(\mathbf{r}, E)$ by an operator that depends solely on the energy variable, if the slowing-down density is separable then the flux is also.† It should be noted that the thermalization kernel $P_T(\mathbf{r}' \to \mathbf{r})$ used earlier is simply the slowing-down density kernel evaluated at the thermal cutoff energy E_T, i.e., $P^*(\mathbf{r}' \to \mathbf{r}, E_T)$.

The ratio of the number of neutrons slowing down past energy E to the number of U-235 fission neutrons born at energies above

† The separability of the flux can also be demonstrated directly by arguments similar to those that led to the conclusion that the slowing-down density is separable, i.e. by Fourier transforming the equation for φ $(\mathbf{r}, E)$.

E is:

$$\frac{\int_R q(\mathbf{r}, E)\, \mathrm{d}^3 r}{\int_E^\infty \mathrm{d}E' \int_R S(\mathbf{r}, E')\, \mathrm{d}^3 r} = \frac{\bar{P}^*_\infty(B_g^2, E) \int_R S(\mathbf{r})\, \mathrm{d}^3 r}{\int_E^\infty \mathrm{d}E' \int_R S(\mathbf{r}, E')\, \mathrm{d}^3 r} = \frac{\bar{P}^*_\infty(B_g^2, E)}{\int_E^\infty f(E)\, \mathrm{d}E} \qquad \text{(I.36)}$$

where $S(\mathbf{r}, E)$ is the source of neutrons due to U-235 fission.† In particular, if E is below the energy region in which the fission spectrum is appreciable, the escape probability becomes simply $\bar{P}^*(B_g^2, E)$, a result which is frequently called the "Second Fundamental Theorem of Reactor Theory" [14]. Formally, this theorem states that the escape probability in a critical system is the Fourier transform of the slowing-down density kernel, evaluated at the geometric buckling of the system. This explains the previously used fact that $\bar{P}_\infty(k^2)$ is a monotonically decreasing function of k^2. As a reactor becomes larger, k^2 decreases and the escape probability clearly increases.

The separability of the flux and slowing-down density, the critical condition, the fact that the flux obeys the Helmholtz equation, and the expression for the escape probability, are the basic results of asymptotic reactor theory.

17. The Six Factor Formula

The escape probability, as defined in the last section, includes effects of fast fission and absorption of fast neutrons, as well as fast leakage. These effects are usually separated in a manner convenient for calculation. Consider fission neutrons starting to slow down from high energies. In the process of slowing down they may induce fast fission, thus increasing their number by the factor ε. Alternatively, they may be absorbed without inducing fission before reaching thermal energies. Here we define the *resonance escape probability*,

† Strictly speaking, Equation (I.35c) is valid only below the U-238 fission threshold since the fast fission factor (ε) has been lumped into the definition of the slowing-down density kernel. This difficulty can be avoided by either introducing an energy dependent fast fission factor (which is undesirable) or by treating U-238 fissions independently. When ε is near unity this difficulty does not arise.

[14] A. M. WEINBERG and E. P. WIGNER, *The Physical Theory of Neutron Chain Reactors*, University of Chicago Press (1959), p. 420.

$p(E' \to E)$, as the probability that a neutron born at energy E' will not be captured before reaching energy E ($p(E)$ will denote the resonance escape probability averaged over a fission spectrum). Finally, neutrons may leak out of the system before reaching thermal energies. To allow for this, the non-leakage probability, $\lambda(B_g^2, E' \to E)$, is defined as the probability that a neutron born at energy E' will not leak out of a system of geometric buckling B_g^2 before reaching energy E. Its fission spectrum average is denoted by $\lambda(B_g^2, E)$. Here B_g^2 has been written explicitly as an argument of λ to call attention to the fact that λ depends very strongly on the geometry of the system. The implication that λ depends only on B_g^2 is accurate only within the limits of asymptotic theory as, of course, are most of the other results of this chapter.

Frequently, ε, p, and λ are assumed entirely independent of one another; this is only true within a certain limit. For example $p(E)$ generally depends on the geometry of the system (and hence on B_g^2) while $\lambda(B_g^2, E)$ may depend to a rather great extent on the amount of absorption in the system. However the contribution to ε comes entirely from very high energy neutrons and hence occurs during only a small fraction of the slowing-down process. Fast fission then occurs before much resonance absorption or leakage has taken place (except in water moderated reactors) and ε is effectively independent of p and λ.

On the other hand, the statement that $p(E)$ is independent of geometry is equivalent to saying that increasing the leakage has no effect on the fast absorption. But this is clearly untrue, for a neutron which leaks out at energy E has zero probability of being captured at energies below E, and vice versa. Hence, as long as the two processes occur simultaneously, the probabilities of their occurring cannot be truly independent. If, however, the probabilities are each close to unity, the interaction will be small, and in such a case for every neutron born due to U-235 fission

$$\varepsilon p(E_T)\, \lambda(B_g^2, E_T) \tag{I.37}$$

will reach thermal energies. Also, in many systems, such as light water assemblies, resonance absorption sets in at energies below those at which most of the fast leakage takes place.

After reaching thermal energies the neutrons may either leak or be absorbed. We denote the *thermal non-leakage probability* by

$\lambda_T(B_g^2)$, i.e., $\lambda_T(B_g^2)$ is the fraction of neutrons which are thermalized that get absorbed. Of the neutrons that suffer thermal absorption, a fraction f (the thermal utilization) is absorbed in uranium and each absorption by a uranium atom produces η neutrons on the average. Hence a total of

$$k_{\text{eff}} = \eta f \varepsilon p(E_T)\, \lambda_T(B_g^2)\, \lambda(B_g, E_T) \tag{I.38a}$$

are produced on the average, for every neutron that started to slow down. This, the famous six factor formula, defines k_{eff}, which must evidently equal unity for criticality. Comparison of Equation (I.38a) with Equation (I.25e) shows that

$$\varepsilon p(E_T)\, \lambda(B_g^2, E_T)\, \lambda_T(B_g^2) = \frac{\bar{P}_{T\infty}(B_g^2)}{1 + L_T^2 B_g^2} \tag{I.38b}$$

One may note that if neutrons did not diffuse after thermalization there would be no thermal leakage, i.e., λ_T would be unity. This limit also obtains by setting $D_T = 0$ (and hence, $L_T^2 = 0$). Hence

$$\varepsilon p(E_T)\, \lambda(B_g^2, E_T) = \bar{P}_{T\infty}(B_g^2) \tag{I.38c}$$

Further, in an infinite reactor there can be no leakage, and the geometric buckling, which is related to the inverse square of a dimension (see Table 1), goes to zero. Setting $B_g^2 = 0$ in Equation (I.38c) gives, therefore:

$$\varepsilon p(E_T) = \bar{P}_{T\infty}(0) \tag{I.38d}$$

Thus, the fast non-leakage probability is the ratio of Equations (I.38c) and (I.38d):

$$\lambda(B_g^2, E) = \frac{\bar{P}_{T\infty}(B_g^2)}{\bar{P}_{T\infty}(0)} \tag{I.38e}$$

The arguments leading from Equation (I.38b) to (I.38c) further show that:

$$\lambda_T(B_g^2) = \frac{1}{1 + L_T^2 B_g^2} \tag{I.38f}$$

a result which may also be derived from the Second Fundamental Theorem as follows. The Green's function for the thermal diffusion equation (Eqn. (I.21j) with D_T spatially constant) in an infinite medium is

$$\varphi_T(\mathbf{r}, \mathbf{r}') = \frac{e^{-|\mathbf{r}-\mathbf{r}'|/L_T}}{4\pi D_T |\mathbf{r} - \mathbf{r}'|} \tag{I.39}$$

The kernel from which the non-leakage probability may be obtained is $\Sigma_{aT}\varphi_T(\mathbf{r}', \mathbf{r})$, whose Fourier transform is indeed the right-hand side of Equation (I.38f).

18. The Non-leakage Probability—Non-critical Systems

The fast non-leakage probability can be interpreted in terms of $\overline{P}(B_g^2, E)$ even for non-critical systems. Consider a subcritical system $(k < 1)$; such a system will have a steady neutron population only if a non-fission source is placed in it. This steady flux will obey the thermal diffusion equation with an "extraneous" source $S(r)$:

$$-D_T\nabla^2\Phi_T(\mathbf{r}) + \Sigma_{\partial T}\Phi_T(\mathbf{r}) = \int d^3\mathbf{r}' P_T(|\mathbf{r} - \mathbf{r}'|)\{\nu\Sigma_{fT}\Phi_T(\mathbf{r}') + S(\mathbf{r}')\} \tag{I.40}$$

where it is assumed that the "extraneous" source energy distribution is sufficiently similar to that of the fission spectrum to allow use of the same slowing-down kernel for both. The flux $\varphi_T(r)$ of Equation (I.40) must obey the usual boundary condition that it vanish on an extrapolated surface. Now let the flux and source be expanded in terms of the eigenfunctions of the Helmholtz equation (note that one eigenfunction will not suffice in this case since criticality does not obtain):

$$\Phi_T(\mathbf{r}) = \sum_{n=0}^{\infty} \varphi_n \chi_n(\mathbf{r})$$

$$S(\mathbf{r}) = \sum_{n=0}^{\infty} s_n \chi_n(\mathbf{r}) \tag{I.41 a}$$

Substituting these expansions into Equation (I.40), applying Lemma (I.34) and making use of the linear independence of the $\chi_n(\mathbf{r})$, we have:

$$\varphi_n = \frac{S_n \overline{P}_{T\infty}(B_n^2)}{\nu\Sigma_{fT} \overline{P}_{T\infty}(B_n^2) - (1 + L^2 B_n^2)} = \frac{S_n \overline{P}_{T\infty}(B_n^2)/\Sigma_{aT}(1 + L^2 B_n^2)}{1 - k_{\text{eff}\,n}} \tag{I.41 b}$$

where, by definition:

$$k_{\text{eff}\,n} = \frac{\eta f \overline{P}(B_n^2)}{1 + L^2 B_n^2} \tag{I.41 c}$$

is the *effective multiplication for the nth mode*, now shown to be the number of nth mode neutrons (the coefficient of the nth eigenfunction in the flux expansion) absorbed per nth mode neutron absorbed in the preceding generation.

The physical meaning of Equation (I.4b) is as follows. If S neutrons per second are emitted by a source into a non-multiplying system the neutron density will be S multiplied by the neutron lifetime. This lifetime must be

$$l = (\Sigma_a + DB^2)^{-1}\, v^{-1}\, \bar{P}(B^2) \tag{I.41 d}$$

since the mean free path of the neutron is $(\Sigma_a + DB^2)^{-1}\, \bar{P}_{T\infty}(B^2)$. Thus we would have

$$n = \frac{S\, \bar{P}_{T\infty}(B^2)}{(\Sigma_a + DB^2)\, v} \tag{I.42 a}$$

which agrees with Equation (I.41 b) except for the factor $(1-k_{\text{eff}})^{-1}$. This factor arises from the fact that in a multiplying medium each source neutron produces, on the average, k_{eff} second generation neutrons, which produce k_{eff}^2 third generation neutrons, etc. The total number of neutrons in the system is thus increased by the factor

$$1 + k_{\text{eff}} + k_{\text{eff}}^2 + k_{\text{eff}}^3 + - \ldots = \frac{1}{1 - k_{\text{eff}}} \tag{I.42 b}$$

19. Some Results of Asymptotic Reactor Theory

The development of time-independent asymptotic reactor theory is now essentially complete. We have demonstrated that within the limitations of asymptotic reactor theory, the problem of criticality may be formulated in terms of a problem in slowing-down theory, namely that of determining the slowing-down kernel. In so doing, several assumptions and approximations have been used, the validity of which is, to say the least, open to question.

One of the foremost of these, that of the one group treatment of all thermal neutrons, has been under considerable study in recent years [15]. The work being done in this area has implications not only for reactor theory but also for the study of the structure of matter. These calculations provide better means of describing

[15] Proceedings of the Brookhaven Conference on Neutron Thermalization, BNL 719, 1962.

thermal neutrons but, as they are somewhat outside the range of a work on slowing down, we shall give no detailed discussion of "thermalization" here.

In evaluating the principal results of asymptotic reactor theory, namely the space–energy separability of the flux (often called the "First Fundamental Theorem") [16] and the Second Fundamental Theorem mentioned earlier, one should try to separate the effects of the various assumptions and approximations.

İnönü [17] and Dresner [18] and Yip and Zweifel [19] have investigated the validity of the Second Fundamental Theorem. Their results indicate that asymptotic reactor theory is applicable (in one speed theory (even to systems whose linear dimensions are of the order of a neutron mean free path. Dresner's results for the non-leakage probability are in better agreement with the rigorous theory than İnönü's due to the fact that he used more nearly correct boundary conditions. For small systems, in which the neutron leakage is great, the boundary conditions discussed earlier are no longer valid and asymptotic reactor theory breaks down. A further discussion of the validity of asymptotic reactor theory is given in Appendix C [20].

D. Calculation of the Elastic Scattering Frequency

20. Problems Associated with the Calculation of $F(E' \to E, \mu_0)$

In the previous section, we have derived the various results of asymptotic theory in terms of a "slowing-down kernel" which still remains to be calculated. We have also obtained the basic means for obtaining that kernel, i.e., the Boltzmann equation. In the remainder of this chapter we begin a discussion of the methods for obtaining the slowing down kernel from the Boltzmann equation. The first part of this problem is to derive the elastic scattering frequency $F(E' \to E, \mu_0)$.

A rigorous treatment of slowing down should, of course, include the effects of inelastic scattering. However, it is difficult to arrive at a satisfactory mathematical development of the theory if

[16] A. M. Weinberg and E. P. Wigner, *op. cit.*
[17] E. İnönü, *Nucl. Sci. Eng.* **5**, 248 (1959).
[18] L. Dresner, *Nucl. Sci. Eng.* **7**, 419 (1960).
[19] S. Yip and P. F. Zweifel, *Nucl. Sci. Eng.* **10**, 362 (1961).
[20] B. Davison, *op. cit.*

inelastic scattering effects are included. Hence, the discussion of the effect of inelastic scattering on reactor calculations is generally made only in the context of multigroup theory; our discussion of this effect is therefore deferred until Chapter IV.

The problem of low-energy molecular effects is also quite difficult to treat and is a topic which is more properly discussed in connection with the problem of calculating the thermal neutron spectrum [21]. The slowing-down calculations described here will be valid only above the energies associated with the chemical bond in molecules, or approximately 1 eV [22].

For many purposes it is sufficient to assume that the elastic scattering is spherically symmetric in the center-of-mass system of neutron and scattering nucleus (*s*-wave scattering), but restriction to this special case is not necessary in general. It is well known that *s*-wave scattering (zero relative angular momentum) is isotropic in the center of mass while *p*-wave scattering (angular momentum of one unit) is strongly peaked in the forward direction [23]. Thus, anisotropic corrections are important at energies at which *p*-wave scattering comes into prominence. This energy can be estimated by the following simple quasi-classical argument.

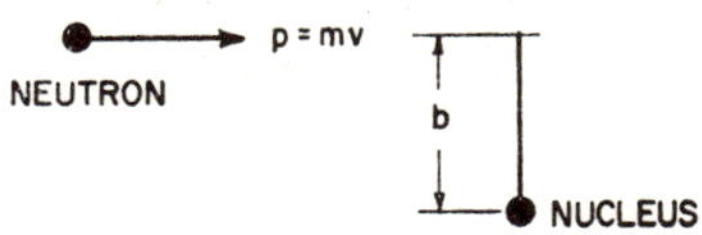

FIG. I.9.

[21] See T. J. KRIEGER and M. S. NELKIN, *Phys. Rev.* **106**, 290 (1957) for a discussion of this problem. These authors give a list of further references for the reader interested in this problem.

[22] Classically, if the energy that would be gained by a free atom is much greater than the chemical binding energy of the molecule, the atom acts as if it were originally free. At lower energies the effects of binding make the atom appear heavier than it actually is. A useful approximation at very low energies is that the atom acts as if it had the mass of the entire molecule in which it is bound. Quantum mechanically, however, an atom may act as if it were free even if it does not gain enough energy to break the chemical bond so long as the energy levels are sufficiently close together to be representable by a continuum. This was first pointed out by R, E. MARSHAK.

[23] L. I. SCHIFF, *Quantum Mechanics*, New York, McGraw-Hill, 1957.

Consider a neutron with momentum p, traveling toward a nucleus. It is assumed that scattering will occur when the impact parameter b (see Fig. I.9) is of the order of the nuclear radius. Then for collision to occur the angular momentum must be of the order:

$$pb = [l(l + 1)]^{\frac{1}{2}} \hbar$$

$[l(l + 1)]^{\frac{1}{2}} \hbar$ being the quantum mechanical angular momentum. For p-wave scattering $l = 1$, and the nuclear radius is approximately $1.5 \times 10^{-13} A^{1/3}$ cm, so we have $E \sim 20\, A^{-2/3}$ MeV in the center-of-mass system. Thus, p-wave scattering tends to become important at lower energies for heavy nuclei and is negligible for many moderators. This argument applies only to potential scattering.

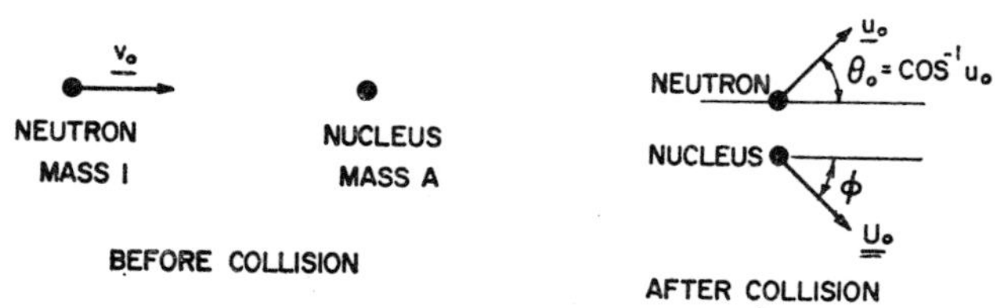

FIG. I.10. Kinematics of a neutron–nuclear collision. (Laboratory system)

21. Kinematics of Elastic Collisions

It is helpful to describe the collision of a neutron with a nucleus of mass number A in both the center-of-mass and laboratory coordinate systems. In particular, the relationships among the scattering angle in the laboratory system, $\cos^{-1} \mu_0$, the scattering angle in the center-of-mass system, $\cos^{-1} \mu_c$, and the energy transfer in the collision are to be found. The laboratory system (L) is that in which measurement can be made, that is, the system in which the nucleus may be assumed at rest while the neutron approaches it with a velocity $v_0 = (2E)^{\frac{1}{2}}$, if the neutron mass is taken to be unity (see Fig. I.10). After the scattering event, the neutron goes off at an angle $\cos^{-1}\mu_0 = \theta_0$ (the scattering angle in L) with speed u_0, after having imparted a speed U_0 to the nucleus, which goes off at angle φ.

In the center-of-mass system (C), on the other hand, the center of mass of nucleus–neutron system is assumed to be at rest (see Fig. I.11). Before the collision, the neutron is moving to the right with a speed $v_0 - V_M$, where V_M is the speed of the center-of-mass of the neutron–nuclear system relative to the laboratory system:

$$V_M = \frac{v_0}{A+1} \tag{I.43a}$$

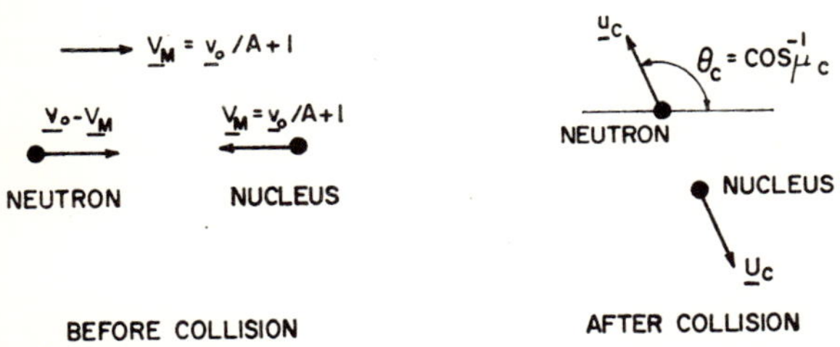

FIG. I.11. Kinematics of a neutron–nuclear collision. (Center-of-mass system)

The neutron speed in the center-of-mass system is thus:

$$v_0 - V_M = \frac{Av_0}{A+1} = v_c \tag{I.43b}$$

The nucleus moves to the left with a speed V_M. In the center-of-mass system the neutron and nucleus appear to scatter each other since they are both moving before the collision. After the collision the neutron has speed u_c and the nucleus speed U_c. The scattering angle in this coordinate system is $\cos^{-1}\mu_c = \theta_c$; the neutron and nucleus come off in opposite directions since the total momentum before the collision in C is

$$1 \cdot (v_0 - V_M) - AV_M = v_0\left[\left(1 - \frac{1}{A+1}\right) - \frac{A}{A+1}\right] = 0 \tag{I.43c}$$

and must, because of momentum conservation, remain zero after the collision. Clearly, this is possible only if the neutron and nucleus are moving in opposite directions after the collision, and furthermore only if

$$u_c = AU_c \tag{I.43d}$$

Energy conservation in C provides another relation between u_c and U_c:

$$\frac{1}{2}A\left(\frac{v_0}{A+1}\right)^2 + \frac{1}{2}\left(\frac{A\,v_0}{A+1}\right)^2 = \frac{1}{2}u_c^2 + \frac{1}{2}AU_c^2 \qquad \text{(I.43e)}$$

Upon solving the last two equations for u_c and U_c in terms of v_0 one finds that *in the center-of-mass system the speeds of neutron and nucleus are unchanged* by an elastic collision, that is

$$v_0 - V_M = u_c = \frac{Av_0}{A+1} \qquad \text{(I.43f)}$$

and

$$V_M = U_c = \frac{v_0}{A+1} \qquad \text{(I.43g)}$$

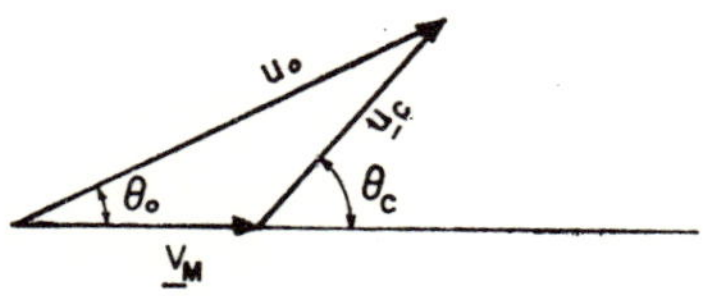

FIG. I.12. Relation between laboratory and center-of-mass system velocities.

We are now in a position to derive a relation between μ_0 and μ_c. Noting that the velocity of the neutron in L is the vector sum of its velocity in C and the velocity of the center-of-mass in L, one can write (see Fig. I.12)

$$u_0\mu_0 = V_M + u_c\mu_c \qquad \text{(I.44a)}$$

Substituting from Equations (I.43f) and (I.43g) for u_c and V_M:

$$\left(\frac{u_0}{v_0}\right)\mu_0 = \frac{1}{A+1} + \frac{A}{A+1}\mu_c \qquad \text{(I.44b)}$$

The ratio u_0/v_0 may also be obtained by applying the law of cosines to Fig. I.12:

$$u_0^2 = u_c^2 + V_M^2 + 2u_c\,V_M\,\mu_c \qquad \text{(I.44c)}$$

or, using Equations (I.43f) and (I.43g) again:

$$\frac{u_0^2}{v_0^2} = \frac{A^2 + 2A\mu_c + 1}{(A + 1)^2} \tag{I.44d}$$

Equations (I.44b) and (I.44d) yield the desired relation between μ_0 and μ_c:

$$\mu_0 = \frac{A\mu_c + 1}{(A^2 + 2A\mu_c + 1)^{\frac{1}{2}}} \tag{I.44e}$$

or, in terms of $\gamma = A^{-1}$,

$$\mu_0 = \frac{\mu_c + \gamma}{(\gamma^2 + 2\gamma\mu_c + 1)^{\frac{1}{2}}} \tag{I.44f}$$

Next, consider the energy change in scattering. In L, the ratio of the energy after collision, E, to that before the collision, E', is just u_0^2/v_0^2, already given by Equation (I.44d):

$$\frac{E}{E'} = \frac{\gamma^2 + 2\gamma\,\mu_c + 1}{(\gamma + 1)^2} \tag{I.44g}$$

or, in terms of a more convenient parameter:

$$\alpha \triangleq \left(\frac{1 - \gamma}{1 + \gamma}\right)^2 = \left(\frac{A - 1}{A + 1}\right)^2 \tag{I.44h}$$

the ratio of the energies is

$$\frac{E}{E'} = \frac{1}{2}\,[(1 + \alpha) + (1 - \alpha)\mu_c] \tag{I.44i}$$

Note that Equation (I.44i) implies that $\alpha E' \leqq E \leqq E'$, since $-1 \leqq \mu_c \leqq 1$. Thus, a neutron cannot be speeded up and cannot lose more than a fraction $(1 - \alpha)$ of its energy in a single elastic collision. The parameter α is tabulated in Table 1, Chapter II.

It is convenient to introduce a new variable, the *lethargy* (logarithmic energy), u, defined by

$$u = \ln(E_0/E); \quad E = E_0 e^{-u} \tag{I.45a}$$

(E_0 is some convenient reference energy, generally chosen to be 10 MeV), for the above results become somewhat simpler. For example, Equation (I.44i) becomes:

$$e^{-U} = \tfrac{1}{2}[(1 + \alpha) + (1 - \alpha)\mu_c] \tag{I.45b}$$

where

$$U = u - u' \tag{I.45c}$$

is the lethargy gained by the neutron in the collision. Inverting and writing the result in terms of γ, we have

$$\mu_c(U) = 1 - \frac{(1 + \gamma)^2}{2\gamma}(1 - e^{-U}) \tag{I.45d}$$

Combining this with Equation (I.44f), one finds

$$\mu_0(U) = \frac{1}{2\gamma}\left[(1 + \gamma)\,e^{-U/2} - (1 - \gamma)\,e^{U/2}\right] \tag{I.45e}$$

Here, μ_0 and μ_c have been written explicitly as functions of the lethargy change U to call attention to the fact that *the angle of scattering is uniquely related to the lethargy change* (gain) in a collision. For some purposes later in the development it will be convenient to consider μ_0 as the independent variable; at other times the choice of U will be more convenient. The maximum and minimum possible neutron energy transfer place restrictions on minimum and maximum lethargy change:

$$0 \leqq U \leqq ln\, 1/\alpha \tag{I.45f}$$

or

$$u - ln\, 1/\alpha \leqq u' \leqq u \tag{I.45g}$$

22. Transformation of Scattering Cross Sections

In deriving the scattering frequency, it is, of course, highly desirable to express the result in terms of the most basic data available, namely, the differential scattering cross section.† The relationship between the energy transfer in a scattering collision and the scattering angles was derived in the last section; the probability of a scattering angle occurring, which is proportional to the differential cross section, is now needed. However, cross sections are frequently presented in the center-of-mass system, whereas one

† The differential microscopic cross section, defined only for scattering, is the area presented by a nucleus to a neutron for scattering into an element of solid angle $d\boldsymbol{\Omega}$ located about $\boldsymbol{\Omega}$, i.e., $d\sigma/d\boldsymbol{\Omega}$.

would like to solve neutron moderation problems in the laboratory system. The conversion from one system to another can be carried out by the following method [24].

First, we define the $\sigma_s^0(u, \mu_0)\, d\mu_0$ as the cross section for scattering of neutrons into solid angle $2\pi\, d\mu_0$ about μ_0 in the L system, and $\sigma_s(u, \mu_c)\, d\mu_c$ as the cross section for scattering into $2\pi\, d\mu_c$ in the C system. Since the number of neutrons scattered into a unit solid angle cannot depend on the choice of coordinate system:

$$\sigma_s^0(u, \mu_0)\, d\mu_0 = \sigma_s^c(u, \mu_c)\, d\mu_c \qquad \text{(I.46a)}$$

where μ_0 and μ_c are related by Equation (I.44e). The quantity previously called the microscopic scattering cross section, $\sigma_s(u)$, is the integral of $\sigma_s^c(u', \mu_c)$ over all solid angle or, equivalently, the integral of $\sigma_c^0(u', \mu_0)$ over solid angle:

$$\sigma_s(u) = \int d\mu_c\, \sigma_s^c(u, \mu_c) = \int d\mu_0\, \sigma_s^0(u, \mu_0) \qquad \text{(I.46b)}$$

Both σ_s^c and σ_s^0 may be expanded in terms of Legendre polynomials of the cosines of the scattering angles. For the L system:

$$\sigma_s^0(u, \mu_0) = \sum_{L=0}^{\infty} \frac{2L+1}{2} B_L^0(u)\, P_L(\mu_0) \qquad \text{(I.46c)}$$

while for the C system

$$\sigma_s^c(u, \mu_c) = \sum_{L=0}^{\infty} \frac{2L+1}{2} B_L^c(u)\, P_L(\mu_c) \qquad \text{(I.46d)}$$

The expansion coefficients are readily found from the orthogonality properties of the Legendre polynomials:

$$\int_{-1}^{1} P_L(\mu)\, P_{L'}(\mu)\, d\mu = \frac{2}{2L+1}\, \delta_{LL'} \qquad \text{(I.46e)}$$

Multiplying Equations (I.46c) and (I.46d) by $P_L(\mu)$ and integrating over μ from -1 to 1, we have:

$$B_L^0(u) = \int_{-1}^{1} d\mu_0\, \sigma_s^0(u, \mu_0)\, P_L(\mu_0) \qquad \text{(I.46f)}$$

and

$$B_L^c(u) = \int_{-1}^{1} d\mu_c\, \sigma_s^c(u, \mu_c)\, P_L(\mu_c) \qquad \text{(I.46g)}$$

[24] The ensuing development follows P. F. Zweifel and H. Hurwitz, Jr., *J. Appl. Phys.* **25**, 1241 (1954).

Since $P_0(\mu) = 1$, it is also evident that

$$\sigma_s(u) = B_0^0(u) = B_0^c(u)$$

It is possible to interrelate these expansion coefficients by use of the transformation law (I.46a). Insertion of (I.46a) into (I.46f) yields

$$B_L^0(u) = \int_{-1}^{1} \mathrm{d}\mu_c\, \sigma_s^c(u, \mu_c)\, P_L(\mu_0) \tag{I.47a}$$

which, upon insertion of the Legendre polynomial expansion of $\sigma_s^c(u, \mu_c)$ (I.46d), becomes

$$B_L^0(u) = \int_{-1}^{1} \mathrm{d}\mu_c \left[\sum_{L'=0}^{\infty} \frac{2L'+1}{2} B_{L'}^c(u)\, P_{L'}(\mu_c)\right] P_L(\mu_0) \tag{I.47b}$$

This equation assumes a simpler form if one defines a matrix $||T||$, with elements $T_{LL'}$:

$$T_{LL'} = \frac{2L'+1}{2} \int_{-1}^{1} \mathrm{d}\mu_c\, P_{L'}(\mu_c)\, P_L(\mu_0) \tag{I.47c}$$

for then we have

$$B_L^0(u) = \sum_{L'=0}^{\infty} T_{LL'} B_{L'}^c(u) \tag{I.47d}$$

That is, the expansion coefficients of the differential cross section in L are the elements of the column vector resulting from multiplication of the matrix $||T||$ and the column vector formed from the expansion coefficients of the differential cross section in C. In an analogous manner, one may obtain the inverse transformation, i.e., the one which gives the $B_L^c(u)$ in terms of the $B_L^0(u)$. To do this one need only substitute (I.64a) and (I.46c) successively into (I.46f) to obtain the results

$$B_L^c(u) = \sum_{L'=0}^{\infty} T_{LL'}^{-1} B_L^0(u) \tag{I.47e}$$

with the elements of the matrix $||T||^{-1}$ given by†

$$T_{LL'}^{-1} = \frac{2L'+1}{2} \int_{-1}^{1} P_L(\mu_c)\, P_L'(\mu_0)\, \mathrm{d}\mu_0 \tag{I.47f}$$

† The superscript "-1" will always denote "inverse" and is not to be confused with the $T_{L'L}^n (n \geqq 0)$ introduced later.

The matrices $\|T\|$ and $\|T\|^{-1}$ are clearly inverses since successive applications of them to a cross section produces the original cross section.† Thus:

$$\sum_{L'=0}^{\infty} T_{LL'}\, T_{L'M}^{-1} = \delta_{LM} \tag{I.47g}$$

It might be noted that these results may be extended from the case of elastic scattering of neutrons from nuclei, discussed here, to the case of any nuclear reaction (including inelastic scattering) by redefinition of γ and U [25]:

$$\gamma = \left(\frac{m_1 m_3}{m_2 m_4}\, \frac{E_c}{E_c + Q}\right)^{\frac{1}{2}}$$

and U is taken to be the logarithm of the ratio of the energy of particle m_3 to the energy it would have if it were ejected in the forward direction. Here, a particle of mass m_1 strikes a particle of mass m_2 yielding products of mass m_3 and m_4 with a reaction energy Q (which may be either positive or negative). E_c, the initial total kinetic energy of the particles in the center-of-mass system, is related to E, the laboratory energy of particle 1, by

$$E_c = m_1 E/(m_1 + m_2)$$

if particle 2 is initially at rest. Note that for elastic scattering $Q = 0$ and no mass changes occur, so γ becomes the ratio of the neutron mass to the mass of the scattering nucleus.

23. Derivation of the Elastic Scattering Frequency

Having developed the basic materials, we now attempt to exhibit the elastic scattering frequency $F(E' \to E, \mathbf{\Omega}' \to \mathbf{\Omega})$ or $F(E' \to E, \mu_0)$ explicitly in terms of basic cross sections. A slight extension of the work of the previous section is needed as lethargy changes, which are not described by the formalism just developed, must be taken into account. We have seen that the variable U is spurious if μ_0 is specified since its value is uniquely determined by μ_0. It is, however, convenient to use both U and μ_0 as variables (one of them being a dummy) in order to carry out the transformation from μ_0, an

† This result also follows directly from the definitions of T_{LL} and T_{LL}^{-1} and the orthogonality properties of the Legendre polynomials.

[25] See SCHIFF, *op. cit.*

observable, to U, the variable required for slowing-down calculations.

One can write

$$\sigma_s(\boldsymbol{\Omega}' \to \boldsymbol{\Omega}, u' \to u) = \sigma_s(\mu_0, u', U) \tag{I.48a}$$

or

$$F(\boldsymbol{\Omega}' \to \boldsymbol{\Omega}, u' \to u) = \frac{\sum N_i \sigma_s^{0i}(\mu_0, u', U)}{\Sigma_s(u')} \tag{I.48b}$$

where N_i is the number of nuclei of species i per cm^3 and $\sigma_s^i(\mu_0, u', U)$ is the differential scattering cross section in L for the ith species and the spurious variable U has been introduced. Since U is spurious, $\sigma_s^{0i}(\mu_0, u', U)$ may be written

$$\sigma_s^{0i}(\mu_0, u', U) = \sigma_s^{0i}(\mu_0, u')\, \delta[\mu_0(U) - \mu_0]\left(-\frac{\partial \mu_0(U)}{\partial U}\right) \tag{I.48c}$$

Here $\delta[\mu_0(U) - \mu_0]$ is the Dirac delta function [26]. Equation (I.48c) expresses the fact that $\sigma_s^{0i}(\mu_0, u', U)$ vanishes unless μ_0 and U are related by Equation (I.45e). Then the relation:

$$\sigma_s^{0i}(\mu_0, u') = \int_0^{\ln 1/\alpha} \sigma_s^{0i}(\mu_0, u', U)\, \mathrm{d}U \tag{I.48d}$$

obtains. The negative sign in front of $\partial\mu_0(U)/\partial U$ in Equation (I.48c) arises from the fact that the derivative [see Eqn. (I.45e)] is always negative.

Next, we introduce the Legendre polynomial expansion of $\sigma_s^0(\mu_0, u, U)$, which is similar to Equation (I.46c) but contains the spurious variable U. For convenience, the superscript 'i' is dropped for the time being, but it is to be remembered that σ^0, σ^c, $B_L^0(u')$, $B_L^0(u', U)$, α, γ, and the $\|T\|$ matrices are all functions of the species of scattering nucleus being considered:

$$\sigma_s^0(\mu_0, u', U) = \sum_{L=0}^{\infty} \frac{2L+1}{2} B_L^0(u', U)\, P_L(\mu_0) \tag{I.49a}$$

where the orthogonality of the $P_L(\mu_0)$ gives, as before:

$$B_L^0(u', U) = \int \mathrm{d}\mu_0\, \sigma_s^0(\mu_0, u', U)\, P_L(\mu_0) \tag{I.49b}$$

[26] For properties of this function see, for example, P. A. M. Dirac, *The Principles of Quantum Mechanics*, Oxford, 1957.

The above expression for $B_L^0(u, U)$ can be evaluated by use of Equation (I.48c):

$$B_{L'}^0(u', U) = \int_{-1}^{1} d\mu_0\, \sigma_s^0(\mu_0, u')\, P_{L'}(\mu_0)\, \delta(\mu_0(U) - \mu_0) \left(\frac{-\partial\mu_0(U)}{\partial U}\right)$$

$$= \sigma_s^0(\mu_0(U), u')\, P_{L'}(\mu_0(U)) \left(\frac{-\partial\mu_0(U)}{\partial U}\right) \qquad \text{(I.49c)}$$

Finally, the $B_L^0(u', U)$ can be found in terms of the $B_L^c(u')$. First, substitute the expansion (I.46d) into Equation (I.46a) and express μ_0 and μ_c in terms of U to obtain

$$\sigma_s^0(\mu_0(U), u') = \sum_{L=0}^{\infty} \frac{2L+1}{2} B_L^c(u') P_L(\mu_c(U)) \left(\frac{\partial\mu_c}{\partial\mu_0}\right) \qquad \text{(I.49d)}$$

Then substitute Equation (I.49d) into Equation (I.49c) and write the results as

$$B_L^0(u', U) = \sum_{L'=0}^{\infty} T_{LL'}(U)\, B_{L'}^c(u') \qquad \text{(I.50a)}$$

where the $T_{LL'}(U)$ are the elements of a new transformation matrix given by

$$T_{LL'}(U) = \frac{2L'+1}{2} P_L(\mu_0(U)) P_{L'}(\mu_c(U)) \left(\frac{-\partial\mu_c(U)}{\partial\mu}\right) \qquad \text{(I.50b)}$$

Finally, substitution of Equation (I.50a) into Equation (I.49a) gives the desired laboratory system scattering function. Explicitly,

$$\sigma_s^0(\mu_0, u', U) = \sum_{L, L'=0}^{\infty} T_{LL'}(U) \frac{2L+1}{2} P_L(\mu_0)\, B_{L'}^c(u')$$

$$= \sum_{L, L'=0}^{\infty} \frac{2L+1}{2} \frac{2L'+1}{2} P_L(\mu_0)\, P_L\left[\frac{1}{2\gamma}[(1+\gamma)e^{-U/2} - (1-\gamma)e^{U/2}]\right]$$

$$P_{L'} \times \left[1 - \frac{(1+\gamma)^2}{2\gamma}(1 - e^{-U})\right] \left[\frac{(1+\gamma)^2}{2\gamma} e^{-U}\right] B_{L'}^c(u') \qquad \text{(I.50c)}$$

The last square bracket is equal to $(-\partial\mu_c/\partial U)$ and was obtained by differentiating Equation (I.45d). It is interesting to calculate the angularly independent scattering function. Integrating Equation (I.50c) over all μ_0, we find:

$$\sigma_s(u' \to u) = \sum_{L'=0}^{\infty} T_{0L'}(U)\, B_{L'}^c(u') \qquad \text{(I.50d)}$$

The $B_L^c(u')$, it may be remembered, are the Legendre coefficients of the differential cross section in the center-of-mass system. Since in the present section it has been shown that the lethargy change and the scattering angle are uniquely related by the kinematics of the collision, one might ask why it was necessary to introduce the angular dependent cross section at all. The reason is that the angular differential cross section is simpler to measure (and, in cases in which no measurements are available, to calculate) than is the energy or lethargy change differential cross section. But, because the Boltzmann equation requires that we deal with energy or lethargy variables rather than angular variables, the formalism of the present section has been introduced.

It may be noted that the relation between the $T_{LL'}(U)$ of this section and the $T_{LL'}$ of the previous section is:

$$T_{LL'} = \int_0^{\ln\frac{1}{\alpha}} T_{LL'}(U)\,\mathrm{d}U \tag{I.50e}$$

An inverse matrix $||T(U)||^{-1}$ is of little value and has, therefore, not been defined.

24. The Case of Isotropic Scattering

In many cases, elastic scattering is entirely *s*-wave and is therefore isotropic in the center-of-mass system. For this case, $\sigma_s^c(u', \mu_c)$ reduces to a single term, namely the coefficient of $P_0(\mu_c) = 1$, i.e.,

$$B_{L'}^c(u') = \sigma_s(u')\delta_{L'0} \tag{I.51a}$$

so that from Equation (I.50a) it is seen that

$$B_L^0(u', U) = T_{L0}(U)\,\sigma_s(u') \tag{I.51b}$$

where the T_{L0} (U), as given by Equation (I.50b) are

$$T_{L0}(U) = \tfrac{1}{2}\,P_L(\mu_0(U))\frac{(1+\gamma)^2}{2\gamma}\,\mathrm{e}^{-U} \tag{I.51c}$$

Combining the last two equations, one obtains

$$B_L^0(u', U) = \frac{(1+\gamma)^2}{4\gamma}\mathrm{e}^{-U}\,P_L\left[\frac{1}{2\gamma}[(1+\gamma)\,\mathrm{e}^{-U/2} - (1-\gamma)\,\mathrm{e}^{U/2}]\right]\sigma_s(u') \tag{I.51d}$$

and finally, the scattering frequency is

$$\sigma_s^0(\mu_0, u', U) = \sigma_s(u') \sum_{L=0}^{\infty} \frac{2L+1}{2} \frac{(1+\gamma)^2}{4\gamma} e^{-U} \times$$

$$P_L \times \left[\frac{1}{2\gamma}[(1+\gamma)\,e^{-U/2} - (1-\gamma)\,e^{U/2}]\right] P_L(\mu_0) \quad \text{(I.51e)}$$

The function, $\sigma_s(u' \to u)$ becomes, for isotropic scattering,

$$\sigma_s(u' \to u) = \sigma_s(u') \frac{(1+\gamma)}{4\gamma} e^{-U} = \frac{\sigma_s(u')\,e^{-U}}{1-\alpha} \quad \text{(I.51f)}$$

or, in terms of energy,

$$\sigma(E' \to E) = \frac{\sigma_s(E')}{(1-\alpha)\,E'} \quad \text{(I.51g)}$$

Note that in this case, and this case only,

$$F(u' \to u) = \frac{\sigma_s(u' \to u)}{\sigma_s(u')} \quad \text{(I.51h)}$$

is a function of $(u - u')$ only. Also note that

$$F(E' \to E) = \frac{\sigma_s(E' \to E)}{\sigma_s(E')}$$

is independent of E in this case, i.e., the energy distribution is flat in the energy range $\alpha E' < E < E'$.

E. Evaluation and Significance of the Transformation Matrices

25. Evaluation of some of the Matrix Elements

Before attempting to discuss the physical significance of some of the matrix elements derived in the previous section, some of these elements will be evaluated explicitly so that they will be available when their significance is discussed.

The simplest way to evaluate the integrals for the $T_{LL'}$ and $T_{LL'}^{-1}$ is to substitute the expressions for $\mu_0(U)$ and $\mu_c(U)$, given by Equations (I.45d, e) into Equation (I.47c) or Equation (I.47f) directly, and then perform the integrations over U. This reduces all the integrands to sums of exponential functions, that is,

$$T_{LL'} = \frac{2L'+1}{2} \int_0^{\ln 1/\alpha} P_L(\mu_0(U))\, P_{L'}(\mu_c(U)) \left|\frac{\partial \mu_c}{\partial U}\right| \mathrm{d}U \quad \text{(I.52a)}$$

$$T^{-1}_{LL'} = \frac{2L'+1}{2} \int_0^{\ln 1/\alpha} P_L(\mu_c(U))\, P_{L'}(\mu_0(U)) \left|\frac{\partial \mu_0}{\partial U}\right| \mathrm{d}U \quad \text{(I.52b)}$$

where the arguments of the Legendre polynomials are given by Equations (I.45d) and (I.45e). The results of some of these integrations are given below [27].

$$T_{0L'} = \delta_{0L'} \qquad T_{20} = \frac{5\gamma^2 - 3}{8\gamma^2} - \frac{3 - 6\gamma^2 + 3\gamma^4}{32\gamma^3} \ln\alpha$$

$$T_{10} = \frac{2\gamma}{3}$$

$$T_{11} = \frac{1}{5}(5 - 3\gamma^2) \quad T_{21} = \frac{9 - 6\gamma^2 + 9\gamma^4}{16\gamma^3} + \frac{9 - 9\gamma^2 - 9\gamma^4 + 9\gamma^6}{6 + \gamma^4} \ln\alpha \quad \text{(I.53a)}$$

Some of the elements of the inverse matrix $\|T\|^{-1}$ are

$$T^{-1}_{0L'} = \delta_{0L'}$$

$$T^{-1}_{10} = -2\gamma/3 \quad \text{(I.53b)}$$

$$T^{-1}_{11} = \frac{3}{8\gamma^2}(1 + \gamma^2) - \frac{3(1+\gamma)^2(1-\gamma)^2}{32\gamma^3} \ln\alpha$$

For small values of γ, i.e., for a heavy scattering nucleus, $T_{LL'}$ and $T^{-1}_{LL'}$ may be represented by the following expansions, correct to order γ:

$$T_{LL'} = \delta_{LL'} - \gamma \left[\frac{L'(L'-1)}{2L'-1} \delta_{L',L+1} - \frac{(L'+1)(L'+2)}{2L'+3} \delta_{L',L-1}\right] \quad \text{(I.54a)}$$

and

$$T^{-1}_{LL'} = \delta_{LL'} + \gamma \left[\frac{L'(L'-1)}{2L'-1} \delta_{L',L+1} - \frac{(L'+1)(L'+2)}{2L'+3} \delta_{L',L-1}\right] \quad \text{(I.54b)}$$

[27] These results are largely taken from Zweifel and Hurwitz, *op. cit.*

Amster [28] has derived a general expression for $T_{1L'}$

$$T_{1L'} = \frac{L'}{2L'-1}(-\gamma)^{L'-1} - \frac{L'+2}{2L'+3}(-\gamma)^{L'+1} \tag{I.55a}$$

and a recursion relation:

$$T_{L+1,L'} = (2L'+1)\left(\frac{2L+1}{L+1}\right)\sum_{j=0}^{\infty} T_{1j} \sum_{m=|j-L'|}^{|j+L'|} (jL'\,00|j\,L'm0)^2 \times \\ \times \frac{T_{Lm}}{2m+1} - \frac{L}{L+1}T_{L-1,L'} \tag{I.55b}$$

where the $(jL'00|jL'm0)$ are Clebsch–Gordon coefficients [29].†

For later applications, in Chapter III, it will be convenient to define still another set of matrices,

$$T^n_{LL'} = \frac{1}{n!}\int_0^{\ln 1/\alpha} (-U)^n\, T_{LL'}(U)\, dU \tag{I.56}$$

Note that $T^0_{LL'} = T_{LL'}$, according to Equation (I.50e). Using straightforward integration with $T_{LL'}(U)$ written in terms of exponentials, Equation (I.50b), one obtains:

$$T^1_{00} = -\left(1 - \frac{(1-\gamma)^2}{4\gamma}\ln\alpha\right) \to -2\gamma \tag{I.57a}$$

$$T^1_{01} = \frac{3+3\gamma^2}{4\gamma} - \frac{3-6\gamma^2+3\gamma^4}{16\gamma^2}\ln\alpha \to 2\gamma \tag{I.57b}$$

where the arrows indicate limits valid for small γ. $T^1_{0L'}$, for $L' > 1$, vanishes to order γ.

Amster [30] has also derived expansions for $T^1_{0L'}$ and $T^2_{0L'}$:

$$T^1_{0L'} = (-)^{L'+1}(2L'+1)\sum_{p=L'+\delta_{0L'}}^{\infty} A_{pL'}(1-\alpha)^p \tag{I.58a}$$

$$T^2_{0L'} = (-)^{L'}(2L'+1)\sum_{p=L'+\delta_{1L'}+2\delta_{0L'}} A_{pL'}\left[\sum_{j=1}^{p}\frac{1}{j}\right](1-\alpha)^p$$

† This relation also holds for $T^n_{LL'}$ if all the $T_{LL'}$ except T_{1j} in Equation (I.55 b) are replaced by $T^n_{LL'}$.

[28] H. AMSTER, *J. Appl. Phys.* **27**, 307 (1956).
[29] E. U. CONDON and C. SHORTLEY, *Theory of Atomic Spectra*, Cambridge Univ. Press.
[30] H. AMSTER, *op. cit.*

where
$$A_{pL'} = \frac{(p!)^2}{p(p-L')!\,(p+L'+1)!} \tag{I.58b}$$

and α is the usual
$$\alpha = \left(\frac{1-\gamma}{1+\gamma}\right)^2$$

Furthermore, it was shown by Amster [31] that the T^n_{LL} approach zero as γ goes to zero and $T^n_{LL'}$ is bounded by the lesser of γ_n and $\gamma^{|L-L'|}$ for small γ.

26. Moments of the Scattering Frequency

The significance of the matrix elements is readily determined. For example, consider $\xi_i = \overline{U}_i$, the average lethargy gain per collision of a neutron with a nucleus of species i, which is given by

$$\xi_i = \frac{\int d\mu_0 \int dU\,\sigma_s^{0i}(\mu_0, u', U)\,U}{\int d\mu_0 \int dU\,\sigma_s^{0i}(\mu_0, u', U)} \tag{I.59a}$$

From the expansion (I.42c) and the orthogonality properties of the Legendre polynomials, this reduces to

$$\xi_i = \frac{\int dU\,UB_0^0(u', U)}{\int dU\,B_0^0(u', U)}$$

which can, in turn, be written in terms of the expansion coefficients in the center-of-mass system by use of the transformation (I.50a)

$$\xi_i = \frac{\sum\limits_{L=0}^{\infty} B_L^{ci}(u') \int\limits_0^{\ln 1/\alpha} dU\,UT_{0L}(U)}{\sum\limits_{L=0}^{\infty} B_L^{ci}(u') \int\limits_0^{\ln 1/\alpha} dU\,T_{0L}(U)}$$

The integral in the numerator is just $-T_{0Li}$ while that in the denominator is $T_{0Li} = \delta_{0L}$. Remembering that $B_0^c(u) = \sigma_s(u)$ we have finally:

$$\xi_i = \frac{-\sum\limits_{L=0}^{\infty} T^1_{0Li}\,B_L^{ci}(u)}{\sigma_s^i(u)} \tag{I.59b}$$

[31] H. Amster, *J. Appl. Phys.* **29**, 623 (1958).

For isotropic scattering in C, an interesting result is that ξ_i is not a function of lethargy. The sum in the numerator of (I.59b) reduces to a single term, $T^1_{00}B^c_0(u')$, so ξ_i becomes simply

$$\xi_{iso\,i} = -T'_{00_i} = 1 - \frac{(1-\gamma)^2}{4\gamma}\ln\alpha \rightarrow 2\gamma \qquad \text{(I.59c)}$$

which shows that ξ_i is independent of neutron energy in this case. For non-isotropic scattering, however, the average lethargy change generally depends on u. The limit 2γ for ξ_{iso} given above is correct to order γ, while to order γ^2

$$\xi_{iso\,i} = \frac{2\gamma}{1 + 2/3\gamma} = \frac{2}{A + 2/3} \qquad \text{(I.59d)}$$

For hydrogen $\gamma = 1$ and $\alpha = 0$ and in this limit Equation (I.59c) yields $\xi = 1$, a useful result, as scattering from hydrogen is isotropic at energies of interest in reactor calculations.

Another useful quantity is the average value of the cosine of the laboratory scattering angle,

$$\langle\mu_0(u')\rangle = \frac{\int \mathrm{d}\mu_0 \int \mathrm{d}U\, \mu_0\, \sigma^0_s(\mu_0, u', U)}{\int \mathrm{d}\mu_0 \int \mathrm{d}U\, \sigma^0_s(\mu_0, u', U)} \qquad \text{(I.60a)}$$

By following a procedure similar to that above, this can be shown to be

$$\langle\mu_0(u')\rangle_i = \frac{\int \mathrm{d}U\, B^{0i}_1(u', U)}{\sigma^i_s(u')} \qquad \text{(I.60b)}$$

or in terms of the expansion coefficients in C:

$$\langle\mu_0(u')\rangle_i = \frac{\sum\limits_{L'=0}^{\infty} T_{1L'_i}\, B^{ci}_{L'}(u')}{\sigma^i_s(u')} \qquad \text{(I.60c)}$$

For isotropic scattering in C, $\langle\mu\rangle$ is also not a function of neutron energy, and is exactly

$$\langle\mu_0(u)\rangle_i = T_{10_i} = \frac{2\gamma}{3} \qquad \text{(I.60d)}$$

A relation between $\xi(u')$ and $\langle\mu(u')\rangle$ may be found to some approximation. By inverting Equation (I.45e) and expanding in a power series, one has:

$$U = 2\gamma\left[1 - \mu_0 + \frac{\gamma^2}{2}(1 - \mu_0)^2 + \cdots\right] \qquad \text{(I.61a)}$$

which, upon averaging, yields:

$$\xi_i = 2\gamma\left[1 - \langle\mu_0\rangle_i + \frac{\gamma^2}{2}\langle(1 - \mu_0)^2\rangle + \cdots\right] \qquad \text{(I.61b)}$$

Hence, to second order in γ:

$$T^1_{0L} = -2\gamma\,[T_{0L} - T_{1L}] = -2\gamma\,[\delta_{0L} - T_{1L}] \qquad \text{(I.61c)}$$

Expansions of $\xi_i\sigma^i_s$ and $\langle\mu_i\rangle\ \sigma^s_i$, to first order in γ (two quantities which will be very important later),

$$\xi_i\sigma^i_s = \sum_{L'=0}^{\infty} T^1_{0L'}\,B^c_{L'}(u') \approx 2\gamma\,(B^c_0 - B^c_1) \approx \xi_{iso}\sigma^i_{tr} \qquad \text{(I.62a)}$$

$$\langle\mu_i\rangle\sigma^i_s = \sum_{L'=0}^{\infty} T_{1L'}\,B^c_{L'}(u') \approx B^c_1 + \frac{2\gamma}{3}(B^c_0 - B^c_2) \qquad \text{(I.62b)}$$

where

$$\sigma^i_{tr} = \sigma^i_t - \sigma^i_s\langle\mu_i\rangle$$

Alternatively, ξ and $\langle\mu_i\rangle$ could be expressed in terms of the B^0_L rather than the B^c_L. For example, from Equation (I.60b) and Equation (I.50a) it follows immediately that

$$\langle\mu_i\rangle = \frac{B^{0i}_1(u)}{\sigma_{s_i}(u)} \qquad \text{(I.63a)}$$

which, when substituted into Equation (I.61b), yields

$$\xi_i = 2\gamma\left[1 - \frac{B^{0i}_1(u)}{\sigma_s(u)} + 0(\gamma^2)\right] \qquad \text{(I.63b)}$$

By the method employed for ξ and $\langle\mu\rangle$ above it is possible to find the more general moment:

$$\langle U^n P_L(\mu_0)\rangle_i = \frac{\int d\mu_0 \int dU\, U^n P_L(\mu_0)\, \sigma_s^{0i}(\mu_0, u', U)}{\int d\mu_0 \int dU\, \sigma_s^{0i}(\mu_0, u', U)} \tag{I.64a}$$

$$= \sum_{L'=0}^{\infty} (-)^n\, n!\, T^n_{LL'}\, B^{ci}_{L'}(u)/\sigma_{s_i}(u)$$

For isotropic scattering this becomes simply

$$\langle U^n P_L(\mu_0)\rangle_i = T^n_{L0_i} \tag{I.64b}$$

CHAPTER II

SLOWING DOWN IN AN INFINITE MEDIUM

A. Introductory Remarks

1. Importance of the Infinite Medium Problem

In this chapter, slowing down in an infinite medium is discussed. It was stated in Chapter I that if sources are uniformly distributed in space and are isotropic, then the neutron angular flux is a function only of energy, $\varphi(E)$. The proof follows from various invariance properties of the Boltzmann equation [1]. Such a case is not as academic as one might at first think. Many reactors are so large that geometric effects are relatively unimportant; these systems display many of the characteristics of infinite systems. For example, in natural uranium–graphite reactors, which must be large to obtain criticality, leakage is of relatively minor importance and the major problems involve spectral calculations (treated in this chapter) and the effects of inhomogeneities, treated only briefly here. The same is true of large power reactors for which engineering considerations dictate large cores.

For other systems, the results derived here will be useful mainly for providing insight into the physical processes occurring. Most systems may be treated accurately only by the numerical methods discussed in Chapter IV. Here, the results obtained are useful aids for making reasonable first guesses as to the flux. Accurate guesses may greatly reduce the numerical work required and thus result in considerable savings in calculational time.

Finally, in some systems, the escape probability separates, to very good approximation, into the product of a non-leakage probability and a non-absorption (resonance escape) probability as

[1] See K. M. Case and P. F. Zweifel, *Neutron Transport Theory* (to be published).

was discussed in Chapter I. In such cases, the infinite medium problem is a significant aid in obtaining the resonance escape probablity.

2. The Boltzmann Equation for an Infinite Medium

The basic equation that we shall deal with in this chapter is, from the arguments cited in the previous section,

$$\Sigma_t(E)\,\varphi(E) = S(E) + \int_E^\infty \Sigma_s(E')\,\varphi(E')\,F(E' \to E)\,\mathrm{d}E' \qquad \text{(II.1)}$$

In terms of lethargy it becomes

$$\Sigma_t(u)\,\varphi(u) = S(u) + \int_0^u \Sigma_s(u')\,\varphi(u')\,F(u' \to u)\,\mathrm{d}u' \qquad \text{(II.2)}$$

Equations (II.1) and (II.2) are quite difficult to solve except for the simplest cases. Generally, the solution requires the use of approximate and/or numerical techniques. The difficulties may be circumvented in many cases of interest by noticing that elastic scattering is the dominant means of slowing down neutrons with energies of less than 100 keV. This, however, is precisely the range of energies through which the major part of the slowing-down process takes place and it is here that resonance absorption is most important. Then, the elastic scattering frequency developed in Chapter I will be applicable to the solution of Equations (II.2). Specification of individual scattering frequencies will be delayed, however, until such time as they are required.

B. Slowing Down in Hydrogeneous Media

3. A Rigorous Solution

A simple rigorous solution for Equation (II.2a) is possible when hydrogen is the only moderating material present. This does not preclude the possibility that absorbing materials which do not moderate ("infinitely heavy absorbers") may be present. This may seem an oversimplification, but is a reasonable approximation to light water–uranium systems. An important point is that scattering from hydrogen is isotropic in the center-of-mass systems up to several MeV, which includes almost all of the energy region of interest in reactors.

Assuming only a single heavy element, which for convenience is assumed to scatter isotropically (these conditions are not essential but simplify the calculations), Equation (II.2) may be written:

$$(\Sigma_a^M + \Sigma_s^M + \Sigma_a^H + \Sigma_s^H)_u \varphi(u) = \int_0^u \varphi(u')\, \Sigma_s^H(u')\, \mathrm{e}^{-(u-u')}\, \mathrm{d}u'$$

$$+ \int_{u-\ln 1/\alpha_M}^{u} \mathrm{d}u' \left[\Sigma_s^M(u')\, \varphi(u') \frac{\mathrm{e}^{-(u-u')}}{1-\alpha_M}\right] + S(u) \qquad \text{(II.3a)}$$

where the superscripts H and M refer to hydrogen and the heavy absorbing element respectively. As the atomic weight of the absorber approaches infinity, $\alpha = [(A-1)/(A+1)]^2$ approaches unity and

$$\lim_{\alpha \to 1} \int_{u-\ln 1/\alpha}^{u} \Sigma_s^M(u')\, \varphi(u') \frac{\mathrm{e}^{-(u-u')}}{1-\alpha}\, \mathrm{d}u' = \Sigma_s^M(u)\, \varphi(u)$$

Thus for the case of "infinitely heavy absorber" one has

$$(\Sigma_a^M + \Sigma_a^H + \Sigma_s^H)_u \varphi(u) = \int_0^u \Sigma_s(u')\, \varphi(u')\, \mathrm{e}^{-(u-u')} \mathrm{d}u' + S(u) \qquad \text{(II.3b)}$$

It was shown, by physical arguments, in Section I.15 that the slowing-down density $q(\mathbf{r}, E)$ is more slowly varying in energy than the flux. Therefore, much of the development in this and the following chapter will utilize the slowing-down density. From Equation (I.32b), remembering that the flux is independent of position, we have for the case of slowing down by hydrogen alone:

$$q(u) = \int_0^u \Sigma_s^H(u')\, \varphi(u')\, \mathrm{e}^{-(u-u')}\, \mathrm{d}u' \qquad \text{(II.4a)}$$

which, accidentally, happens also to be the integral term on the right-hand side of Equation (II.3b), the "degradation integral". A useful relation may be obtained by differentiating Equation (II.4a) and adding the result to itself:

$$q(u) + \frac{\partial q}{\partial u} = \Sigma_s^H(u)\, \varphi(u) \qquad \text{(II.4b)}$$

Equations (II.4a) and (II.4b), which are peculiar to hydrogen moderation, hold also in the space dependent case, since none of the arguments leading to them are peculiar to the infinite medium problem. Eliminating the degradation integral between Equations (II.3b) and (II.4a), one finds

$$q(u) = \Sigma_t(u)\,\varphi(u) - S(u) \tag{II.4c}$$

where

$$\Sigma_t(u) \triangleq \Sigma_s^H(u) + \Sigma_s^H(u) + \Sigma_a^M(u) \tag{II.4d}$$

We next define the *collision density*, $\chi(u)$, the number of collisions per unit lethargy involving neutrons with lethargies about u per second.

$$\chi(u) \triangleq \Sigma_t(u)\,\varphi(u) \tag{II.4e}$$

Elimination of $q(u)$ between Equations (II.4b) and (II.4c) yields a differential equation for $\chi(u)$:

$$\frac{d\chi(u)}{du} + c_0(u)\,\chi(u) = S(u) + \frac{dS}{du} \tag{II.5a}$$

where

$$c_0(u) \triangleq 1 - \frac{\Sigma_s^H(u)}{\Sigma_t(u)} = \frac{\Sigma_a(u)}{\Sigma_t(u)} \tag{II.5b}$$

Let $u = 0$ correspond to some energy above which no neutrons are born, i.e., $S(u) = 0$ for $u < 0$. Then integration of Equation (II.5a) is straightforward and yields

$$\chi(u) = \int_0^u S(u)\,c(u')\,e^{-\int_{u'}^{u} c_0(u'')du''}\,du' + S(u) \tag{II.6a}$$

where

$$c(u) \triangleq 1 - c_0(u) = \frac{\Sigma_s^H(u)}{\Sigma_t(u)} \tag{II.6b}$$

From the definition of the resonance escape probability (see Section I.17) $p(u)$ is the ratio of the number of neutrons reaching lethargy u to the total number which would reach it if no absorption took place:

$$p(u) = \frac{q(u)}{\int_0^u S(u')\,du'} = 1 - \frac{\int_0^u \Sigma_a(u')\,\varphi(u')\,du'}{\int_0^u S(u')\,du'} \tag{II.7a}$$

Or, for hydrogen

$$p(u) = 1 - \frac{\int_0^u c_0(u') \left[\int_0^u S(u'')\, c(u'')\, e^{-\int_{u''}^{u} c_0(u''')\, du} + S(u') \right] du'}{\int_0^u S(u')\, du'} \qquad \text{(II.7b)}$$

An often quoted form for p may be obtained if we let $S(u'')$ equa $\delta(u'' - u')$, a monoenergetic source of one neutron per second. The integral over u'' is then trivial and yields the result:

$$p(u' \to u) = c(u')\, e^{-\int_{u''}^{u} c_0(u'')du''} \qquad \text{(II.7c)}$$

In the limit that $c_0(u)$ vanishes $p(u' \to u) = 1$, as expected. At energies below the source energies, the collision density, slowing-down density and resonance escape probability are all identical:

$$\chi(u) = q(u) = p(u) \qquad \text{(II.8a)}$$

(assuming a unit source). Furthermore, at these energies:

$$\varphi(u) = \frac{q(u)}{\Sigma_t(u)} \qquad \text{(II.8b)}$$

The simplicity of the results, e.g., a collision density that is independent of u, is one reason for introducing the lethargy variable. To obtain φ in terms of the energy one sets

$$\varphi(E)\, dE = \varphi(u)\, du \qquad \text{(II.9a)}$$

so that

$$\varphi(E) = \frac{p(E)}{E\Sigma_t(E)} \qquad \text{(II.9b)}$$

In the limit of zero absorption, noting that the scattering cross section of hydrogen is constant between about 0.1 eV and 100 keV, one has

$$\varphi(E) = \frac{1}{E\Sigma_s} \qquad \text{(II.9c)}$$

the well known "$1/E$" spectrum.

An important equation may be derived by differentiating Equation (II.7a)

$$\frac{dq}{du} = S(u) - \Sigma_a(u)\, \varphi(u) \qquad \text{(II.9d)}$$

This equation and Equations (II.7a) from which it was derived, are valid in the space-independent case for any moderator, not just for the case of hydrogen moderation. Equation (II.9d) is a simple statement of neutron conservation, i.e., the slowing-down density is increased by the source and decreased by neutron absorption. For the space-dependent case, by the way, we must also subtract from the right-hand side of Equation (II.9d) the rate at which neutrons leak, which is simply the divergence of the current **J**:

$$\frac{\partial q(\mathbf{r}, u)}{\partial u} = S(\mathbf{r}, u) - \Sigma_a(\mathbf{r}, u)\,\varphi(\mathbf{r}, u) - \nabla \cdot \mathbf{J}(\mathbf{r}, u) \quad \text{(II.9e)}$$

We shall return to this point in Chapter III.

C. Slowing Down in Media Containing Heavy Elements: I. Non-absorbing Media

4. Neutron Slowing-down Spectra

For media containing materials of mass number greater than unity, exact solutions of Equation (II.2) cannot be obtained in terms of a function which is continuous at all energies. It can, in fact, be shown [2] that the only solution of Equation (II.2) that is continuous at all energies is a constant in lethargy, which is the asymptotic solution (far from sources) as we shall soon see. Considering the case of a single non-hydrogenous moderator, we note that a monoenergetic neutron source gives rise to once collided neutrons with energies between αE_0 and E_0; neutrons of energy less than αE_0 can arise only from multiple collisions. The first collision distribution is indeed continuous for $\alpha E_0 < E < E_0$; for isotropic scattering in the center-of-mass system, it was seen in Chapter I that this distribution is actually constant (in energy) over this energy range and zero elsewhere. Similarly, the twice-collided distribution is continuous and non-zero for $\alpha^2 E_0 < E < E_0$ and zero everywhere else. A careful analysis will show that the twice-collided distribution actually goes to zero at $\alpha^2 E_0$ and is therefore continuous there. However, its first derivative is readily shown to be discontinuous there.

[2] V. C. Boffi, *Annals of Physics* **9**, 435 (1960).

From the above argument, we see that the flux at energies just above αE_0 will be greater than that at energies just below αE_0 by the contribution to the flux of the first collided neutrons reaching energies just above αE_0, i.e., the flux is discontinuous at αE_0. Similarly, the nature of the twice-collided flux creates a discontinuity in the first derivative of the flux at $\alpha^2 E_0$. Continuation of this analysis shows that in general the nth derivative is discontinuous at $\alpha^{n+1} E_0$. Far from the source, this transient behavior damps out and the flux approaches a smooth asymptotic function (the $1/E$ spectrum). Hydrogen, of course, displays none of this behavior since a neutron may lose all of its energy in one collision. Finally, we remark that these discontinuities are properties of a monoenergetic source; for distributed sources, integration over source energies tends to remove the discontinuities, but there are still oscillations.

A monoenergetic sink, such as might arise from an idealized absorption resonance, will produce analogous results, except, of course, the transient will have to be subtracted from rather than added to the flux due to other sources and/or sinks. Fortunately, in most cases of practical interest, resonance absorption does not set in until some lethargy from the source. In these cases it is sufficient simply to study the effect of capture upon the asymptotic energy distribution arising from the sources. The fact that the rapidity with which the asymptotic distribution is reached decreases sharply with decrease in the mass number of the moderator (except in the anomalous case of hydrogen) makes deuterium moderation problems particularly difficult to treat.

5. Transient Solutions—Placzek Functions [3]

First, we shall derive solutions to Equation (II.2) near the source, i.e., the transient solutions, which, despite their rigor, have found relatively small practical application. This is largely due to the fact that there are no transients when hydrogen is the only moderator present, and that the transients usually die out before resonance absorption sets in beryllium or graphite moderated reactors. It is possible that these solutions will find increased application if heavy water reactors become more common. One application of

[3] G. Placzek, *Phys. Rev.* **69**, 423 (1946).

these transient solutions, called the "Plazcek functions", will be given later in this chapter.

Consider the case of pure heavy element moderation. Mixtures of elements may be treated by this method, but it usually suffices to take only the longest lived transient, i.e., that due to the lightest element (other than hydrogen) that may be present. A monoenergetic source is considered, but it is again remarked that the solution for this case is also the Green's function of the more general problem. Since the functional behavior of the flux is not continuous to all orders at the points $\alpha^n E_0$, $n = 1, 2, \ldots$, the flux itself must be represented by a different smooth function, which we denote by $\varphi_n(E)$ or $\varphi_n(u)$, on each energy interval $\alpha^{n-1}E_0 < E < \alpha^n E_0$ (or in lethargy $(n+1) \ln 1/\alpha > u > n \ln 1/\alpha$).

On the interval $0 < u < \ln 1/\alpha$ the equation will be slightly different from that on the succeeding intervals. Source neutrons collide to give a first collision lethargy distribution $e^{-u}/(1-\alpha)$ for $u < \ln 1/\alpha$; this must be added to the multiple collision contribution† so that Equation (II.2) becomes, on the first collision interval, ignoring absorption,

$$\chi_0(u) = \Sigma_s(u)\,\varphi_0(u) = \frac{e^{-u}}{1-\alpha} + \int_0^u \frac{e^{-(u-u')}}{1-\alpha}\,\chi_0(u')\,du' \qquad \text{(II.10a)}$$

where $\chi_n(u)$ is to represent the collision density for lethargies in the nth interval. Equation (II.10a), a linear Volterra Integral Equation of the second kind, can be solved by any of several methods, among them Laplace transforms and the iteration method used earlier. The simplest attack is probably to multiply Equation (II.10a) by e^u and differentiate, to obtain the differential equation

$$\frac{d\chi_0(u)}{du} - \frac{\alpha}{1-\alpha}\,\chi_0(u) = 0 \qquad \text{(II.10b)}$$

A suitable boundary condition is obtained from Equation (II.10a) by taking the limit as $u \to 0$:

$$\lim_{u\to 0} \chi_0(u) = \frac{1}{1-\alpha} \qquad \text{(II.10c)}$$

† The source could also be treated by substituting $\delta(u)$ for $e^{-u}/(1-\alpha)$ in Equation (II.10a). The solution is the same except for a term $\delta(u)$ appearing in $\chi_0(u)$.

The solution of (II.10b) with boundary condition (2.10c) is relatively straightforward:

$$\chi_0(u) = \frac{e^{(\alpha/1-\alpha)u}}{1-\alpha} \tag{II.10d}$$

Using the transformation law (II.9a), we have in terms of the energy:

$$\chi_0(E) = \frac{1}{(1-\alpha)E}\left(\frac{E_0}{E}\right)^{\alpha/1-\alpha} \tag{II.10e}$$

Note that this reduces to the exact solution for slowing down without absorption in hydrogen. Further, it shows that the deviation from the "$1/E$" spectrum in the first collision interval increases with increasing α or A but, of course, the first collision interval does not extend far in lethargy for heavy elements.

For the other $\chi_n(u)$ one must separate the contributions from the lethargy intervals $n \ln 1/\alpha < u' < u$ and $u - \ln 1/\alpha < u' < n \ln 1/\alpha$, thus arriving at the integral equation:

$$\chi_n(u) = \int_{n \ln 1/\alpha}^{u} \chi_n(u')\frac{e^{-(u-u')}}{1-\alpha}\,du' + \int_{n-\ln 1/\alpha}^{n \ln 1/\alpha} \chi_{n-1}(u')\frac{e^{-(u-u')}}{1-\alpha}\,du' \tag{II.11a}$$

which may be solved by the method used to find $\chi_0(u)$. Again multiplying be e^u and differentiating with respect to u, we have

$$\frac{d\chi_n(u)}{du} - \frac{\alpha}{1-\alpha}\chi_n(u) = \frac{-\chi_{n-1}(u-\ln 1/\alpha)}{1-\alpha} \tag{II.11b}$$

On introducing the integrating factor $e^{-(\alpha/1-\alpha)u}$, one has

$$e^{(\alpha/1-\alpha)u}\frac{d}{du}[\chi_n(u)\,e^{-(\alpha/1-\alpha)u}] = \frac{-\chi_{n-1}(u-\ln 1/\alpha)}{1-\alpha} \tag{II.11c}$$

which is readily integrated to yield

$$\chi_n(u) = \chi_n(n \ln 1/\alpha)\,e^{(\alpha/1-\alpha)(u-n\ln\alpha)} - \int_{n \ln 1/\alpha}^{u} \chi_{n-1}(u'-\ln 1/\alpha)\frac{e^{(\alpha/1-\alpha)(u-u')}}{1-\alpha}\,du' \tag{II.11d}$$

The value of $\chi_n(n \ln 1/a)$ may be obtained by setting $u = n \ln a$ in Equation (II.11a):

$$\chi_n(n \ln 1/\alpha) = \int_{(n-1)\ln 1/\alpha}^{n \ln 1/\alpha} \chi_{n-1}(u') \frac{e^{-(n \ln 1/\alpha - u')}}{1-\alpha} du' \equiv \chi_{n-1}(n \ln 1/\alpha) \quad \text{(II.11e)}$$

except when $n = 1$. For $n = 1$, the integral in Equation (II.11e) may be evaluated by means of Equation (II.10a) to give

$$\chi_1(\ln 1/\alpha) = \int_0^{\ln 1/\alpha} \chi_0(u') \frac{e^{-(\ln 1/\alpha - u')}}{1-\alpha} du' = \chi_0(\ln 1/\alpha) - \frac{\alpha}{1-\alpha} \quad \text{(II.11f)}$$

Thus, one has for $\chi_1(u)$:

$$\chi_1(u) = \frac{1-\alpha^{1/1-\alpha}}{1-\alpha} e^{(\alpha/1-\alpha)u} - \frac{\alpha^{(\alpha/1-\alpha)}}{(1-\alpha)^2}(u - \ln 1/\alpha)\, e^{(\alpha/1-\alpha)u} \quad \text{(II.11g)}$$

The functions $\chi_n(u)$ for $n > 1$ are readily obtained by iterating with Equation (II.11d) using the boundary condition (II.11e). They increase rapidly in complexity and are plotted as $(1 - \alpha)\chi(u)$ in Fig. II.1 for moderators of $A = 4$ (beryllium) and $A \to \infty$. One notes that this function has almost reached its asymptotic value at $E = \alpha^3 E_0$ which is rather close to the source energy except for deuterium. It is possible to calculate the asymptote using this formulation, but another, simpler, method can be used, and will be displayed in the next section. The deviation of $\chi(u)$ from the asymptotic value is shown in Fig. II.2.

6. Asymptotic Energy Distribution

For a non-absorbing medium the asymptotic energy distribution may be found quite straightforwardly by a method due to Marshak [4], which has the advantages of greater flexibility and, hence, wider applicability. First, the case of a single moderator will be considered as an illustration of the method and then the physically more interesting case of mixture will be discussed. Later, it will be applied to a simple case that includes capture. The method applies to hydrogen as well as any other moderator.

[4] R. E. Marshak, *Rev. Mod. Phys.* **19**, 185 (1947). That the Placzek functions can also be derived in this way was shown by T. Teichmann, *Nucl. Sci. Eng.* **7**, 292 (1960).

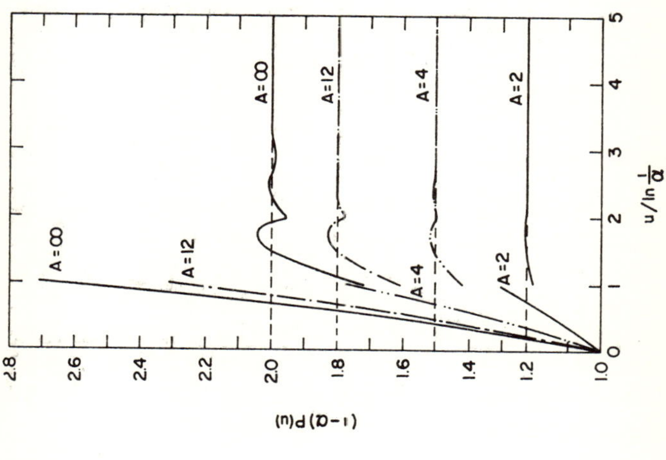

FIG. II.1. The Placzek function $P(u)$ as a function of lethargy. After Weinberg and Wigner, *op. cit.*

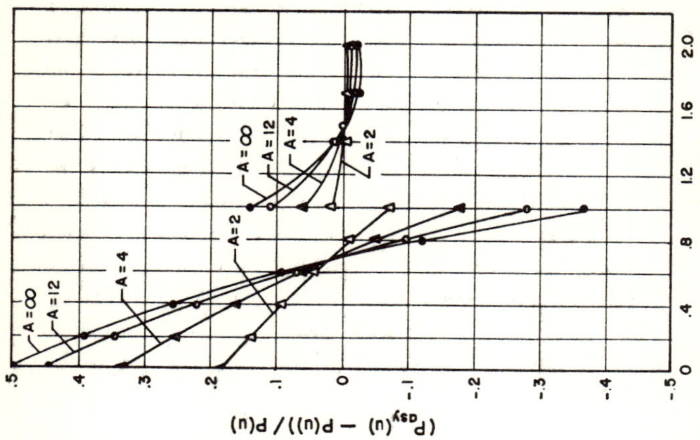

FIG. II.2. The deviation of the Placzek function from its asymptotic value. After Weinberg and Wigner, *op. cit.*

Again we consider only a non-absorbing medium in which isotropic elastic scattering is the only means of slowing down. (Anisotropic scattering is allowed by this method only if the anisotropy is the same at all energies.) Since we will be interested in the behavior of χ far from the source, we let $S(u) = \delta(u)$. Thus the equation:

$$\Sigma_s(u)\,\varphi(u) = \int\limits_{u-\ln 1/\alpha}^{u} \Sigma_s(u')\,\varphi(u')\,F(u' - u)\,\mathrm{d}u' + \delta(u) \quad \text{(II.12a)}$$

is the starting point. The fact that $F(u' \to u)$ is a function only of $(u - u')$ has been used explicitly. The Laplace transform method [5] may be applied when it noted that the lower limit of the integral in Equation (II.12a) could be taken as zero since $F(u' - u)$ is defined as zero outside the interval $u' < u < u' + \ln 1/\alpha$. Then the integral is a convolution and the Laplace transform of Equation (II.12a) is simply:

$$\psi(s) = \psi(s) f(s) + 1 \quad \text{(II.12b)}$$

where

$$\psi(s) = \mathscr{L}\{\chi(u)\} = \int\limits_0^\infty \chi(u)\,\mathrm{e}^{-su}\,\mathrm{d}u \quad \text{(II.12c)}$$

is the Laplace transform of $\chi(u)$ and

$$f(s) = \mathscr{L}\{F(u)\} = \int\limits_0^\infty F(u)\,\mathrm{e}^{-su}\,\mathrm{d}u \quad \text{(II.12d)}$$

$$= \int\limits_0^{\ln 1/\alpha} \frac{\mathrm{e}^{-(s+1)u}}{1-\alpha}\,\mathrm{d}u = \frac{1-\alpha^{s+1}}{(1+\alpha)(s+1)} \quad \text{(II.12e)}$$

is the Laplace transform of $F(u)$ for elastic scattering, isotropic in the center-of-mass system. Solving Equation (II.12b) for $\psi(s)$, we have:

$$\psi(s) = \frac{1}{1-f(s)} = 1 + \frac{f(s)}{1-f(s)}$$

$$= 1 + \frac{1-\alpha^{s+1}}{(1-\alpha)(s+1) - (1-\alpha^{s+1})} \quad \text{(II.13a)}$$

[5] For a review of the properties of Laplace transforms see, for example, R. V. CHURCHILL, *Operational Mathematics*, New York, McGraw-Hill, 1958.

The unity term has been separated from the remainder of the right-hand side of this equation since it is the Laplace transform of the source term, $\delta(u)$. Applying the inversion theorem for Laplace transforms to $\psi(s)$, we have:

$$\chi(u) = \delta(u) + \frac{1}{2\pi i}\int_{b-i\infty}^{b+i\infty} \frac{(1 - \alpha^{s+1})\, e^{su}}{(1 - \alpha)(s + 1) - (1 - \alpha^{s+1})}\, ds \quad \text{(II.13b)}$$

where $\delta(u)$ represents the direct source contribution and the line of integration in the complex plane, $Re(s) = b$, is to the right of all poles of the integrand. These poles are at the zeros of the denominator, i.e., at the roots of

$$(1 - \alpha)(s + 1) = 1 - \alpha^{s+1} \quad \text{(II.13c)}$$

since all of the functions appearing in the integrand are, by themselves, analytic everywhere in the finite complex plane. By closing the contour of integration around the left half plane and noting that the integral on the semicircle (see Fig. II.3) is zero, we can evaluate $\chi(u)$ as a sum of residues:

$$\chi(u) = \delta(u) + \sum_{i=0}^{\infty} \frac{(1 - \alpha^{s_i+1})\, e^{s_i u}}{(1 - \alpha) + \alpha^{s_i+1} \ln \alpha} \quad \text{(II.13d)}$$

where the s_i are the roots of Equation (II.13b) and each term represents the contribution of the residue of the pole at s_i. Thus, the flux may be expressed as a sum of exponentials. It may be shown with a little algebraic manipulation that Equation (II.13c) has only one solution with a non-negative real part, for $0 < \alpha < 1$, namely $s_0 = 0$. All other terms of the series (II.13d) therefore die out for large u so the asymptotic solution (i.e., the solution at energies far below the source energy) becomes:

$$\chi_{asy}(u) = \Sigma_s(u)\, \varphi_{asy}(u) = \frac{1 - \alpha}{1 - \alpha + \alpha \ln \alpha} = \frac{1}{1 + \dfrac{\alpha}{1 - \alpha} \ln \alpha} = \frac{1}{\xi_{iso}} \quad \text{(II.14a)}$$

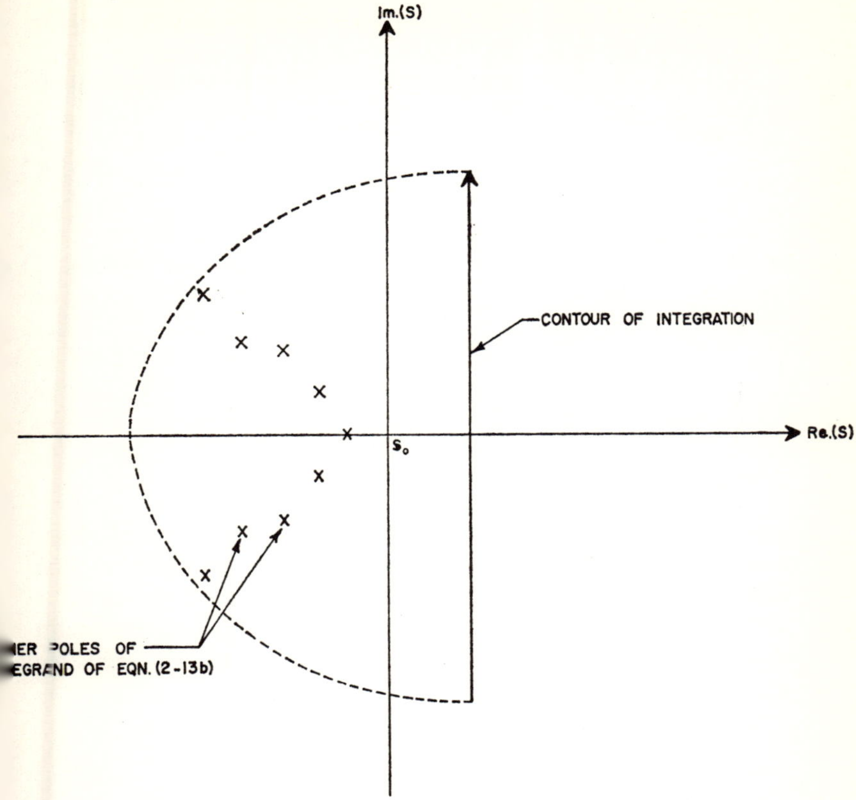

FIG. II.3. Contour of integration for Equation (II.13b).

where ξ_{iso} is the average logarithmic energy loss per collision defined by Equation (I.59c). Marshak [6] has determined that the poles with the second largest real part are

$$s = -1.55 \pm 3.37i \quad \text{for deuterium} \quad A = 2 \qquad \text{(II.14b)}$$

$$s = -0.52A \pm 1.87Ai \quad \text{for} \quad A \gg 1 \qquad \text{(II.14c)}$$

Therefore, terms of order $e^{-1.55u}$ for deuterium and $e^{-0.52Au}$ for heavy elements, are neglected in the asymptotic solution (2.14a).

For hydrogen, $\alpha = 0$, and Equation (II.13b) has only a single root, that at $s = 0$, and the rigorous result again follows. Again,

[6] R. E. MARSHAK, *op. cit.*

deuterium moderation is the worst case, as the series (II.13c) converges most slowly for this case.

For constant Σ_s, the asymptotic flux spectrum is again constant in lethargy, and therefore yields the "$1/E$" spectrum

$$\varphi_{asy}(E) = \frac{1}{E\xi_{iso}\,\Sigma_s} \tag{II.14d}$$

Clearly, the series (II.13c) must converge to the Placzek function. However, the rapid variation of this function near the source (see Fig. II.1 and II.2) would require a large number of terms for reasonable convergence, particularly near the discontinuity at αE_0, where the Gibbs phenomenon will occur. In this region, an expansion proposed by Teichmann (see page 13, ref. [4]) can be used to derive the Placzek functions.

7. Mixtures of Moderators

A procedure analogous to that of Section 6 may be followed for a mixture of elements, provided the scattering cross sections of all elements vary in the same way with energy, and scattering is isotropic. A solution has not been found for the more general case. The development will be sketchy as it is essentially the same as the one just given.

Define

$$\beta_j = \frac{\Sigma_s^j}{\Sigma_s} \tag{II.15a}$$

the fraction of scattering from nuclei of the jth type. Then Equation (II.2b) becomes

$$\chi(u) = \sum_j \beta_j \int_{u\ln 1/\alpha_j}^{u} \chi(u')\,F_j(u'-u)\,\mathrm{d}u' + \delta(u) \tag{II.15b}$$

Proceeding as before, the Laplace transform of $\chi(u)$ is found to be

$$\psi(s) = \frac{1}{1-\sum_j \beta_j f_j(s)} = 1 + \frac{\sum_j \beta_j f_j(s)}{1-\sum_j \beta_j f_j(s)} \tag{II.15c}$$

where $f_j(s)$ is defined exactly as before:

$$f_j(s) = \frac{1-\alpha_j^{s+1}}{(1-\alpha_j)(s+1)} \tag{II.15d}$$

The inversion integral is:

$$\chi(u) = \delta(u) + \frac{1}{2\pi i}\int_{b-i\infty}^{b+i\infty} \frac{\sum_j \frac{\beta_j(1-\alpha_j^{s+1})}{(1-\alpha_j)(s+1)} e^{su}}{1 - \sum_j \frac{\beta_j(1-\alpha_j^{s+1})}{(1-\alpha_j)(s+1)}}\, ds \quad \text{(II.16a)}$$

and the poles of the integrand are at the zeros of:

$$\sum_j \frac{\beta_j(1-\alpha_j^{s+1})}{(1-\alpha_j)(s+1)} = 1 \quad \text{(II.16b)}$$

Thus

$$\chi(u) = \delta(u) + \sum_{i=0}^{\infty}\left[\frac{\sum_j \frac{\beta_j(1-\alpha^{s_i+1})}{(1-\alpha_j)(s_i+1)} e^{s_i u}}{\sum_j \left[\frac{\beta_j a^{s_i+1}\ln\alpha_j}{(1-\alpha_j)(s_i+1)} + \frac{\beta_j(1-\alpha^{s_i+1})}{(1-\alpha_j)(s_i+1)^2}\right]}\right] \quad \text{(II.16c)}$$

Again, the only root of Equation (II.22a) with non-negative real part is $s_0 = 0$ so that the asymptotic flux is †

$$\chi_{asy}(u) = \Sigma_s(u)\varphi_{asy}(u) = \frac{1}{1 + \sum_j \frac{\beta_j\alpha_j \ln\alpha_j}{(1-\alpha_j)}} = \frac{1}{\xi_{iso}} \quad \text{(II.16d)}$$

where the definition

$$\xi_{iso} = \sum_j \beta_j \xi_{j\,iso} \quad \text{(II.16e)}$$

has been invoked. As one would expect, ξ_{iso} is the average lethargy gain per collision in the mixture. This is readily shown. Since β_j is the fraction of neutrons which scatter off j type nuclei, the scattering frequency is (as in Equation (II.15b):

$$F(u'-u) = \sum_j \beta_j F_j(u'-u)$$

and upon averaging $(u - u')$ over this scattering frequency one finds that $\langle u - u' \rangle = \xi_{iso}$.

In the limit of vanishingly small capture, one might assert that the effect of capture on the neutron energy spectrum would be

† For hydrogenous mixtures, there is a second real root which is negative but smaller in absolute value than the real part of any complex root. See H. Amster, *Nucl. Sci. Eng.* **21**, 206 (1965).

negligible. The resonance escape probability for this limit could then be written:

$$p(u) = 1 - \int_0^u \Sigma_a(u')\,\varphi(u')\,du' = 1 - \int_0^u \frac{\Sigma_a(u')}{\xi_{iso}\,\Sigma_s(u')}\,du' \quad \text{(II.17a)}$$

assuming that capture takes place at lethargies far from the source. Since the integral is, by assumption, small, we may write

$$p(u) \approx \exp\left\{-\int_0^u \frac{c_0(u')}{\xi_{iso}}\,du'\right\} \quad \text{(II.17b)}$$

8. Scattering Resonances

The case just treated is a useful first approximation; it does not, however, successfully account for scattering resonances. In heavy elements, such as the common reactor fuels, absorption resonances are almost always accompanied by scattering resonances.† Furthermore, other materials, for example moderators, structural materials and coolants, generally display strong scattering resonances in the energy region traversed by a neutron in slowing down. Thus a more detailed treatment of the problem must be made. This problem is of very great significance in fast reactors, in which most neutron absorption takes place in the resonance region.

Qualitatively, a change in the scattering cross section of one element changes ξ, sometimes by quite a significant amount. For the sake of argument, let us say that the scattering resonance occurs in a light element and that the resonance width is of the order of at least some of the $\ln 1/\alpha$, a not unreasonable case. Now, most of the neutrons arriving at energies near the high-energy end of the resonance will come from energies at which the effect of the resonance is unimportant. Hence the collision density $\Sigma_s(u)\,\varphi(u)$ will still behave like $1/\xi_0$, where ξ_0 is the non-resonance value of ξ. In the resonance, a greater proportion of the neutrons will be scattered by the light nucleus displaying the resonance and on the average they will lose more energy per collision than they would outside

† Generally speaking, resonances at low energies tend to absorb more heavily (i.e., σ_a/σ_s is higher) than high energy resonances. This is due to the behavior of the Breit–Wigner parameters Γ_n and Γ_γ discussed later.

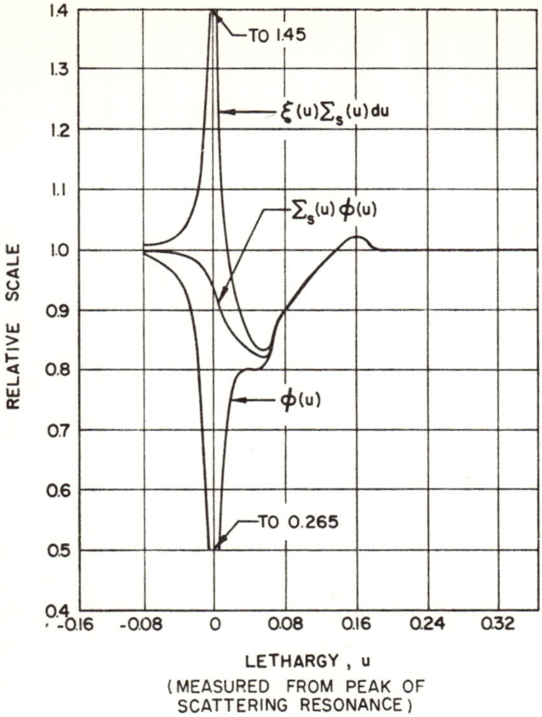

FIG. II. 4. The flux and collision density through a sodium resonance in a sodium–iron mixture. The resonance is assumed at $u = 0$ with a lethargy width $\Delta u = 0.01$ (Breit–Wigner width) and the parameters were $\Sigma_p^{\mathrm{Na}} = 1$, $\Sigma_p^{\mathrm{Fe}} = 3$, Σ_0 (resonance height) $= 10$.

the resonance. As a result, the collision density at the low-energy end of the resonance will be less than $1/\xi_0$. Still later, at energies just below the resonance, the collision density will tend to return to its $1/\xi_0$ behavior. However, some of the neutrons scattered in the resonance may reach these energies and the collision density may be greater than $1/\xi_0$. Eventually, of course $\Sigma_s(u)\,\varphi(u)$ does resume its asymptotic value, assuming that no other resonances are found below the resonance in question. The important point is that the behavior of the collision density is not at all like $1/\xi(u)$, as might be inferred from the previous section. For sufficiently narrow resonances, the approximation $\Sigma_s \varphi = 1/\xi_0$ is valid throughout the resonance.

A complete description of these processes can only be obtained numerically. The results of such a calculation [7], which displays the features described above, is given in Fig. II.4.

D. Slowing Down in Media Containing Heavy Elements: II. Absorbing Media

9. The Case of Constant c

Now consider the more general case in which the effects of neutron absorption are no longer necessarily small. Absorption primarily affects the asymptotic spectrum, since absorption usually does not become important until after the Placzek oscillatons have died out. A solution for the case in which absorption takes place close to the source is possible but will be delayed until the more important case of the effect on the asymptotic flux has been presented.

For a single moderating element, with isotropic scattering, and monoenergetic source, Equation (II.2b) becomes:

$$\chi(u) = \Sigma_t(u)\,\varphi(u) = \int_{u-\ln 1/\alpha}^{u} c(u')\,\chi(u')\,\frac{e^{-(u-u')}}{1-\alpha}\,du' + \delta(u) \qquad \text{(II.18a)}$$

with

$$c(u) \triangleq \frac{\Sigma_s(u)}{\Sigma_t(u)} \qquad \text{(II.18b)}$$

This equation is not amenable to solution by the previous method as it stands. If, however, $c(u)$ is constant, i.e., the ratio of capture and scattering cross sections is independent of energy,† the mathematical solution is the same as in the no-capture case but with $f(s)$ replaced by $cf(s)$. The case of a single element will be illustrated for simplicity; mixtures are, in principle, no more difficult to treat but lead to more complicated expressions.

† As the scattering cross section is usually nearly constant, this amounts to constant capture in most cases.

[7] R. B. Nicholson, P. F. Zweifel and J. H. Ferziger, *Trans. Amer. Nuc. Soc.* **4**, 19 (1961).

In analogy to Equation (II.13c) one obtains:

$$\chi(u) = \delta(u) + \sum_i \frac{c\,(1 - \alpha^{\nu_i + 1})\,e^{\nu_j u}}{(1 - \alpha) + c\alpha^{\nu_i + 1} \ln \alpha} \qquad \text{(II.19a)}$$

where the ν_i are the roots of the transcendental equation:

$$(1 - \alpha)\,(\nu_i + 1) = c(1 - \alpha^{\nu_i + 1}) \qquad \text{(II.19b)}$$

Again it may be shown (by separating the real and imaginary parts of the equation) that there are no ν_i with real parts greater than zero. For $c = 1$, it is clear that the ν_i are identically the roots of Equation (II.13b). For c slightly less than unity one expects that the roots will change only slightly. In particular, the root at zero for $c = 1$ is a single root and can be expected to remain real since complex roots must occur in conjugate pairs. This root can be expected to move to the left along the real axis in the complex plane, but should still be the root with largest real part for c near unity. The asymptotic distribution will then be that due to the first term of the series (II.19a):

$$\chi_{asy}(u) = \Sigma_t(u)\,\varphi_{asy}(u) = \frac{c(1 - \alpha^{\nu_0 + 1})\,e^{\nu_0 u}}{(1 - \alpha) + c\alpha^{\nu_0 + 1} \ln \alpha} \qquad \text{(II.20a)}$$

For c near unity, and therefore small ν_0, the expansion

$$\alpha^{\nu_0} \approx 1 + \nu_0 \ln \alpha \qquad \text{(II.20b)}$$

is valid and

$$\nu_0 \approx \frac{1 - c}{1 + \dfrac{c\alpha}{1 - \alpha} \ln \alpha} \approx \frac{c_0}{\xi_{iso}} \qquad \text{(II.20c)}$$

So that the asymptotic flux is approximately

$$\varphi_{asy}(u) = \frac{1}{\Sigma_t(u)} \left[\frac{c(1 - \nu_0(\xi_{iso} - 1))}{1 + c(1 + \nu_0 \ln \alpha)\,(\xi_{iso} - 1)} \right] e^{c_0 - u/\xi_{iso}}$$

$$\approx \frac{e^{-c_0 u/\xi_{iso}}}{\xi_{iso} \Sigma_t(u)} \qquad \text{(II.20d)}$$

The asymptotic flux is only roughly proportional to $1/E$. The resonance escape probability for this situation is approximately given by:

$$p(u) = 1 - \int_0^u \Sigma_a(u')\,\varphi_{asy}(u')\,du'$$

$$= 1 - \int_0^u \frac{c_0}{\xi_{iso}}\,e^{-c_0 u'/\xi_{iso}}\,du' = e^{-c_0 u/\xi_{iso}} \qquad \text{(II.20e)}$$

10. Slowly Varying Capture [8]

If c is not a constant, but is a function of lethargy, the Laplace transform method breaks down. Of most practical interest is the case in which absorption is negligible up to a considerable lethargy from the source. Then as we have seen in Section 5 source transients will have died out and only the effect of absorption on the asymptotic flux need be considered. This requires solution of Equation (II.18a) subject to the conditions:

$$c(u) = 1 \qquad u < u_0$$
$$\chi(u_0) = \frac{1}{\xi} \tag{II.21a}$$

where u_0 is the smallest lethargy at which Σ_a is non-zero.

To facilitate solution for this case, a transformation will be applied to Equation (II.18a). First it is rewritten (noting that sources are absent):

$$e^u \chi(u) = \int_{u - \ln 1/\alpha}^{u} \frac{c(u')\,\chi(u')}{1-\alpha} e^{u'}\, du' \tag{II.21b}$$

Differentiation with respect to u and subsequent integration over u (changed to u' for convenience) from u_0 to u gives

$$\int_{u_0}^{u} \chi(u')\, du' + \chi(u) - \chi(u_0) = \int_{u_0}^{u} \frac{c(u')\,\chi(u')}{1-\alpha}\, du'$$
$$- \frac{\alpha}{1-\alpha} \int_{u_0}^{u} c\left(u' - \ln\frac{1}{\alpha}\right) \chi\left(u' - \ln\frac{1}{\alpha}\right) du' \tag{II.21c}$$

Now change variables in the last term on the right side of this equation to $w = u' - \ln 1/\alpha$. The resulting integral, which runs between the limits of $u_0 - \ln 1/\alpha$ to $u - \ln 1/\alpha$, can be broken into three integrals. Symbolically:

$$\int_{u_0 - \ln 1/\alpha}^{u - \ln 1/\alpha} = \int_{u_0 - \ln 1/\alpha}^{u_0} + \int_{u_0}^{u} - \int_{u - \ln 1/\alpha}^{u} \tag{II.21d}$$

[8] This treatment follows G. PLACZEK, *op. cit.*

Now, noting that the boundary conditions (II.21 a) state that $c = 1$ for $u < u_0$ and further assuming that $\chi(u) = 1/\xi$ for $u > u_0 - \ln 1/\alpha$, we can evaluate the integral from $u_0 - \ln 1/\alpha$ to u_0. Combining the resulting terms leads to

$$\chi(u) = 1 - \int_{u_0}^{u} c_0(u')\,\chi(u')\,du' + \frac{\alpha}{1-\alpha}\int_{u-\ln 1/\alpha}^{u} c(u')\,\chi(u')\,du' \tag{II.21 e}$$

If $(\xi - 1)\,\chi(u)$ is now added to both sides of this equation, and it is noted that $(\xi - 1)\,\chi(u)$ can be written as

$$(\xi - 1)\,\chi(u) = -\frac{\alpha}{1-\alpha}\int_{u-\ln 1/\alpha}^{u} \chi(u)\,du' \tag{II.21 f}$$

we obtain finally the following equation, which is equivalent to Equation (II.18) under the assumptions made:

$$\xi\chi(u) = 1 - \int_{u_0}^{u} c_0(u')\,\chi(u')\,du' + \frac{\alpha}{1-\alpha}\int_{u-\ln 1/\alpha}^{u} du'[c(u')\,\chi(u') - \chi(u)] \tag{II.21 g}$$

Or, defining $\xi\chi(u) = q_0(u)$, we have

$$q_0(u) = 1 - \frac{1}{\xi}\int_{u_0}^{u} c_0(u')\,q_0(u')\,du' + \frac{\alpha}{1-\alpha}\,\frac{1}{\xi}\int_{u-\ln 1/\alpha}^{u} du'[c(u')\,q_0(u') - q_0(u)] \tag{II.21 h}$$

Note the restriction implicit in the above derivation, that the asymptotic distribution be attained at least one collision interval before the absorption sets in (i.e., the absorption is assumed to begin at u_0 but the asymptotic flux is assumed to exist from $u_0 - \ln 1/\alpha$ on).

For hydrogen moderation, $\alpha = 0$, and the last term on the right-hand side of Equation (II.21 h) drops out. Thus, this term distinguishes between hydrogen and heavy element slowing down. Note that $\dfrac{\alpha}{1-\alpha} = \dfrac{(A-1)^2}{4A} \approx \dfrac{A}{4}$ and $\xi \approx 2/3A$ for large A so that the coefficient of the last term of Equation (II.21 h) is proportional to A^2 and is important for heavy moderators.

For slowly varying capture, the effect of changes of $c(u)$ in one collision interval may be neglected. Then, if we try to satisfy Equation (II.21h) with a function of the form [9] $q_0(u) = e^{s(u)}$ and approximate $s(u - \ln 1/\alpha)$ by a two-term Taylor series expansion about u we find that, on keeping only the first two terms, $s'(u)$ satisfies a relation exactly like (II.20a):

$$(1 - \alpha)(s'(u) + 1) = c(u)(1 - \alpha^{s'(u)+1}) \tag{II.22a}$$

where

$$s'(u) = \mathrm{d}s/\mathrm{d}u$$

In analogy to our previous notation we denote the solution of Equation (II.22a) with largest real part by $\nu_0(u)$, so that

$$s_0'(u) = \nu_0(u) \tag{II.22b}$$

or

$$s_0(u) = \int_{u_0}^{u} \nu_0(u')\,\mathrm{d}u' \tag{II.22c}$$

Thus we arrive at the result

$$q_0(u) = \xi\Sigma_t(u)\,\varphi(u) = e^{\int_{u_0}^{u} \nu_0(u')\,du'} \tag{II.22d}$$

which is analogous to Equation (II.20d) except that ν_0 is now a function of lethargy. This result can be generalized to mixtures but it would be necessary to derive an equation similar to (II.21h) for this case and it probably is not worth while to carry out the procedure here. It is perhaps intuitively obvious, and can be shown [10], that the results of such a calculation are that $\nu_0(u)$ is the root of

$$1 = \sum_j \frac{\beta_j(u)(1 - \alpha_j^{\nu_0(u)-1})}{(1 - \alpha_j)(1 + \nu_0(u))} \tag{II.23}$$

with largest real part. Here β_j is the fraction of neutrons of lethargy u that collide with j type nuclei [see Eq. (II.15a)]. In the limit of small $c_0(u)$ an approximation to $\nu_0(u)$ is given by Equation (II.20c), and the resonance escape probability becomes:

$$p(u) = \exp\left\{-\int_{u_0}^{u} \frac{c_0(u')}{\xi(u')}\,\mathrm{d}u'\right\} \tag{II.24a}$$

[9] This is essentially the well known WKB approximation of quantum mechanics. See any text in quantum mechanics, e.g., SCHIFF, *op. cit.* For a slightly different application of the WKB method of this problem see L. DRESNER, *Resonance Absorption of Neutrons in Nuclear Reactors*, Pergamon Press, 1961.

[10] R. E. MARSHAK, *op. cit.*

where $\xi(u)$ is a function of lethargy by virtue of the fact that $\beta_j(u)$ is a function of lethargy.

In terms of energy, the flux can be written

$$\varphi(E) = \frac{1}{\xi \Sigma_t(E)\, E} \exp\left\{-\int_E^{E_0} \frac{c_0(E')}{\xi(E')} \frac{\mathrm{d}E'}{E'}\right\} \tag{II.24b}$$

E_0 here is the energy corresponding to lethargy u_0. In this case, the flux drops off somewhat slower than $1/E$. These formulae may be applied to non-resonance capture and, in particular, they would be used to determine the resonance escape probability in a medium with "$1/v$" absorption, i.e., a material whose absorption cross section varies with the reciprocal of the neutron velocity, a situation which obtains in many cases of interest.

11. Rapidly Varying Capture—Strong Resonances [11]

The problem with $c_0(u)$ varying rapidly with energy is important as it covers the case of strong absorption resonances. A rigorous solution, by a method similar to that used in deriving the Placzek function, is possible, but the calculations reach high orders of complexity very rapidly. We shall display the calculation and then derive a useful formula for widely spaced narrow resonances.

Again it is assumed that the asymptotic flux is established one collision interval prior to the onset of absorption, so that we may solve Equation (II.18a) subject to the conditions that $\chi(u) = 1/\xi$ and $c(u) = 1$ for $u < u_0$. For $u_0 + \ln 1/\alpha > u > u_0$, Equation (II.18a) may then be written

$$\chi(u) = \int_{u-\ln 1/\alpha}^{u_0} c(u')\,\chi(u')\,\frac{\mathrm{e}^{-(u-u')}}{1-\alpha}\,\mathrm{d}u' + \int_{u_0}^{u} c(u')\,\chi(u')\,\frac{\mathrm{e}^{-(u-u')}}{1-\alpha}\,\mathrm{d}u'$$

$$= \frac{\mathrm{e}^{u_0-u} - \alpha}{\xi(1-\alpha)} + \int_{u_0}^{u} c(u')\,\chi(u')\,\frac{\mathrm{e}^{-(u-u')}}{1-\alpha}\,\mathrm{d}u' \tag{II.25a}$$

which is of the same form as some of the integral equations we have met previously. Differentiation of Equation (II.25a) yields a simple linear first-order differential equation whose solution is

[11] This treatment closely follows G. Placzek, *op. cit.*

readily found to be:

$$\xi\chi(u) = q_0(u) = e^{\gamma(u)}\left[1 - \frac{\alpha}{1-\alpha}\int_{u_0}^{u} e^{-\gamma(u')}du'\right] \quad \text{(II.25b)}$$

where the exponent of the integrating factor, $\gamma(u)$, is

$$\gamma(u) = \int_{u_0}^{u}\left[\frac{c(u')}{1-\alpha} - 1\right] du' \quad \text{(II.25c)}$$

Following the derivation of the Placzek function, this process may be repeated iteratively on the intervals $u_0 + n \ln 1/\alpha < u < u_0 + (n+1) \ln 1/\alpha$ to obtain the complete energy spectrum. The formulae thus obtained grow in complexity even more rapidly than the Placzek functions but they are rarely used even when resonances are closely spaced.

An important case in the one in which there are sharp resonances widely separated from each other. Since absorption in such resonances does not disturb the flux distribution far from the resonances very greatly the oscillations can be ignored and the problem is greatly simplified. In this region the last term of Equation (II.21h) vanishes and the flux can be calculated quite simply:

$$q_0(u) = \xi\Sigma_t(u)\,\varphi(u) = 1 - \int_{u_0}^{u} c_0(u')\,\chi(u')\,du' \quad \text{(II.26a)}$$

For a very sharp resonance, in which practically all of the absorption takes place in a lethargy interval much smaller than $\ln 1/\alpha$, the collision density in the resonance will be nearly that of the asymptotic distribution so that $\chi(u')$ in the integrand of Equation (II.26a) may be replaced by $1/\xi$.† Furthermore, most of the contribution to the integral will come from the resonance. Thus:

$$q_0(u) \approx 1 - \frac{1}{\xi}\int_{\text{Res}} c_0(u')\,du' \quad \text{(II.26b)}$$

where the subscript "Res" on the integral denotes integration only over the lethargy region for which Σ_a is large. If the integral is

† i.e., a scattering collision within the resonance is assumed to remove the neutron from the resonance energy. This is the basic assumption of the narrow resonance (NR) approximation which is discussed more fully later.

small:

$$q_0(u) \approx \exp\left\{-\frac{1}{\xi}\int\limits_{\text{Res}} c_0(u')\,du'\right\} \tag{II.26c}$$

The resonance escape probability for this case is obtained from Equation (II.7a) and is simply:

$$p(u) \approx \exp\left\{-\frac{1}{\xi}\int\limits_{\text{Res}} c_0(u')\,du'\right\} \tag{II.26d}$$

which is the well known "narrow resonance" (NR) formula. In Section E absorption by individual resonances will be considered in more detail.

12. Anisotropic Scattering

These results are readily generalized to non-isotropic scattering by means of the formalism of Chapter I. For non-isotropic scattering, it is necessary to employ the more general scattering frequency (I 50d) which, when substituted into Equation (II.18a), yields

$$\Sigma_t(u)\,\varphi(u) = \sum_i N_i \int\limits_{u-\ln 1/\alpha_i}^{u} \varphi(u') \sum_{L'=0}^{\infty} B_{L'}^{c}(u')\,T_{0L'}(u-u')\,du' + S(u) \tag{II.27a}$$

where it is again understood that α, $B_L^c(u')$ and $T_{LL'}(U)$ are properties of particular nuclear types. Unfortunately, Equation (II.27a) cannot be solved in general, firstly, because the $T_{0L'}(U)$ are rather complex functions of their argument, and secondly because the $B_L^c(u)$ are generally supplied as empirical data, which cannot be easily fitted to analytic functions. To simplify matters, one frequently expands the functions $B_{L'}^c(u')\,\varphi(u')$ in a Taylor series about u. Doing so, and noting that the necessary convergence for interchanging the sum and integral exists, we can write

$$\Sigma_t(u)\,\varphi(u) = \sum_i N_i \sum_{L'=0}^{\infty} \sum_{\nu=0}^{\infty} \frac{1}{\nu!} \int\limits_{u-\ln 1/\alpha}^{u} (u'-u)^{\nu} \frac{\partial^{\nu}}{\partial u^{\nu}}\left[B_{L'}^c(u)\,\varphi(u)\right] \times$$
$$\times\, T_{0L'}(U)\,du' + S(u) \tag{II.27b}$$

The derivative terms may be taken outside outside the integral and the remaining integral (with the $\nu!$) is just the definition of $T_{0L'}^{\nu}$ [see

Eq. (I.56)]. Thus

$$\Sigma_t(u)\,\varphi(u) = \sum_i N_i \sum_{\nu=0}^{\infty} \sum_{L'=0}^{\infty} T^{\nu}_{0L'} \frac{\partial^{\nu}}{\partial u^{\nu}} [B^c_{L'}(u)\,\varphi(u)] + S(u) \tag{II.27c}$$

But if consideration is restricted to terms of order γ or greater only,† from Equations (I.53a) and (I.57a, b) we have

$$\Sigma_t(u)\,\varphi(u) = \sum_i N_i \left[B^c_0(u)\,\varphi(u) + 2\gamma \frac{\partial}{\partial u} [(B^c_1(u) - B^c_0(u))\,\varphi(u)] \right] + S(u) \tag{II.27d}$$

or, since $B^c_0(u) = \sigma_s(u)$ and

$$\Sigma_s(u) = \sum_i N_i \sigma^i_s(u) \tag{II.27e}$$

Equation (II.27d) becomes

$$\Sigma_a(u)\,\varphi(u) = \sum_i \frac{\partial}{\partial u} [2\gamma_i (N_i\, B^{ci}_1(u) - \Sigma_s(u))\,\varphi(u)] + S(u) \tag{II.27f}$$

But, from Equations (I.62a) and (II.16c),

$$\sum_i 2\gamma_i \left(\frac{N_i B^{ci}_1(u)}{\Sigma_s(u)} - 1 \right) = -\,\xi(u) \tag{II.27g}$$

is seen to be the mean lethargy change per collision. Thus we arrive at the differential equation:

$$\frac{\partial}{\partial u} (\xi(u)\,\Sigma_s(u)\,\varphi(u)) + \frac{\Sigma_a(u)}{\xi(u)\,\Sigma_s(u)} [\xi(u)\,\Sigma_s(u)\,\varphi(u)] = S(u) \tag{II.28a}$$

for which the solution, found by methods used previously in this chapter, is:

$$\xi(u)\,\Sigma_s(u)\,\varphi(u) = \int_0^u S(u')\, e^{-\int_{u'}^{u} \frac{\Sigma_a(u'')}{\xi(u'')\,\Sigma_s(u'')} du''} du' \tag{II.28b}$$

The resonance escape probability is easily extracted and is [setting $S(u) = \delta(u)$ again]:

$$p(u' \to u) = e^{-\int_{u'}^{u} \frac{\Sigma_a(u'')}{\xi(u'')\Sigma_s(u'')} du''} \tag{II.28c}$$

which is the extension of Equation (II.26d) to the more general case, subject to the approximations made above.

† This is closely related to the Age-Diffusion Approximation of Chapter III and in fact the results derived here are the Age-Diffusion results.

13. Resonance Escape Probability in the Goertzel—Greuling Approximation

A slightly more accurate formula for the resonance escape probability can be developed from an approximation due to Goertzel and Greuling [12]. In Chapter III we shall derive an approximation to the Boltzmann equation which reduces to

$$\left[\Sigma_a(u) + \frac{\xi}{\lambda}\Sigma_s(u)\right]\varphi(u) = S(u) + \frac{\xi}{\lambda^2}\int_0^u e^{-(u-u')/\lambda}\,\Sigma_s(u')\,\varphi(u')\,du' \tag{II.29a}$$

for an infinite medium. Here,

$$\lambda = \frac{\langle U^2\rangle}{2\xi} = 1 - \frac{\alpha\left[1 + \ln\dfrac{1}{\alpha} + 2\ln^2\dfrac{1}{\alpha}\right]}{1 - \alpha\left(1 + \ln\dfrac{1}{\alpha}\right)} \tag{II.29b}$$

the last equality holding for isotropic scattering.

Let the function on the left-hand side of Equation (II.29a) be defined as $\chi(u)$. The solution is similar to those of previous cases; differentiation of Equation (II.29a) yields

$$\frac{\partial\chi}{\partial u} + \frac{\Sigma_a(u)}{\lambda\Sigma_a(u) + \xi\Sigma_s(u)}\chi(u) = \frac{S(u)}{\lambda} + \frac{\partial S}{\partial u} \tag{II.29c}$$

which may be solved to yield a solution analogous to Equation (II.6a):

$$\chi(u) = \frac{\lambda\Sigma_a(u) + \xi\Sigma_s(u)}{\lambda}\varphi(u) = S(u)$$
$$+ \frac{1}{\lambda}\int_0^u \frac{\xi\Sigma_s(u')}{\lambda\Sigma_a(u') + \xi\Sigma_s(u')} S(u')\exp\left\{-\int_{u'}^u \frac{\Sigma_a(u'')\,du''}{\lambda\Sigma_a(u'') + \xi\Sigma_s(u'')}\right\}du' \tag{II.29d}$$

The resonance escape probability for this case is given by

$$p = 1 - \int_0^u du'\,\frac{\Sigma_a(u')}{\lambda\Sigma_a(u') + \xi\Sigma_s(u')}\,S(u') +$$
$$\frac{1}{\lambda}\int_0^{u'} S(u'')\frac{\xi\Sigma_s(u'')}{\lambda\Sigma_a(u'') + \xi\Sigma_s(u'')}\exp\left\{-\int_{u''}^{u'}\frac{\Sigma_a(u''')\,du''}{\lambda\Sigma_a(u''') + \xi\Sigma_s(u''')}\right\}du''' \tag{II.30a}$$

[12] G. Goertzel and E. Greuling, *Nucl. Sci. Eng.* 7, 69 (1960).

which, for a monoenergetic source, becomes:

$$p = \frac{\xi \Sigma_s(0)}{\lambda \Sigma_a(0) + \xi \Sigma_s(0)} \exp\left\{-\int_0^u \frac{\Sigma_a(u'')\,\mathrm{d}u''}{\lambda \Sigma_a(u'') + \xi \Sigma_s(u'')}\right\} \quad \text{(II.30b)}$$

Values of ξ, λ and α for common moderators are given in Table 1. Since ξ and λ are not greatly different, it will usually suffice to use Equation (II.24a) if absorption is not too large.

TABLE 1

Element	ξ	λ	α
Hydrogen	1.000	1.000	0
Deuterium	0.725	0.583	0.111
Beryllium	0.209	0.149	0.640
Carbon	0.158	0.116	0.716

E. Capture by Individual Resonances

14. The Narrow Resonance Approximations

We now turn our interest to some approximation methods which are frequently used to calculate the resonance escape probability for individual resonances. Consider a homogeneous mixture of moderator and a heavy absorbing element. Most heavy-element cross sections are a superposition of resonances on a constant potential scattering base. The absorption cross section is entirely due to the resonances; their width is generally (but not always) small compared with the average energy loss suffered by a neutron in a single collision. The approximations introduced will be based on assumptions concerning resonance widths.

A quantity which can be used as a yardstick for describing resonances, the *practical width* [13], is defined as that energy (or lethargy) interval for which the macroscopic *capture* cross section is greater than the macroscopic *non-resonance total* cross section (note that this definition makes the practical width dependent on the composition of the system). For resonances encountered in reactors, the practical width is generally much smaller than the average energy a neutron at the resonance energy would lose in a collision with a moderator atom. There is, however, another means

[13] The notion of "practical width" is due to E. P. WIGNER.

of distinguishing between resonances. Some resonances are so narrow that even the energy loss suffered in an elastic collision with an absorber nucleus (e.g., U-238) removes neutrons from the resonance; other resonances are so broad that a single collision between a neutron with an energy near the resonance peak and a heavy absorber nucleus could not possibly remove the neutron from the resonance. Whether the quantity $E_R(1 - \alpha_U)$, where E_R is the energy at the resonance peak and $\alpha_U = \left(\frac{A_U - 1}{A_U + 1}\right)^2$ ($\approx$ 0.98 for U-238), is greater or less than the practical width of the resonance is frequently [14] taken as a means of distinguishing the first case from the second [15]. The quantity $E_R(1 - \alpha_U)$ is, of course, the maximum energy loss in an elastic collision with a heavy nucleus. Thus, to a reasonable approximation, if the maximum energy loss suffered by a neutron in a collision with a uranium nucleus is greater than the width of the resonance, elastic scattering in uranium can be considered to remove neutrons from the resonance; this is the narrow resonance (NR) approximation. Similarly, if the inequality is reversed, elastic scattering with uranium atoms can be assumed not to remove neutrons from the resonance; this is the narrow resonance–infinitely heavy absorber (NRIA) approximation. If, in a particular resonance, $\Sigma_{au} \gg \Sigma_{su}$, the effect of scattering from uranium may be ignored entirely. It may be noted that in uranium and thorium most of the resonances are of the NR type. However, as the NRIA resonances are broader and tend to occur at lower energies [16], where the flux is highest, it is this type of resonance which is responsible for most of the resonance absorption (see Table 2).

A simple way to derive the NR formula is to note that $N(u)$, the number of neutrons making collisions within a unit lethargy interval about u within the resonance, is given by

$$N(u) = \sum_i N_i \int_{u - \ln 1/\alpha_i}^{u} \Sigma_s^i(u')\, \varphi(u')\, F_i(u' \to u)\, \mathrm{d}u' \qquad \text{(II.31 a)}$$

[14] K. T. Spinney in BNL- 433, p. 103; L. Dresner, ORNL 2659.

[15] Some resonances fall into neither category; see W. Rothenstein and J. Chernick, *Nucl. Sci. Eng.* **7**, 454 (1960) and R. Goldstein and E. R. Cohen, *Nucl. Sci. Eng.* **13**, 132 (1962).

[16] This is a direct result of the fact that *s*-state scattering resonance widths vary as $E^{\frac{1}{2}}$ while the energy loss in a collision is directly proportional to E. For a discussion of resonance widths see J. M. Blatt and V. F. Weisskopf, *Theoretical Nuclear Physics*, Wiley, 1952.

TABLE 2†

1 Resonance energy E_R	2 Natural width Γ	3 Practical with ($kT = 0.025$)	4 Maximum energy loss in collision with U-238	5 Resonance escape probability
418	0.105	2.56	7.00	0.001950
278	0.065	1.87	4.65	0.003495
212	0.085	2.63	3.55	0.005021
192	0.165	5.66	3.21	0.007119
117.5	0.040	1.32	1.97	0.009170
90	0.0251	0.13	1.51	0.001104
81.3	0.0271	0.76	1.36	0.009596
36.9	0.0575	3.65	0.62	0.05820
21.0	0.0339	1.95	0.35	0.06755

† From K. T. Spinney, in BNL 433, p. 105. The practical width includes the effect of Doppler broadening (see Section 19). These are only a few (typical) U-238 resonances.

Since the resonance is assumed narrow, most of the contribution to the integral comes from lethargies outside the resonance, and we can neglect the difference between the integral in (II.31 a) and that obtained by replacing $\Sigma_s\varphi$ by its value outside the resonance, $1/\xi$. It is assumed that the non-resonance or potential scattering is constant, and that the resonance under consideration is far enough removed in energy from any neutron sources or other resonances so that φ has attained its asymptotic value. In this case, Equation (II.16d) applies.

Under these assumptions, Equation (II.31 a) becomes

$$N(u) = 1/\xi \tag{II.31 b}$$

since

$$\int F(u' \to u)\, \mathrm{d}u' = 1$$

The number of neutrons absorbed in du will be $N(u)\,\dfrac{\Sigma_a(u)}{\Sigma_t(u)}\,\mathrm{d}u$ since every scattering event is assumed to degrade the neutron to below the resonance. Then, the total number of neutrons absorbed in the resonance will be

$$\int \frac{\Sigma_a(u)}{\xi\Sigma_t(u)}\,\mathrm{d}u \tag{II.31 c}$$

and p, the resonance escape probability in the narrow resonance (NR) approximation is simply (since $\xi\Sigma_s\varphi = 1$ implies unit slowing down density)

$$p_{\mathrm{NR}} = 1 - \int \frac{\Sigma_a(u)}{\xi\Sigma_t(u)} \mathrm{d}u \approx \mathrm{e}^{-\int_{\mathrm{Res}} \frac{\Sigma_a(u)}{\xi\Sigma_t(u)} \mathrm{d}u} \tag{II.31d}$$

In the NRIA approximation it is assumed that a scattering event with an absorber atom does not change the neutron energy. Thus Equation (II.31c) must be modified by subtracting $\Sigma_s^U(u)$, the absorber scattering cross section, from $\Sigma_t(u)$. In this approximation we find:

$$p_{\mathrm{NRIA}} = 1 - \int \frac{\Sigma_a(u)}{\xi(\Sigma_t(u) - \Sigma_s^U(u))} \mathrm{d}u \approx \mathrm{e}^{-\int_{\mathrm{Res}} \frac{\Sigma_a(u)}{\xi(\Sigma_t(u) - \Sigma_s^U(u))} \mathrm{d}u} \tag{II.31e}$$

Note that here ξ is defined only in terms of potential scattering cross sections

$$\xi = \frac{\sum_{i \neq U} \Sigma_{pi} \xi_i}{\sum_{i \neq U} \Sigma_{pi}} \tag{II.31f}$$

Equations (II.31d) and (II.31e) are frequently applied to the computation of the resonance escape probabilities for individual resonances. For the case of many resonances, we have seen earlier that to a good approximation the flux recovers its asymptotic "$1/E$" spectrum between resonances and that the resonance escape probabilities of the individual resonances are independent. Thus the overall resonance escape probability is the product of the individual probabilities, i.e., it is obtained by summing the exponents in Equations (II.31d) or (II.31e) over all resonances, using the more appropriate of the two formulae for each resonance. The utility of these formulae lies not so much in their ability to predict the resonance escape probability of a given reactor (although for simple systems this is possible) as in their application to the development of "effective" absorption cross sections for numerical calculations. The means of doing this is outlined in the next section and is discussed more fully in Chapter IV.

15. Resonance Integrals

Examination of Equations (II.31d) and (II.31e) shows that for the case of a single absorber the resonance escape probability depends only on the properties of the absorber (including its concentration) and the properties ξ and Σ_p of the system. Since ξ is, by assumption, a constant, it may be taken out of the integrals, which are now parameterized only by Σ_p. This suggests that we write Equations (II.31d) and (II.31e) in the form

$$p = \exp\left[-\frac{N_u}{\xi \Sigma_p} I_{\text{eff}}\right] \tag{II.32a}$$

for I_{eff} is characterized simply by the microscopic cross section of the absorber and the quantity Σ_p/N_u. I_{eff} has the dimensions of a microscopic cross section and is called the *effective resonance integral.* Its physical interpretation may be obtained by comparison with Equation (II.20e), the constant capture result; it is the constant absorption cross section which would give the same resonance escape probability as the actual cross section.

Using Equation (II.32a) we see that for the narrow resonance (NR) case, I_{eff} is given by:

$$I_{\text{eff}}^{\text{NR}} = \Sigma_p \int \frac{\sigma_a^U(u)}{\Sigma_t(u)}\,\mathrm{d}u \tag{II.32b}$$

while in the NRIA case,

$$I_{\text{eff}}^{\text{NRIA}} = \Sigma_p \int \frac{\sigma_a^U(u)}{\Sigma_t(u) - \Sigma_s^U(u)}\,\mathrm{d}u \tag{II.32c}$$

If the absorber concentration is very small, $\Sigma_t(u) \approx \Sigma_p$ and both (II.32b) and (II.32c) approach the "infinitely dilute resonance integral"

$$I_\infty = \int \sigma_a(u)\,\mathrm{d}u \tag{II.32d}$$

which is larger than either the NR or NRIA effective resonance integrals. The effect which causes the resonance integrals (II.32b) and (II.32c) to be smaller than the infinitely dilute value is sometimes called "energy self-shielding" [17] and is simply the depletion of the resonance flux.

Again, we note that the resonance integrals can be written in terms of $\sigma_p = \dfrac{\Sigma_p}{N_u}$, the total potential scattering cross section per absorber atom and $\sigma_s(u)$, the resonance scattering cross section. In

[17] G. Goertzel, H. Feshbach and J. Yamanchi, *Nucl. Sci. Eng.* **1**, 4 (1956).

this notation Equations (II.32b) and (II.32c) become†

$$I_{\text{eff}}^{\text{NR}} = \int \frac{\sigma_a(u)}{1 + \dfrac{\sigma_a(u) + \sigma_s(u)}{\sigma_p'}} \, du \tag{II.32e}$$

$$I_{\text{eff}}^{\text{NRIA}} = \int \frac{\sigma_a(u)}{1 + \dfrac{\sigma_a(u)}{\sigma_p'}} \, du \tag{II.32f}$$

The distinction between NR and NRIA resonances can be quite important as scattering and absorption resonances always occur together. If the scattering cross section of the absorber in the resonance is comparable to or larger than the absorption cross section, it is also likely to be a large part of total scattering cross section. Then it is extremely important to choose the correct approximation, a point clearly illustrated by the results given in Table 3.

TABLE 3‡

1 Resonance energy E_R	Values of $(1 - p)$		
	2 Exact	3 NR	4 NRIA
418	0.001950	0.001816 (− 6.9%)	0.003855 (+97.7%)
278	0.003495	0.003318 (− 5.1%)	0.004936 (+41.2%)
212	0.005021	0.004440 (−11.6%)	0.007689 (+53.1%)
192	0.007119	0.005086 (−28.6%)	0.01228 (+73.0%)
117.5	0.009170	0.009035 (− 1.5%)	0.009500 (+ 3.6%)
90	0.001104	0.001114 (+ 0.9%)	0.000998 (− 9.6%)
81.3	0.009596	0.010163 (+ 5.9%)	0.008109 (−15.5%)
36.9	0.05820	0.04739 (−18.6%)	0.06110 (+ 5.0%)
21.0	0.06755	0.07455 (+10.4%)	0.07059 (+ 4.5%)

‡ From Spiney, BNL-433, p. 105. These resonances are the ones for which resonance parameters are given in Table 2.

Spinney [18] and others have done calculations for resonances which fit neither the NR nor the NRIA description. In addition, Spinney has compared the results predicted by the formulae just given with exact results obtained by numerical integration of the

† In the NRIA case scattering by the absorber is to be ignored in computing σ_p'.

[18] K. T. SPINNEY, *op. cit.*; ROTHENSTEIN and CHERNICK, *op. cit.*

slowing-down equations for the case of a uranium-hydrogen mixture; this comparison is shown in Table 3. Figure II.5, also taken from Spinney's work, compares the NR and NRIA calculations.

In some resonances, interference between the resonance scattering and potential cross sections reduces the absorber scattering cross section very sharply just below the resonance. Neumann [19] has made some calculations of the effect of this phenomenon on resonance absorption, but the reader is referred to the original work for greater detail.

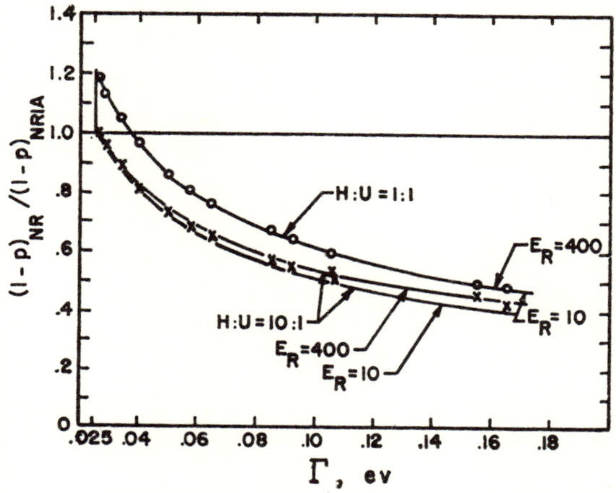

FIG. II.5. Comparison of narrow resonance approximation with narrow resonance with infinitely heavy absorber approximation for mixtures of hydrogen and uranium. After K. Spinney, BNL-433, p. 107.

The narrow resonance approximation can also be used for scattering resonances. Here, the concept of a resonance integral no longer has any meaning and effective cross sections for particular elements are calculated directly. The effective scattering cross section of element "i" is defined by

$$\Sigma^i_{s\,\mathrm{eff}} = \frac{\int \Sigma^i_s(u)\,\varphi(u)\,\mathrm{d}u}{\int \varphi(u)\,\mathrm{d}u} \tag{II.32g}$$

where the integrals extend over an arbitrary lethargy interval. In the narrow resonance approximation $\varphi = 1/\xi_0\Sigma_t$, where ξ_0 is the

[19] H. NEUMANN, BNL-433, p. 110.

non-resonant value of ξ and so we have

$$\sigma^i_{seff} = \frac{\displaystyle\int \frac{\sigma^i_s(u)\,du}{\sigma^i_{sr}(u) + \sigma'_p}}{\displaystyle\int \frac{du}{\sigma^i_{sr} + \sigma'_p}} \tag{II.32h}$$

where σ^i_{sr} is the resonance portion of the scattering cross section and σ'_p is the potential scattering cross section per "i" atom.

16. Corrections to the Six Factor Formula

It was mentioned earlier that although the narrow resonance formulae are valid only for individual resonances, the overall resonance escape probability can be obtained by multiplying the individual resonance escape probabilities, i.e., by summing the individual resonance integrals.

This method has the advantage of being readily adaptable to multi-group procedures, as will be seen in Chapter IV. As a simple example, suppose one wishes to obtain a better estimate of k_{eff} without greatly increasing the complexity of the formulae obtained in Chapter I. One procedure might be to break the fast group arbitrarily into two smaller groups and assign to each a resonance escape probability, say

$$\begin{aligned} p_1 &= p(E') \\ p_2 &= p(E' \rightarrow E_T) \end{aligned} \tag{II.33a}$$

where p_1 is averaged over a fission spectrum. In a similar fashion one could break up the non-leakage probability into $\lambda_1(E')$ and $\lambda_2(E' \rightarrow E_T)$. The fast effect is, as we have seen, usually independent of both p and λ. Using the resonance formulae just developed, it is a relatively simple matter (in principle) to calculate p_1 and p_2; the calculation of λ_1 and λ_2 will be indicated in Chapter III. Then, one could write for the multiplication factor

$$k = \frac{\eta f p_1 p_2 \lambda_1 \lambda_2 \varepsilon}{1 + L^2 B^2} \tag{II.33b}$$

which is presumably more accurate than the formulae of Chapter I as it allows for an interaction between the resonance escape and non-leakage probabilities neglected in the simple six factor formula.

By further breaking up the fast group one could obtain still more accurate expressions for k_{eff}, which in the limit of infinite subdivision must yield the exact result.

17. Closely Spaced Resonances

The assumption of wide separation of resonances is frequently not valid, e.g., U-238 has at least 20 strong resonances below 200 eV. However, the method just described is usually sufficiently accurate since even the strongest resonances capture only a small fraction of the neutrons passing through them. Thus the transients they introduced into the neutron spectrum cannot be more than a few per cent of the asymptotic flux in amplitude and may be ignored to good approximation.

The resonance escape probability for closely spaced resonances can be calculated more nearly rigorously by using the rapidly-varying capture solution already obtained in Section 11. The solution is, however, not readily adaptable to simple calculations. Another method of calculating the interference effect in resonance capture is due to Weinberg and Wigner [20].

Absorption in a sharp resonance may be considered a monoenergetic neutron sink, or, at worst, a superposition of such sinks. The effect of such sinks on the neutron flux is the introduction of transients of the Placzek function type. These transients must, of course, be subtracted from the asymptotic flux that would exist if the resonance were not present. One can write

$$\chi(u) = \frac{1}{\xi} - \int_0^u c_0(u')\,\chi(u')\,\phi_p(u', u)\,\mathrm{d}u' \qquad \text{(II.34a)}$$

where $\phi_p(u, u')$ is the Placzek function—the collision density at lethargy u due to a unit source of neutrons at u'. The Placzek function may be decomposed into an asymptotic $(1/E)$ part and a

[20] A. M. WEINBERG and E. P. WIGNER, BNL-433, p. 125. N. CORNGOLD has derived a variational principle for this problem and has also shown that the interference between a scattering resonance and an absorption resonance may be even larger than the type of interference considered here. See *Proc. Phys. Soc.* (London), **A70**, 793 (1958).

difference term:

$$\phi_p(u', u) = \frac{1 - \varphi_p(u', u)}{\xi} \tag{II.34b}$$

When this is substituted into Equation (II.34a) one finds

$$\xi\chi(u) = 1 - \int_0^u c_0(u')\,\chi(u')\,\mathrm{d}u' + \int_0^u c_0(u')\,\chi(u')\,\varphi_p(u, u')\,\mathrm{d}u' \tag{II.34c}$$

In Equation (II.34c), the last term represents the correction sought; omission of it leads once again to the narrow resonance formula. However, if the correction is small, as expected, the narrow resonance formula may be used to evaluate the last term of Equation (II.34c). That is, we set

$$\chi(u') = \chi(u) \exp\left\{\int_{u'}^{u} c(u'')\,\mathrm{d}u''\right\} \tag{II.34d}$$

where the positive exponent is chosen because $u' < u$. If one further defines

$$\int_0^u c_0(u')\,\varphi_p(u, u') \exp\left\{\int_{u'}^{u} c_0(u'')\,\mathrm{d}u''\right\} = \xi_1(u) \tag{II.34e}$$

the integral equation for $\chi(u)$ simplifies to

$$[\xi - \xi_1(u)]\,\chi(u) = 1 - \int_0^u c_0(u')\,\chi(u')\,\mathrm{d}u' \tag{II.34f}$$

This equation is exactly that involved in the narrow resonance problem except that ξ has been replaced by $[\xi - \xi_1(u)]$. Proceeding in the same manner as that previously followed, we obtain:

$$\chi(u) = \frac{1}{\xi - \xi_1(u)} \exp\left\{-\int_0^u \frac{c_0(u'')}{\xi - \xi_1(u'')}\,\mathrm{d}u''\right\} \tag{II.35a}$$

which yields the resonance escape probability:

$$p(u) = \exp\left\{-\int_0^u \frac{c_0(u')}{\xi - \xi_1(u')}\,\mathrm{d}u'\right\} \tag{II.35b}$$

This procedure is successful in obtaining the correction in a relatively simple form. However, the difficulty in calculating $\xi_1(u)$ and in evaluating Equation (II.34b) has rendered this method relatively inutilitarian.

F. Calculation of Resonance Integrals

18. Natural Line Shape

For a single isolated resonance, the Breit–Wigner formulae give good approximations to the line shape. Within an absorption resonance, the resonance portion of cross section is well fitted by [21]:

$$\sigma_a(E) = \sigma_0 \frac{\Gamma_\gamma}{\Gamma} \frac{\Gamma^2/4}{(E - E_0)^2 + (\Gamma/2)^2} \tag{II.36a}$$

where σ_0 is the maximum total cross section (at energy E_0), Γ_γ is the level width for gamma emission and Γ is the total level width. In terms of a new variable

$$x = \frac{E - E_0}{\Gamma/2} \tag{II.36b}$$

Equation (II.36a) becomes:

$$\sigma_a(E) = \frac{\sigma_0 \Gamma_\gamma/\Gamma}{1 + x^2} \tag{II.36c}$$

The infinitely dilute resonance integral for a single resonance of this type [see Eqn. (II.32d)] is:

$$I_\infty = \int_{-\infty}^{\infty} \frac{\sigma_0 \Gamma_\gamma/\Gamma}{1 + x^2} \frac{\Gamma}{2E} \mathrm{d}x \tag{II.37a}$$

Little error is introduced by extending the integral to $(-\infty, \infty)$ since the integrand is sharply peaked near $x = 0$ $(E = E_0)$. Furthermore, this enables us to set $E = E_0$ in the integrand without making a serious error and we may take it and Γ_γ, which is essen-

[21] See J. M. Blatt and V. F. Weisskopf, *op. cit.* Briefly, the behavior of the parameter is roughly: $\sigma_0 \sim \Gamma_n/E_0\Gamma_1$; $\Gamma = \Gamma_\gamma + \Gamma_n$, where Γ_n is the neutron level width; Γ_0 is the capture width and is nearly constant from resonance to resonance; $\Gamma_n = \Gamma_n^0 E^{\frac{1}{2}}$ where Γ_n^0 fluctuates from resonance to resonance but has an average value independent of E.

tially constant in energy, out of the integral. Thus,

$$I_\infty = \int_{\text{Res}} \sigma_a(E)\frac{dE}{E} = \frac{\sigma_0\Gamma_\gamma}{2E_0}\int_{-\infty}^{\infty}\frac{dx}{1+x^2} \tag{II.37b}$$

The integration is straightforward, and yields:

$$I_\infty = \frac{\pi}{2}\frac{\sigma_0\Gamma_\gamma}{E_0} \tag{II.37c}$$

This is the contribution to I_∞ from a single resonance line and can be evaluated if Γ_γ, σ_0, and E_0 are known. Since these quantities have to be determined experimentally, Equation (II.37c) can be applied only to those resonances which have been resolved.

Resonance integrals for the self-shielded case are also readily calculated. If we ignore interference between resonance and potential scattering the resonance scattering cross section is given by

$$\sigma_s(u) = \frac{\sigma_0\Gamma_n/\Gamma}{1+x^2} \tag{II.37d}$$

where Γ_n is the neutron width and

$$\Gamma = \Gamma_n + \Gamma_\gamma$$

Inserting Equations (II.36c) and (II.37d) into Equation (II.32e) we find

$$I_{\text{eff}}^{\text{NR}} = \frac{\sigma_0\Gamma_\gamma}{2E_0}\int_{-\infty}^{\infty}\frac{dx}{1+x^2+\dfrac{\sigma_0}{\sigma_p'}} \tag{II.37e}$$

Or, defining

$$\beta = \frac{\sigma_p'}{\sigma_0} \tag{II.37f}$$

we have

$$I_{\text{eff}}^{\text{NR}} = \frac{\pi}{2}\frac{\sigma_p'\Gamma_\gamma}{E_0}[\beta(1+\beta)]^{-\frac{1}{2}} = I_\infty\sqrt{\frac{\beta}{1+\beta}} \tag{II.37g}$$

For the NRIA case, a similar calculation gives

$$I_{\text{eff}}^{\text{NRIA}} = \frac{\pi}{2}\frac{\sigma_p'\Gamma}{E_0}[\beta'(1+\beta')]^{-\frac{1}{2}} \tag{II.37h}$$

where

$$\beta' = \frac{\sigma_p'}{\sigma_0}\frac{\Gamma_\gamma}{\Gamma} \tag{II.37i}$$

and absorber potential scattering is to be neglected in computing σ_p'.

The resonance integral cannot be evaluated using Equations (II.37c) or (II.37d) alone. At low energies the absorption cross section is approximately "$1/v$" and its contribution must be included:

$$\int_{E_T}^{\infty} \sigma_{a1/v}(E)\frac{dE}{E} = \sigma_{aT}(0.025)^{\frac{1}{2}} \int_{E_T}^{\infty} \frac{dE}{E^{3/2}} = \frac{\sigma_{aT}}{2} \qquad \text{(II.37j)}$$

where E_T is taken as 0.4 eV, the "cadmium cutoff" energy, which is convenient for experimental work; σ_{aT} is actually σ_a (0.025 eV) and not an averaged thermal value.

Finally, at high energies the resonances can no longer be resolved. Here one has to make some guess, on statistical grounds, as to the resonance parameters. Hence, this energy range is called the statistical region. Rather than enter into the details of these calculations, which are rather involved, we shall quote only the results [22]. They are given in Tables 4 and 5 for U-238 and Th-232 and include the effects of temperature on resonances, dissussed in the next section.

In order to evaluate effective scattering cross sections one may substitute Equation (II.37d) into Equation (II.32h). Care must be

TABLE 4. DRESNER'S CALCULATION OF THE EFFECTIVE RESONANCE INTEGRAL OF HOMOGENEOUS MIXTURES OF URANIUM

Temperature	σ_p (barns)	Contribution of resolved resonances (barns)	Contribution of unresolved resonances (barns)	Total (barns)
0°K	20	7.9	1.7	9.6
	200	24	4.8	28.8
	2000	74	11	85
300°K	20	8.1	3.0	11.1
	80	17	6.6	23.6
	200	27	9.4	36.4
	2000	86	12	98

[22] For a discussion of this and other fine points not fully discussed here see L. DRESNER, *Resonance Absorption of Neutrons in Nuclear Reactors*, Pergamon Press, 1960.

TABLE 5. DRESNER'S CALCULATION OF THE EFFECTIVE RESONANCE INTEGRAL OF HOMOGENEOUS MIXTURES OF THORIUM

Temperature	σ_p (barns)	Contribution of resolved resonances (barns)	Contribution of unresolved resonances (barns)	Total (barns)
0°K	20	3.7	2.2	5.9
	200	12	6.3	18.3
	2000	33	14	47
300°K	20	4.0	4.0	8.0
	80	8.9	8.6	17.5
	200	15	12	27
	2000	42	20	62

TABLE 6. MEASURED RESONANCE INTEGRALS I_∞ (10^{-24} cm^2)

Nuclide	Activation	Absorption
$_{24}Cr$		1.9
$_{26}Fe$		2.1
$_{40}Zr$		3
$_{47}Ag^{107}$	74	
$_{47}Ag^{109}$	1160	
$_{49}In^{113}$	1050	
$_{49}In^{115}$	2640	
$_{62}Sm^{152}$	$>$ 1750	
$_{63}Eu^{153}$	950	
$_{79}Au^{197}$†	1558	
$_{90}Th^{232}$	69.8	
$_{92}U^{238}$	283	

† Adopted as standard, using $\sigma_{th} = 98$ barns.

taken in performing the integrals, however, as they diverge if the limit is on the range $(-\infty, \infty)$. If we take the integral only over an energy range ΔE large compared with a resonance width the result is:

$$\sigma^i_{s\,\mathrm{eff}} = \sigma^i_p + \frac{\pi}{2\Delta E}\sigma_0 \Gamma_n [\beta(1+\beta)]^{-\frac{1}{2}} \qquad \text{(II.37k)}$$

where β is defined by Equation (II.37f). If many resonances are included in the interval ΔE the last term is to be summed over all resonances.

19. Doppler Broadening

The foregoing has been based on the assumption that the absorber nuclei are stationary, which is generally a rather good approximation if cross sections are slowly varying. As we shall soon see, the thermal motion of nuclei results in an apparent broadening of resonance lines and an accompanying increase in absorption; omission of the "Doppler broadening" may lead to underestimation of the effective resonance integral by a large amount.

Clearly, the cross section can depend only on the relative velocity of the nucleus and neutron. Therefore, to take into account the fact that nuclei are in motion one must average the cross section over the velocity distribution of the nuclei. One can thus write:

$$v\sigma_{a\,\mathrm{eff}}(v) = \int |\mathbf{v} - \mathbf{V}|\, \sigma_a(|\mathbf{v} - \mathbf{V}|)\, P(\mathbf{V})\, \mathrm{d}^3 V \qquad \text{(II.38a)}$$

where $\mathbf{v}$ and $\mathbf{V}$ are the laboratory system velocities of the neutron and the absorber respectively, and $P(\mathbf{V})$ is the distribution of absorber velocities. Note that

$$\frac{V}{v} \approx \left(\frac{kT/A}{E}\right)^{\frac{1}{2}} \ll 1 \qquad \text{(II.38b)}$$

at normal temperatures.† Here T is the absolute absorber temperature, A the mass of the absorber, and E the energy of neutron. Then, except for a few very high energy nuclei:

$$|\mathbf{v} - \mathbf{V}| \approx v \qquad \text{(II.38c)}$$

Under this approximation Equation (II.38a) becomes:

$$\sigma_{a\,\mathrm{eff}}(v) = \int \sigma_a(|\mathbf{v} - \mathbf{V}|)\, P(\mathbf{V})\, \mathrm{d}^3 V \qquad \text{(II.38d)}$$

The relative kinetic energy is

$$E' = \tfrac{1}{2} m\, |\mathbf{v} - \mathbf{V}|^2 \qquad \text{(II.38e)}$$

† For U-238 at 300°K, V/v is about 0.004 at the (first) 6.7 eV resonance. At higher energy resonances, the ratio is even smaller. A more careful treatment is necessary for the very low energy ($\lesssim 1\,\mathrm{eV}$) resonances that occur, for example, Pu-240.

where m is the mass of the neutron. Choosing the direction of $\mathbf{v}$ as the z-axis, we observe that for $V/v \ll 1$ (see Fig. II.6):

$$|\mathbf{v} - \mathbf{V}|^2 = v^2 + V^2 - 2vV\cos\theta \approx v^2 - 2vV_z \qquad \text{(II.38f)}$$

so that the relative kinetic energy is

$$E' = \tfrac{1}{2}mv^2 - mvV_z = E-(2mE)^{\frac{1}{2}}\, V_z \qquad \text{(II.38g)}$$

Under the approximations made here, the relative energy E' is therefore a function only of the component of $\mathbf{V}$ in the direction of v. The integration of Equation (II.38e) over V_x and V_y is therefore trivial and we have

$$\sigma_{a\,\mathrm{eff}}(v) = \int \sigma(|\mathbf{v} - \mathbf{V}|)\, P(V_z)\, \mathrm{d}V_z \qquad \text{(II.39a)}$$

where $P(V_z)$ is the one-dimensional velocity distribution of the absorber nuclei.

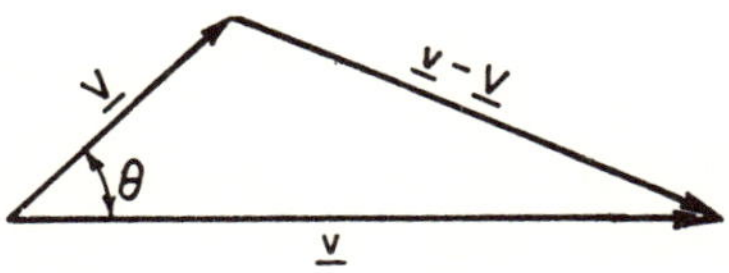

FIG. II.6. Kinematics for computation of the Doppler Effect.

Using Equation (II.38g), we have

$$\sigma_{a\,\mathrm{eff}}(E) \approx \int \sigma(E')\, P\left(\frac{E - E'}{(2mE)^{\frac{1}{2}}}\right) \frac{\mathrm{d}E'}{(2mE)^{\frac{1}{2}}} \qquad \text{(II.39b)}$$

The appropriate distribution of nuclear velocities is the Maxwell–Boltzmann Distribution,† mentioned in Chapter I, Section 1. The distribution of V_z is:

$$P(V_z)\,\mathrm{d}V_z = \left(\frac{M}{2\pi kT}\right)^{\frac{1}{2}} \exp\left\{-\frac{MV_z^2}{2kT}\right\} \mathrm{d}V_z \qquad \text{(II.39c)}$$

where M is the mass of the nucleus. Using the Breit–Wigner line shape (II.30a) for $\sigma_a(E)$ and Equation (II.39c) for $P(V_z)$, the integral

† This distribution is not quite correct in some crystals, but is a sufficiently good approximation for the current application.

in Equation (II.39b) may be written explicitly as

$$\sigma_{a\,\mathrm{eff}}(E) = \int \sigma_0 \frac{\Gamma_\gamma}{\Gamma} \frac{\Gamma^2/4}{(E' - E_0)^2 + (\Gamma/2)^2} \left(\frac{M}{2\pi kT}\right)^{\frac{1}{2}} \times$$

$$\times \exp\left\{\frac{-M(E - E')^2}{2mE\,2kT}\right\} \frac{\mathrm{d}E'}{2mE} \tag{II.40a}$$

Use of the substitutions,

$$y = \frac{E' - E_0}{\Gamma/2} \tag{II.40b}$$

$$\xi = \Gamma/\Delta = \left(\frac{\Gamma^2 A}{4kTE}\right)^{\frac{1}{2}} \tag{II.40c}$$

$$\Delta = \left(\frac{4kTE}{A}\right)^{\frac{1}{2}} \quad A = \frac{M}{m} \tag{II.40d}$$

and the previous definition of x, Equation (II.36b), reduces Equation (II.40a) to

$$\sigma_{a\,\mathrm{eff}}(E) = \sigma_0 \frac{\Gamma_\gamma}{\Gamma} \frac{\xi}{2\sqrt{\pi}} \int_{-\infty}^{\infty} \frac{\exp\left\{-\frac{\xi^2}{4}(x - y)^2\right\}}{1 + y^2} \mathrm{d}y \tag{II.40e}$$

which gives the temperature dependence of the effective absorption cross section in a resonance. The quantity Δ, which has the dimensions of energy, is known as the *Doppler width* of the resonance, as it plays a role similar to that of Γ in the Breit–Wigner formula. As T increases, the Doppler width increases; i.e., the resonance broadens.

$$\psi(\xi, x) = \frac{\xi}{2\sqrt{\pi}} \int_{-\infty}^{\infty} \frac{\exp\left\{-\frac{\xi^2}{4}(x - y)^2\right\}}{1 + y^2} \mathrm{d}y \tag{II.40f}$$

is called the Doppler broadened line shape, and is the temperature-dependent counterpart of the natural line shape $(1 + x^2)^{-1}$ previously considered.

Some properties of the function $\psi(\xi, x)$ may be briefly noted. First,

$$\lim_{\xi \to \infty} \psi(\xi, x) = \int_{-\infty}^{\infty} \frac{\delta(x - y)}{1 + y^2} \mathrm{d}y = (1 + x^2)^{-1} \tag{II.41a}$$

since

$$\lim_{\xi\to\infty}\left[\frac{\xi}{2\sqrt{\pi}}\exp\left\{-\frac{\xi^2}{4}(x-y)^2\right\}\right]=\delta(x-y) \qquad \text{(II.41 b)}$$

The significance of Equation (II.41 a) is somewhat obvious; $\xi \to \infty$ implies $kT \to 0$. Thus, in the limit of stationary absorber nuclei, the Doppler broadened line shape reduces to the natural line shape, as expected. Another property is

$$\int_{-\infty}^{\infty} \psi(\xi, x)\,\mathrm{d}x = \pi \qquad \text{(II.41 c)}$$

which is readily shown by integrating first over x and then over y; this is independent of ξ. Thus, the area under the absorption cross section curve is independent of temperature, but the curve is broader, and not as high at the peak. This is the key to understanding of the phenomenon of the Doppler effect. Absorption resonances are generally strong enough to cause a lump to be "black", i.e., impenetrable, to neutrons with energies near the resonance peak. The Doppler effect does not lower the peak sufficiently to make the lump significantly less "black", but it does increase the energy range over which the lump absorbs. Hence, Doppler broadening increases the neutron absorption in a given resonance.

The Doppler broadened line shape may be substituted into the formulae for I_{NR} and I_{NRIA} to yield the temperature dependent resonance escape probabilities for single resonances.† The results of these calculations are:

$$I_{NR} = \frac{\sigma_p'\Gamma_\gamma}{E_0} J(\xi, \beta) \qquad \text{(II.41 d)}$$

where

$$J(\xi, \beta) = \int_0^{\infty} \frac{\psi(\xi, x)}{\psi(\xi, x) + \beta}\,\mathrm{d}x \qquad \beta = \frac{\sigma_p'}{\sigma_0} \qquad \text{(II.41 e)}$$

is a tabulated function (see p. 104, reference in [22]).

For the NRIA case a similar calculation gives

$$I_{NRIA} = \frac{\sigma_p'\Gamma}{E_0} J(\xi, \beta') \qquad \beta' = \frac{\sigma_p'}{\sigma_0}\frac{\Gamma_\gamma}{\Gamma} \qquad \text{(II.41 f)}$$

where absorber potential scattering is to be neglected in computing σ_p'.

† Note that as a result of Equation (II.41 c), I_∞ is not temperature dependent.

G. Resonance Integrals in Heterogeneous Systems

20. Effects of Heterogeneities

Heretofore, it has been assumed that the reactor under consideration is homogeneous, but it was also mentioned that most reactors are, in the strictly physical sense, heterogeneous. We have indicated that a heterogeneous system might be replaced by a fictitious homogeneous system, if it is composed of repeating substructures or cells. In this section, we will briefly describe some of these cell calculations but great detail will not be given. The reader is once again refered to Dresner's work or to Nordheim's [23] excellent review article.

In the early days of reactor analysis, it was realized that a heterogeneous system could have a higher k_∞ than the same materials in homogeneous form. This results from several phenomena. First, neutrons born in a lump have higher probabilities of colliding with fuel nuclei on their first few collisions and thus increase the fast effect. Then they leave the lump and slow down in the moderator. At resonance energies few neutrons are able to penetrate to the center of the lumps so that the absorber in the center of the lump absorbs fewer neutrons than it would in a homogeneous medium. Thus fewer resonance neutrons are captured and the resonance escape probability increases. This effect also works to reduce the thermal utilization f but the overall effect can be an increase in the infinite multiplication constant k_∞ if the reactor is properly designed. The effect on the resonance escape probability and the fast effect will be briefly indicated here.

21. Effect on Resonance Escape Probability

In order to compute resonance escape probabilities in heterogeneous systems, we shall consider a binary lattice of absorber (region "0") and moderator (region "1"). It has been shown [24] that the flux in such a system is spatially constant at energies far enough removed from neutron sources or sinks. (Close to sources

[23] L. W. Nordheim, Report GA-638 (General Atomics, 1960).

[24] W. T. Hays, M. Luming and P. F. Zweifel, *Trans. Amer. Nuc. Soc.* **3**, 366 (1960).

or sinks, spatial oscillations, similar in nature to the Placzek oscillations in energy, described in Section 5, occur.)

Assuming, then, a spatially constant flux above the resonance energy, we may break the resonance absorption into two parts:

(a) Those neutrons which, before absorption, suffered their last collisions in the absorber (defined as n_0). The volume of the absorber will be denoted by V_0.

(b) Those neutrons which, before absorption, suffered their last collision in the moderator (defined as n_1). The moderator volume is V_1.

Making use of the assumed flat distribution of the flux above the resonance, one can write:†

$$n_0 = V_0 \int_{\text{Res}} \frac{\Sigma_a^0(u)}{\Sigma_t^0(u)} P_{00}(u)\,\mathrm{d}u \int_{u-\ln 1/\alpha_0}^{u} \Sigma_p^0 \varphi(u')\, F_0(u' \to u)\, \mathrm{d}u' \tag{II.42a}$$

Here $\varphi(u)$ is the flux at lethargy u, and $P_{ij}(u)$ is the probability that a neutron, born isotropically and uniformly in region i at lethargy u, will suffer its first collision in region j (i and j being either 0 or 1). The subscript "Res" indicates that the integral is to be taken only over the lethargy range for which the resonance absorption is strong, i.e., approximately over the practical width. Similarly:

$$n_1 = V_1 \int_{\text{Res}} \frac{\Sigma_a^0(u)}{\Sigma_t^0(u)} P_{10}(u)\,\mathrm{d}u \int_{u-\ln 1/\alpha_1}^{u} \Sigma_p^1 \varphi(u')\, F_1(u' \to u)\, \mathrm{d}u' \tag{II.42b}$$

The asymptotic flux in this case is given by:

$$\varphi(u) = \frac{q}{\bar{\xi}\Sigma_p} \tag{II.42c}$$

where

$$V\bar{\xi}\Sigma_p = V_0\xi_0\Sigma_p^0 + V_1\xi_1\Sigma_p^1, \quad V = V_1 + V_0 \tag{II.42d}$$

and q is the slowing-down density in the cell above the resonance. Now, noting that Σ_p is constant (or nearly so) in energy, and neglecting the resonance width compared to both $\ln 1/\alpha_i$ as in

† It is also assumed here that each region contains only one element. This is merely for simplicity and the result is easy to generalize.

Section 14, we have

$$n_0 = \frac{q\Sigma_p^0}{\xi\Sigma_p} V_0 \int_{\text{Res}} \frac{\Sigma_a^0(u)}{\Sigma_t^0(u)} P_{00}(u)\, du \tag{II.43a}$$

and

$$n_1 = \frac{q\Sigma_p^1}{\xi\Sigma_p} V_1 \int_{\text{Res}} \frac{\Sigma_a^0(u)}{\Sigma_t^0(u)} P_{10}(u)\, du \tag{II.43b}$$

Now, defining the resonance escape probability as the ratio of the number of neutrons absorbed to the number slowing down above the resonance, we have:

$$1 - p = \frac{n_0 + n_1}{qV} = \frac{N_u}{\xi\Sigma_p} I_{\text{eff}} \tag{II.43c}$$

Thus

$$I_{\text{eff}}^{\text{NR}} = \frac{V_0\Sigma_p^0}{V} \int_{\text{Res}} \frac{\sigma_a^0(u)}{\Sigma_t^0(u)} P_{00}(u)\, du + \frac{V_1\Sigma_p^1}{V} \int_{\text{Res}} \frac{\sigma_a^0(u)}{\Sigma_t^0(u)} P_{10}(u)\, du \tag{II.43d}$$

Now, it follows from the definitions of the quantities that

$$P_{01}(u) = 1 - P_{00}(u) \tag{II.44a}$$

and, under the assumption of flat flux above the resonance, $P_{01}(u)$, the probability that a neutron born at lethargy u in the absorber suffers its first collision in the moderator is formally identical with the one-velocity escape probability for purely absorbing media, E, defined by Case, DeHoffmann, and Placzek [25]. (Only the first collisions need be considered when the NR approximation is being invoked.) Furthermore, $P_{01}(u)$ is related to $P_{10}(u)$ through the reciprocity relation for one velocity neutron transport [26]:

$$V_0\Sigma_t^0(u)\, P_{01}(u) = V_1\Sigma_p^1 P_{10}(u) \tag{II.44b}$$

[25] CASE, PLACZEK, DEHOFFMANN, *op. cit.* For the *single* pin case or cases in which interactions with neighboring cells may be ignored, one may use the escape probabilities as tabulated in this reference. In the more practical case, escape probabilities for the entire lattice must be computed. See the section on Dancoff corrections.

[26] This result follows quite simply from the more general one velocity reciprocity relation given in Reference [25]. One should also note that Equation (II.44b) is always true, not only for the single pin case.

Thus, the effective resonance integral reduces to

$$I_{\text{eff}}^{\text{NR}} = \frac{V_0}{V}\left\{\int_{\text{Res}} E(u)\,\sigma_a^0(u)\,\frac{(\Sigma_t^0(u) - \Sigma_p^0)}{\Sigma_t^0(u)}\,\mathrm{d}u + \Sigma_p^0 \int \frac{\sigma_a^0(u)}{\Sigma_t^0(u)}\,\mathrm{d}u\right\} \tag{II.44c}$$

In the development above, we have assumed some sort of unit *cell* of volume for which we are computing the resonance escape probability. For a single pin in the midst of an infinite homogeneous medium (a situation roughly approximated in measurements of I_{eff}) note that I_{eff} as we have defined it goes to zero because $V_1 \to \infty$. Such a situation is reasonable physically since we begin with an infinite number of neutrons and only a finite number are captured. Thus p must be unity.

For this reason, some authors lump the volume factor V_0/V into the coefficient of I_{eff}, i.e.,

$$1 - p = \frac{N}{\bar{\xi}\Sigma_p V} R_{\text{eff}}$$

where $N = N_U V_0$ = the total number of absorber atoms (in a cell), and $\xi\Sigma_p V$ is the *cellular* slowing down power. $R_{\text{eff}} = I_{\text{eff}} V/V_0$, and does not vanish as $V_1 \to \infty$. Note that for a homogeneous medium $R_{\text{eff}} = I_{\text{eff}}$. In addition, Equation (II.44c) can be written in terms of the microscopic cross sections rather than macroscopic cross sections so long as we employ the artifice of using $\sigma_p^{0'}$, the potential scattering cross section (in the lump) per absorber atom.

In the more general case in which moderator may be mixed with the absorber (as in UO_2 fuel elements), Σ_p^0 must include the scattering cross section, per absorber atom, of any moderating element mixed with the absorber, e.g., oxygen in uranium oxide fuel elements.

For the NRIA approximation, the situation is somewhat more complicated than in the NR case because collisions with absorber atoms are assumed not to degrade the neutron energy. Such collisions affect the P_{ij} since they change the direction of travel of the neutrons. Thus, the P_{ij} which are required in the NRIA approximation cannot be given in terms of the one velocity escape probability for a purely absorbing medium, E, but are the escape probabilities for absorbing and scattering bodies calculated for slabs by

Stewart *et al.* [27] and for cylinders by Stuart [28] and Stuart and Woodruff [29]. These works calculate the escape probability for bodies which are surrounded by scattering media. To obtain the result required here, we must set

$$\text{“}\Sigma_a^0(u)\text{”} = \Sigma_{au}^0(u) + \Sigma_{pm}^0(u) \tag{II.45a}$$

where $\Sigma_{au}^0(u)$ is the fuel absorption cross section and $\Sigma_{pm}^0(u)$ is the (potential) scattering cross section of any moderator mixed with the fuel. Also

$$\text{“}\Sigma_s^0(u)\text{”} = \Sigma_{su}^0(u) \tag{II.45b}$$

is the "absorber scattering cross section" and

$$\Sigma_s^1(u) = 0 \tag{II.45c}$$

since a collision with the surrounding moderator will degrade the neutron energy to below the resonance and thus can be considered to be an absorption. A first order correction to E for this case is given by

$$P_{01}(u) = \frac{E(u)}{1 - \dfrac{\text{“}\Sigma_s^0(u)\text{”}}{\Sigma_t^0(u)}[1 - E(u)]} \tag{II.45d}$$

a result first derived by Stuart [30]. In addition, $\Sigma_t^0(u)$ must be replaced by $\Sigma_t^0(u) - \Sigma_{su}^0(u)$ in the factor $\sigma_a^0(u)/\Sigma_t^0(u)$ that appears in Equation (II.43d). (The reciprocity relation (II.44b) may, however, be retained intact.) The resonance integral in the NRIA approximation thus becomes:

$$I_{\text{eff}}^{\text{NRIA}} = \frac{\Sigma_p^0 V_0}{V} \int\limits_{\text{Res}} \frac{\sigma_a^0(u)}{\Sigma_t^0(u) - \Sigma_{su}(u)}\, du$$

$$+ \frac{V_0}{V} \int\limits_{\text{Res}} \sigma_a^0(u) \left[\frac{\Sigma_a^0(u)}{\Sigma_t^0(u) - \Sigma_{su}^0(u)}\right] \frac{E(u)}{1 - \dfrac{\text{“}\Sigma_{su}^0(u)\text{”}}{\Sigma_t^0(u)}[1 - E(u)]}\, du \tag{II.45e}$$

These equations have been integrated numerically by Adler, Hinman, and Nordheim [31], who considered the effects of Doppler

[27] N. C. Francis, J. C. Stewart, L. S. Bohl and T. J. Krieger, PICG P/627 (1958).
[28] G. W. Stuart, *Nucl. Sci. Eng.* **2**, 617 (1957).
[29] G. W. Stuart and R. W. Woodruff, *Nucl. Sci. Eng.* **3**, 339 (1958).
[30] G. W. Stuart, *op. cit.*
[31] F. T. Adler, G. W. Hinman, and L. W. Nordheim, PICG P/1988 (1958).

broadening.† Their results display very good agreement with the experimental results of Hellstrand (see Table 7).

It should be noted, however, that Adler's calculations omitted the interference between resonance and potential scattering, an effect studied by Neumann, *op. cit.* Adler also found that the factor $[1 - E(u)$ "$\Sigma_{s_1}^0(u)$"$/\Sigma_t^0(u)]^{-1}$ did not appreciably aid agreement between theory and experiment, i.e., by completely ignoring the effect of multiple collisions in the NRIA case good values for I_{ef} were obtained. This may be due to cancellation of errors introduced by the two neglects, one of interference between resonance and potential scattering, and the other of multiple scattering.

TABLE 7. RESONANCE ESCAPE PROBABILITIES FOR CYLINDRICAL RODS OF U^{238}

Radius	Calculated [31]	Experimental [32]
0.422 cm	15.2	15.21
0.845 cm	11.4	11.26
1.68 cm	8.7	8.72
3.38 cm	7.0	7.21

To apply Adler's results to calculations of R_{eff}, it is necessary to determine first for each resonance whether it falls into the NR or NRIA category. He then writes

$$R_{eff} = R_V + R_S$$

where R_V, the "volume term", is the first term of Equation (II.44c) and R_S, the "surface term", is the second. (The reason for this notation is discussed in the next section.) For the NR case,

$$R_V^{NR} = \frac{\sigma_p^0 \Gamma_\gamma}{E_0} J(\xi, \beta) \tag{II.46a}$$

$$R_S^{NR} = \frac{\sigma_0 \Gamma_\gamma}{E_0} \int_{-\infty}^{\infty} \mathrm{d}x \, P_{01}\left[t\left(\frac{\psi}{\beta} + 1\right)\right] \frac{\psi^2}{\psi + \beta} = \frac{\sigma_0 \Gamma_\gamma}{E_0} L(t, \xi, \beta) \tag{II.46b}$$

† To take Doppler broadening into account, one merely uses the Doppler line shape for σ_a. Otherwise, the formulae given above remain intact.

[32] E. HELLSTRAND, *J. Appl. Phys.* **28**, 1493 (1957).

where σ_0, Γ_γ, β, and E_0 are resonance parameters previously defined, and $t = 4V_0\Sigma_p^0/S_0$ where S_0 is the surface area of the lump. Since $4V_0/S_0$ is the mean chord length of a lump, t is the mean chord length in units of $(\Sigma_p^0)^{-1}$. The function $J(\xi, \beta)$ was defined by Equation (II.41 e) and Equation (II.46 b) defines $L(t, \xi, \beta)$. Extensive tables of these functions are given by Adler and Nordheim [33].

A similar treatment is possible for the NRIA case with some redefinitions:

$$R_V^{\mathrm{NRIA}} = \frac{\sigma_{pm}\Gamma}{E_0} J(\xi, \beta') \qquad \text{(II.47 a)}$$

$$R_S^{\mathrm{NRIA}} = \frac{\sigma_{pm}\Gamma}{\beta' E_0} L(t', \xi, \beta') \qquad \text{(II.47 b)}$$

where σ_{pm} is the scattering cross section per absorber atom of any moderator contained in the lump, β' is defined by Equation (II.41 f) and

$$t' = \frac{4V_0\Sigma_{pm}}{S_0} \qquad \text{(II.47 c)}$$

For the case in which no moderator is mixed with the absorber, $\beta' = 0$ and R_V^{NRIA} becomes zero. However, this is clearly incorrect as some neutrons will always be slowed down into the resonance. We can make an approximate correction for this by saying that all neutrons slowed down into the resonance are captured. Then R_V^{NRIA} becomes simply the slowing-down density and the number of neutrons captured per absorber nucleus will be

$$R_V^{\mathrm{NRIA}} = \xi_u \sigma_{pu} = \frac{2}{A}\sigma_{pu} \qquad \text{(II.48 a)}$$

The surface term may, however, be correctly computed by setting $\sigma_{pm} = 0$ in Equation (II.47 b). Equation (II.47 b) then reduces to

$$R_S^{\mathrm{NRIA}} = \frac{\sigma_0\Gamma_\gamma}{E_0} L'(t'', \xi) \qquad \text{(II.48b)}$$

where

$$L'(t'', \xi) = \int_{-\infty}^{\infty} \mathrm{d}x\, P_{10}(t''\psi(\xi, x))\, \psi(\xi, x) \qquad \text{(II.48 c)}$$

$$t'' = \frac{4V_0\,\Sigma_{a0}}{S_0} \qquad \text{(II.48 d)}$$

[33] F. T. Adler and L. W. Nordheim, *Tables for the Computation of Resonance Integrals*, GA-377 (1958).

and

$$\Sigma_{a0} = \frac{N_u \sigma_0 \Gamma_\gamma}{\Gamma} \tag{II.48e}$$

is the macroscopic absorption cross section at the peak of resonance ignoring Doppler broadening.

22. The Rational Approximation

Wigner noted that for small bodies $E \approx 1$, and for large bodies,

$$E \approx S/4V\Sigma$$

where S is the surface area and V the volume and so devised the well-known rational approximation

$$E = \frac{S_0/4V_0\Sigma_t}{1 + S_0/4V_0\Sigma_t} \tag{II.49a}$$

which gives both limits correctly. Fortunately, by using the Wigner rational approximation, much of the hard work described above can be avoided if extremely accurate answers are not required.

When this approximation is applied to the NR formula (II.44c) one obtains

$$I^{\mathrm{NR}} = \frac{S_0}{4V} \int_{\mathrm{Res}} \frac{\sigma_a^0(u)\,[\Sigma_t^0(u) - \Sigma_p^0]}{[\Sigma_t^0(u)]^2 \left(1 + \dfrac{S}{4V_0\Sigma_t^0(u)}\right)}\,\mathrm{d}u + \frac{V_0\Sigma_p^0}{V} \int_{\mathrm{Res}} \frac{\sigma_a^0(u)}{\Sigma_t^0(u)}\,\mathrm{d}u \tag{II.49b}$$

An interesting result can be arrived at by rearranging Equation (II.49b) and writing the result in terms of R_{eff}:

$$R_{\mathrm{eff}}^{\mathrm{NR}} = \int \frac{\sigma_a(u)\left[\Sigma_p^0 + \dfrac{S_0}{4V_0}\right]}{\Sigma_t(u) + \dfrac{S_0}{4V_0}}\,\mathrm{d}u \tag{II.49}$$

On comparing this result with Equation (II.32b), we see that $R_{\mathrm{eff}}^{\mathrm{NR}}$ is identical to $I_{\mathrm{eff}}^{\mathrm{NR}}$, the resonance integral for a homogeneous system, if Σ_p is replaced by $\Sigma_p^0 + S_0/4V_0$. A similar analysis with respect to the NRIA approximation reveals that $R_{\text{eff}}^{\text{NRIA}}$ becomes identical to the resonance integral $I_{\text{eff}}^{\text{NRIA}}$ for a homogeneous system when the same substitution is made.

The consequences of this result are many-fold. Firstly it gives rise to the equivalence theorem that states that any two systems with the same $(\Sigma_p^0 + S_0/4V_0)$ have the same resonance integral regardless of what the individual quantities Σ_p^0 and $S_0/4V_0$ may be. This result may be used whenever results are desired only to the accuracy of the rational approximation. Since the rational approximation is always valid to within 10%, the results obtained by using this equivalence theorem are generally quite satisfactory. This theorem was first pointed out by Chernick and Vernon [34]. This result also leads to the interpretation of $S_0/4V_0$ as an "escape cross section", i.e., a cross section for removal of a neutron from the lump. In this fashion, the rational approximation may be interpreted as stating that the escape probability, E, is the ratio of the escape cross section to the total (escape plus actual) cross section [35].

The equivalence theorem allows one to write an analytical expression for the resonance integral in the absence of Doppler broadening. Equation (II.37g) may be used provided Σ_p is replaced by $\Sigma_p + S/4V$. Thus, if we redefine:

$$\beta = \frac{\Sigma_p + \dfrac{S_0}{4V_0}}{N^u \sigma_0} \tag{II.50a}$$

we have:

$$R_{\text{eff}}^{\text{NR}} = \frac{\pi}{2} \frac{(\sigma_p' + S_0/4V_0 N_u)\, \Gamma_\gamma}{E_0 \sqrt{\beta(1+\beta)}} \tag{II.50b}$$

Some of the limits of this equation are interesting. When $\sigma_p' + S_0/4VN_u$ is small compared to σ_0 (i.e. for large resonances) β may be neglected compared to unity. Then

$$R_{\text{eff}}^{\text{NR}} = \frac{\pi}{2} \frac{\Gamma_\gamma}{E_0} \left(\sigma_0 \left(\sigma_p' + \frac{S_0}{4V_0 N_u}\right)\right)^{\frac{1}{2}} \tag{II.51a}$$

[34] J. Chernick and E. Vernon, *Nucl. Sci. Eng.* **4**, 649 (1958).
[35] L. W. Nordheim, GA-638 (1959).

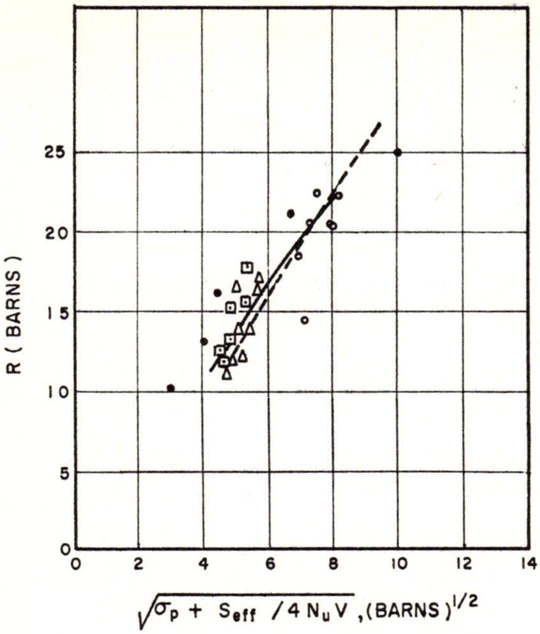

FIG. II.7. U-238 resonance integral.

The experimental data are from the following sources:

● Homogeneous systems; A. M. Weinberg and E. P. Wigner, *op. cit.*, p. 313.
□ Metal rods; H. Kouts *et al.*, PICG P/1841 (1958).
Δ Metal rods; D. Klein *et al.*, *Nucl. Sci. Eng.* **3**, 403 (1958).
⊙ Oxide rods; *Ibid.*
--- Metal rods; E. Hellstrand, *J. Appl. Phys.* **28**, 1443 (1957).
— Oxide rods; *Ibid.*

Hence it has been suggested by Levine [36] that empirical fits to resonance integral data should be given in the form

$$I_0 = f\left(\sigma_p + \frac{S_{\text{eff}}}{4VN_u}\right) \tag{II.51b}$$

where S_{eff} is an effective surface area which includes a correction for interaction between lumps as discussed in the next section. As shown in Figs. II.7 and II.8, this representation is a rather good

[36] M. M. LEVINE, *Nucl. Sci. Eng.* **8**, 363 (1960).

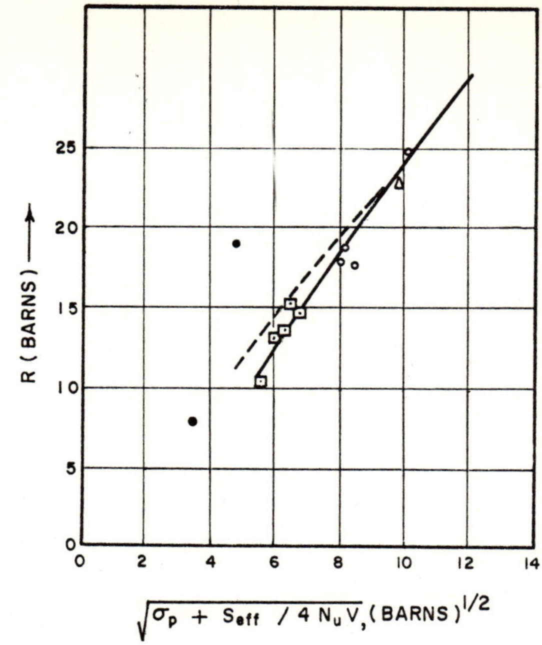

FIG. II.8. Thorium resonance integral.

The experimental data are from the following sources:

● Homogeneous systems; A. M. Weinberg and E. P. Wigner, *op. cit.*, p. 313.

□ Metal rods; J. Phelps and R. Sher, *Nucl. Sci. Eng.* **3**, 403 (1958).

△ Metal rods; M. C. Edlund *et al.*, PICG P/2405 (1958).

○ Oxide rods; *Ibid.*

--- Metal rods; W. G. Pettus, BAW-TM-203 (1959).

— Oxide rods; *Ibid.*

one. If σ_p is small compared with $S_0/4V_0N_u$, as is the case for lumps of pure absorber, Equation (II.51 a) becomes

$$I = \frac{\pi}{2}\frac{\Gamma_\gamma}{E_0}\sigma_0^{1/2}\left(\frac{S}{4VN_u}\right)^{1/2} = c\left(\frac{S}{M}\right)^{1/2} \tag{II.51 c}$$

or

$$I \cong I_V + I_s(S/M)^{1/2} \quad (I_V \approx 0)$$

a form which has frequently been used [37].

[37] This form was first suggested by I. I. GUREVICH and I. Y. POMERANCHOUCK, *PICG* **5**, 466 (1955).

On the other hand, if $(\sigma_p + S_0/4V_0N_u)$ is still small compared with σ_0, but σ_p is large compared to $S_0/4V_0N_u$, Equation (II.51a) can be expanded to give

$$R_{\text{eff}}^{\text{NR}} \approx \frac{\pi}{2}\frac{\Gamma_\gamma}{E_0}\sigma_0^{1/2}\sigma_p'^{1/2}\left[1 + \frac{S_0}{2V_0N_u\sigma_p'}\right] \tag{II.52a}$$

which is similar to the form given originally by Wigner [38]:

$$R_{\text{eff}}^{\text{NR}} = I_V' + I_s'\left(\frac{S}{M}\right) \tag{II.52b}$$

and which was frequently used before Equation (II.51c) was suggested. It is found that Equation (II.51c) is generally more applicable, since $S_0/4V_0N_u$ is generally considerably larger than σ_p.

Finally, if

$$\sigma_p' + \frac{S_0}{4V_0N_u}$$

is large compared with σ_0, $R_{\text{eff}}^{\text{NR}}$ correctly becomes the infinite dilution resonance integral I_∞.

23. Dancoff Corrections; Interactions of Cells

All of the numerical calculations discussed so far have been for the single pin case, i.e., for lattices in which adjacent absorber lumps do not interfere appreciably with each other's resonance absorption. To correct for interference, it is necessary only to use the generalized escape probability in the equations for I_{eff}, i.e., to use the probability that a neutron born uniformly and isotropically in a lump will suffer its first collision neither in that lump nor in any other lump. Otherwise, the formulae may be taken over exactly. Unfortunately, this quantity is quite difficult to calculate so that Monte Carlo calculations are frequently used [39]. Some analytical approaches have also been attempted [40].

The historical approach [41] is based on the following argument. Insofar as absorption is concerned, the effect of a neighboring lump

[38] A. M. Weinberg and E. P. Wigner, *op. cit.*

[39] A short discussion of the Monte Carlo method is given in Chapter III. Examples of its application to the problem of calculating resonance integrals may be found in R. D. Richtmyer, BNL 433, p. 82 (1957).

[40] J. Stewart and P. F. Zweifel, PICG P/631 (1958).

[41] S. M. Dancoff and M. Ginsburg CP-2157 (1944).

is to reduce the number of neutrons incident on a particular lump from the direction of the neighbor. Thus, not as many neutrons will be incident on the part of the surface that faces another lump as on the part that faces only moderator. Hence not all of the surface is equally effective in capturing neutrons. To account for this an effective surface area S_{eff} is introduced and defined as:

$$S_{\text{eff}} = S(1 - c) \qquad \text{(II.53a)}$$

where $(1 - c)$ is the "Dancoff Correction". This approach was mentioned in the previous section with regard to Equation (II.51b) and Figs. II.7 and II.8.

The calculation of c (or of S_{eff}) is of course the most difficult portion of this procedure. Many authors [42] have considered this problem in various approximations and for various geometries but space does not permit inclusion of even a summary of this work here. Rather we shall quote only a very simple result which is quite analogous to the rational approximation [43]:

$$S_{\text{eff}} = \frac{S\Sigma_p^1}{\Sigma_p^1 + S\left(\dfrac{V_0}{V_1}\right)} \qquad \text{(II.53b)}$$

where Σ_p^1 is the moderator (potential) cross section and V_1 is the volume (per cell) of the moderator region. The approximation gives the correct result in the limits $(V_0/V_1) \to 0$ and $(V_0/V_1) \to \infty$ and is sometimes used. Unfortunately, it is not so accurate as the rational approximation [44].

24. The Fast Effect

Earlier, the fast fission factor, ε, was defined as the ratio of the number of neutrons slowing down past the U-238 fission threshold per neutron produced by thermal fission. When defined in this way the concept of a fast fission factor is valid for almost all thermal systems. In low enrichment ($<10\%$ U-235) thermal reactors the fuel is mainly U-238 so that practically all non-thermal fissions

[42] To quote a few: J. A. Thie, *Nucl. Sci. Eng.* **5**, 75 (1959); H. Takahashi, *J. Nucl. Ener.* **12**, 1 (1960); Y. Fuaki, *Trans. Amer. Nucl. Soc.* **2** Paper 8–5 (1960); W. Rothenstein, *Nucl. Sci. Eng.* **7**, 162 (1960).

[43] G. I. Bell, *Nucl. Sci. Eng.* **5**, 75 (1959).

[44] L. W. Nordheim, GA 638.

occur in this isotope. High enrichment systems generally have low fuel concentrations so that ε is very nearly unity. For intermediate and fast reactors the concept of a fast fission factor breaks down and there is, in fact, little advantage in using the six factor formula at all. Thus it is legitimate to consider only U-238 fission in calculating ε.

Since the U-238 fission threshold is at approximately 1 Mev and most fission neutrons are born at energies not much higher than this (in fact, many are born below 1 Mev), an inelastic scattering event is almost certain to degrade the neutron energy to below the U-238 fission threshold. Similarly, a collision with a very light nucleus (hydrogen or deuterium) or several collisions with other common moderators (e.g., beryllium or graphite) will remove the neutron from the energy range of interest. In most systems a neutron that has escaped from a fuel element will make several collisions in the moderator before returning to a fuel element so that a neutron that has escaped from a rod has a very small probability of causing U-238 fission.

These remarks lead to the following model for calculating the fast fission factor. A neutron born in the lump may cause fast fission on its first flight. If it does not cause fission but either scatters inelastically or leaves the lump, it is "dead" insofar as fast fission is concerned. On the other hand, if it scatters elastically within the lump, it may cause fast fission on succeeding flights. By following the neutron through successive flights the number of fast fissions can be obtained.

Thus we consider a single U-235 fission neutron starting out with a fission spectrum in the lump. On its first flight it has probability:

P_1 of escaping from the lump

$(1 - P_1)\,\Sigma_f/\Sigma_t$ of causing fission

$(1 - P_1)\,\Sigma_c/\Sigma_t$ of being captured without causing fission

$(1 - P_1)\,\Sigma_{in}/\Sigma_t$ of scattering inelastically in the lump

$(1 - P_1)\,\Sigma_{el}/\Sigma_t$ of scattering elastically in the lump

where all cross sections are averaged over the fission spectrum. Note that P_1 is not the conventional escape probability, E, used in connection with resonance capture since U-235 fission neutrons are not born uniformly in the lump. Furthermore, the escape probabilities on the various flights, P_n, will all be different as the neutron distribution will be different after each collision.

There are, on the average, $(1 - P_1)\,\Sigma_{el}/\Sigma_t$ neutrons available for causing fissions on the second flight. Of these

$P_2(1 - P_1)\,\Sigma_{el}/\Sigma_t$ escape from the lump

$(1 - P_2)(1 - P_1)(\Sigma_f/\Sigma_t)(\Sigma_{el}/\Sigma_t)$ cause fission

$(1 - P_2)(1 - P_1)(\Sigma_c/\Sigma_t)(\Sigma_{el}/\Sigma_t)$ are captured without causing fission

$(1 - P_2)(1 - P_1)(\Sigma_{in}/\Sigma_t)(\Sigma_{el}/\Sigma_t)$ scatter inelastically

$(1 - P_2)(1 - P_1)(\Sigma_{el}/\Sigma_t)^2$ scatter elastically

This bookkeeping procedure may be carried out indefinitely. However, after two flights very few neutrons are still available for causing fast fission on later flights. Hence it suffices to set P_n, the escape probability for the nth flight, equal to P_2 for $n > 2$. When this is done it is possible to sum the fissions caused by the various flights. Thus, the total number of fast fissions is:

$$(1 - P_1)\left(\frac{\Sigma_f}{\Sigma_t}\right) + (1 - P_1)(1 - P_2)\left(\frac{\Sigma_{el}}{\Sigma_t}\right)\left(\frac{\Sigma_f}{\Sigma_t}\right) + (1 - P_1)(1 - P_2)^2\left(\frac{\Sigma_{el}}{\Sigma_t}\right)^2\left(\frac{\Sigma_f}{\Sigma_t}\right) + \cdots$$

$$= (1 - P_1)\left(\frac{\Sigma_f}{\Sigma_t}\right)\left[1 + (1 - P_2)\left(\frac{\Sigma_{el}}{\Sigma_t}\right) + (1 - P_2)^2\left(\frac{\Sigma_{el}}{\Sigma_t}\right)^2 + \cdots\right]$$

$$= \frac{(1 - P_1)\,\Sigma_f/\Sigma_t}{1 - (1 - P_2)\,\Sigma_{el}/\Sigma_t} \tag{II.54a}$$

Similarly, the number captured is

$$\frac{(1 - P_1)\,\Sigma_c/\Sigma_t}{1 - (1 - P_2)\,\Sigma_{el}/\Sigma_t} \tag{II.54b}$$

Since one neutron is captured for every fission, the net neutron gain by fast fission, i.e., the increase in the number slowing down past the U-238 fission threshold per U-235 fission neutron, is:

$$\frac{(1 - P_1)\,[(\nu - 1)\,\Sigma_f - \Sigma_c]/\Sigma_t}{1 - (1 - P_2)\,\Sigma_{el}/\Sigma_t} = \varepsilon - 1 \tag{II.54c}$$

and is $(\varepsilon - 1)$ by definition.

Equation (II.54c) is the desired result. It has frequently been used for computing ε, with apparently successful results. At least a part of this success must be regarded as fortuitous since some of the approximations made above are clearly not accurate. However, it it obvious that the above procedure actually calculates $\varepsilon - 1$. Using appropriate cross sections Weinberg and Wigner [45] reduce Equation (II.45c) to

$$\varepsilon - 1 = \frac{0.09(1 - P_1)}{1 - 0.52(1 - P_2)} \qquad \text{(II.54d)}$$

from which it is obvious that the maximum value of $(\varepsilon - 1)$ is approximately 1.18; in actual reactors ε is rarely more than $\sim$1.05. Hence relatively large errors in the calculation will not change ε much.

[45] A. M. Weinberg and E. P. Wigner, *op. cit.* p. 706.

CHAPTER III

SLOWING DOWN IN FINITE MEDIA

A. The Slowing-down Kernels

1. Introductory Remarks—Green's Function

In this chapter we will obtain approximations which are applicable to reactors in general. Much of the development is made in the context of asymptotic theory but a more general exposition of the theory, particularly of that portion which is usually applied in reactor calculations, is also given. In the previous chapter it was shown how the resonance escape probability and fast fission factor could be obtained for non-homogeneous systems. These results can be used to define a set of fictitious cross sections which depend on both the material composition and on the geometry and which yield the same resonance escape probability and fast effect when applied to calculations in a homogeneous medium of the same gross dimensions as the actual system.†

In a similar manner, values of other pertinent parameters can be obtained by insisting that the homogenized system have the same overall properties as the actual system. To simplify matters, it will be assumed that this homogenization has been carried out and hence we shall deal only with homogeneous systems in this chapter, except for one method designed specifically for calculations of heterogeneous systems which is discussed near the end of the chapter.

It may be worth while at this point to review the development of asymptotic reactor theory as presented in Chapter I. It was shown that by treating thermal neutrons by one group theory, the calculation of the effective multiplication constant may be reformulated in terms of the calculation of certain slowing-down kernels. In this

† Generally speaking, this homogenization is carried out by regions. Thus, a fuel-moderator region might be homogenized, while a region containing control rods might be homogenized separately.

manner all of the details of slowing down were put aside and hidden within these slowing-down kernels which must still be calculated; this chapter will be devoted largely to the calculation of the required slowing-down kernels. Before attacking this problem directly, several results which will make the task simpler will be introduced.

First, we shall return to the monoenergetic source infinite medium slowing-down kernel† $P_\infty(|\mathbf{r}' - \mathbf{r}|, E' \to E)$, and treat it as a Green's function for the fission-averaged kernel, $P(|\mathbf{r}' - \mathbf{r}|, E)$, discussed in Chapter I. This has the double advantage that:

1. monoenergetic sources are simpler to deal with than are distributed energy sources, and
2. such a slowing-down kernel can be applied to sources other than U-235 fission.

Thus, this kernel can be used in the design of reactors using fuels other than U-235 (e.g., U-233 or Pu-239); it can accurately account for delayed neutrons, neutrons arising from $(n, 2n)$ reactions, or "extraneous" source neutrons which generally do not have the same energy spectrum as fission neutrons.

From the definition of the slowing-down kernel it is seen that in order to obtain the infinite-medium slowing-down kernel $P_\infty(|\mathbf{r} - \mathbf{r}'|, E' \to E)$ it is necessary to solve the Boltzmann equation for an infinite medium with a unit monoenergetic isotropic point source of neutrons at $\mathbf{r}'$. In practice, we will solve the Fourier transformed Boltzmann equation (or, what is the same thing, the Boltzmann equation with a "sinusoidal" source, $e^{-i\mathbf{k}\cdot\mathbf{r}}$). The transform may then be inverted to obtain the spatial distribution or, as we have seen in Chapter I, the tranform itself may be used directly in criticality calculations.

2. The One-dimensional Problem

The solution to the three-dimensional Boltzmann equation in an infinite medium (i.e., the point source problem) is closely related to the solution of the one-dimensional Boltzmann equation (the plane source problem). Also, the Fourier transform of the infinite medium point source slowing-down kernel (the one employed up

† Angular dependence of this kernel, i.e., of the kernel $P(\mathbf{r}' \to \mathbf{r}, E' \to E, \mathbf{\Omega})$ may be described by some of the methods presented.

to now) is precisely the Fourier transform of the kernel defined with respect to the plane source problem. This last kernel, $P(|z' - z|, E' \to E)$ or $P(z, E' \to E)$, is the number of neutrons slowing down into a unit energy range at E at distance z from the source due to a unit isotropic plane source of neutrons (one neutron per cm^2 per sec) of energy E' at $z = 0$.

To prove this, let $P_{pt}(r)$ be the neutron flux† due to a unit point source at $\mathbf{r} = 0$. Then for any source distribution, $S(\mathbf{r})$, the flux distribution $\phi(\mathbf{r})$ is:

$$\phi(\mathbf{r}) = \int S(\mathbf{r}')\, P^*(|\mathbf{r} - \mathbf{r}'|)\, \mathrm{d}^3r' \qquad \text{(III.1 a)}$$

If, in particular, $S(\mathbf{r}) = \delta(z)$, we have, using cylindrical coordinates (see Fig. III.1):

$$P^*_{pl}(z) = \int_0^{2\pi} \mathrm{d}\varphi \int_0^{\infty} \varrho\, \mathrm{d}\varrho \int_{-\infty}^{\infty} \mathrm{d}z'\, \delta(z')\, P^*_{pt}([(z - z')^2 + \varrho^2]^{\frac{1}{2}}) \qquad \text{(III.1 b)}$$

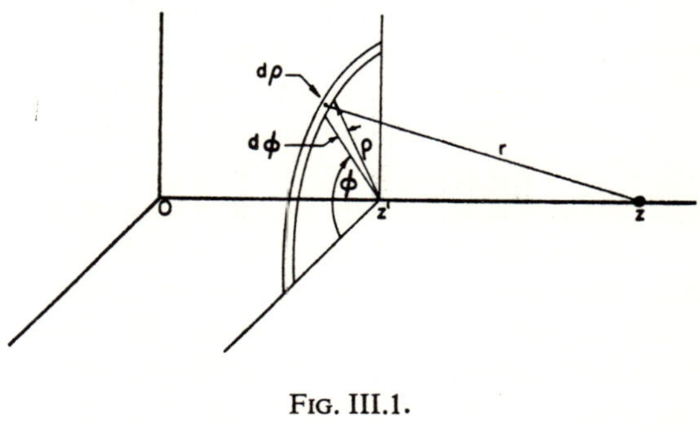

FIG. III.1.

where $P^*_{pl}(z)$ is the flux arising from a unit isotropic plane source. The integration over z' is trivial, as is the azimuthal integration.

$$P^*_{pl}(z) = 2\pi \int_0^{\infty} P^*_{pt}([z^2 + \varrho^2]^{\frac{1}{2}})\varrho\, \mathrm{d}\varrho \qquad \text{(III.1 c)}$$

† The energy variables are temporarily dropped for this argument. Also, note that although the slowing-down kernel is not strictly a flux (it is the integral of the flux multiplied by the transfer cross section $\Sigma(E' \to E, \mathbf{\Omega}'.\mathbf{\Omega})$), it is sufficient to prove the theorem for the flux since the two are related by an operator not involving the spatial variable.

Now, changing the variable of integration from ϱ to $r = \sqrt{(z^2 + \varrho^2)}$, we have

$$P^*_{pl}(z) = 2\pi \int_{|z|}^{\infty} r P^*_{pt}(r)\,\mathrm{d}r \tag{III.1d}$$

which may be differentiated to give

$$P^*_{pt}(r) = -\frac{1}{2\pi r}\left.\frac{\mathrm{d}P^*_{pl}(z)}{\mathrm{d}z}\right|_{z=r} \tag{III.1e}$$

That is, the point source kernel may be found by differentiating the plane source kernel.

To show that the Fourier transforms of the point and plane source kernels are equal, we simply take the Fourier transform of Equation (III.1a). Noting that P_{pl} is a function only of $|z|$ and applying the convolution theorem, one finds:

$$\overline{P}^*_{pl}(k_z^2)\,\delta(k_x)\,\delta(k_y) = \bar{S}(k)\,\overline{P}^*_{pt}(k^2) \tag{III.2a}$$

For a plane source $S(\mathbf{r}) = \delta(z)$ and the Fourier transform of the source is

$$\overline{S}(\mathbf{k}) = \delta(k_x)\,\delta(k_y) \tag{III.2b}$$

whence, we have immediately

$$\overline{P}_{pl}(k_z^2) = \overline{P}_{pt}(k^2) \tag{III.2c}$$

the desired result.

The import of this result is that the entire problem of finding the slowing-down kernels required by asymptotic reactor theory can be formulated and solved in plane geometry. The one-dimensional problem can be solved, Fourier transformed, and inverted to yield the point source.

Having justified the use of the one-dimensional form of the Boltzmann equation, we will use this form whenever asymptotic reactor theory is to be applied. The one-dimensional Boltzmann equation, obtained by using the coordinate system shown in Fig. III.2 and noting the azimuthal symmetry in the problem, is

$$\mu\,\frac{\partial\Phi(z, E, \mu)}{\partial z} + \Sigma_t(E)\,\Phi(z, E, \mu)$$

$$= \int \mathrm{d}\Omega' \int \mathrm{d}E'\,\Phi(z, E', \mu')\,\Sigma_s(\mu_0, E' \to E) + S(z, E, \mu) \tag{III.3a}$$

while its Fourier transform is:

$$(\Sigma_t(E) - ik\mu)\,\bar{\Phi}(k, E, \mu) = \int d\boldsymbol{\Omega}' \int dE'\,\bar{\Phi}(k, E', \mu')\,\Sigma_s(\mu_0, E' \to E) + \bar{S}(k, E, \mu) \quad \text{(III.3b)}$$

This is the basic equation on which much of this chapter will be based. Sources will be assumed isotropic so as to permit transformation to the point source problem, but the restriction is not a serious one for the class of problems under consideration.

Although most of this chapter is based on the one-dimensional Boltzmann equation, the results will be generalized to three dimensions wherever possible.

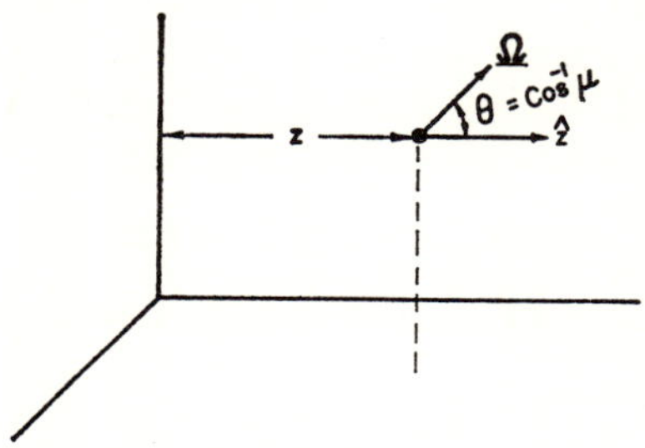

FIG. III.2.

3. Non-leakage Probabilities

Since we obtained expressions for the fast effect and resonance escape probability in the previous chapter, the main concern here will be non-leakage probabilities. Under the assumptions of asymptotic reactor theory the non-leakage probability λ is equal to the Fourier transform of the slowing-down kernel $\bar{P}(B_g^2)$ divided by $\bar{P}(0)$ (see Equation I.38e).

Let us now expand $P(B_g^2)$ in a Maclaurin series in B_g:

$$\bar{P}(B_g^2) \approx \bar{P}(0) + B_g \left.\frac{\partial \bar{P}}{\partial B_g}\right|_{B_g=0} + \frac{1}{2} \left.\frac{\partial^2 \bar{P}}{\partial B_g^2}\right|_{B_g=0} \quad \text{(III.4a)}$$

But from the definition of the Fourier transform one has, for the one-dimensional case:

$$\left.\frac{\partial^n \bar{P}}{\partial B_g^n}\right|_{B_g=0} = \int_{-\infty}^{\infty} (iz)^n\, P(z)\, dz \quad \text{(III.4b)}$$

which is, except for normalization, the *n*th spatial moment of the slowing-down kernel. But we are dealing with the problem of a unit plane isotropic source in an infinite homogeneous medium so that one would physically expect $P(z)$ to be an even function of z,† in which case the odd moments vanish. Frequently, one-half the second moment is defined as the *neutron age* or simply the *age*, $\tau(E)$,

$$\frac{1}{2}\langle z^2(E)\rangle = \tau(E) = \frac{\int_{-\infty}^{\infty} z^2 P(z, E)\, \mathrm{d}z}{\int_{-\infty}^{\infty} P(z, E)\, \mathrm{d}z} \tag{III.4c}$$

and is an important quantity in reactor calculations. Hence, for large reactors B_g^2 is small and the non-leakage probability is approximately

$$\lambda(B_g^2) = 1 - B_g^2\tau \tag{III.4d}$$

For large reactors (small B_g^2) this is equivalent to other frequently encountered formulae:

$$\lambda(B_g^2) = \mathrm{e}^{-B_g^2\tau} \tag{III.4e}$$

and

$$\lambda(B_g^2) = \frac{1}{1 + B_g^2\tau} \tag{III.4f}$$

which are equivalent to (III.4d) for small B_g.

From the identity of the Fourier transform of the point and plane source kernels these results hold in general. However, the second spatial moment for slowing down from a point source $\langle r^2\rangle$ is three times that for a plane source $\langle z^2\rangle$. To prove this, we write

$$\langle r^2\rangle = \frac{\int_0^{\infty} r^2 P_{pt}(r)\, \mathrm{d}^3r}{\int_0^{\infty} P_{pt}(r)\, \mathrm{d}^3r} \tag{III.5a}$$

† Mathematically this can be shown by noting that the one-dimensional Boltzmann equation with an isotropic source and constant cross sections is unchanged under the transformation $z \to -z$ and $\mu \to -\mu$. Therefore, $\Phi(z,\mu) = \Phi(-z, -\mu)$. On integrating over angle we have $\phi(z) = \phi(-z)$ which is the above statement.

or, using Equation (III.1e):

$$\langle r^2 \rangle = \frac{-\dfrac{4\pi}{2\pi}\displaystyle\int_0^\infty r \frac{\mathrm{d}P_{pl}}{\mathrm{d}r}\,\mathrm{d}^3 r}{-\dfrac{4\pi}{2\pi}\displaystyle\int_0^\infty \frac{1}{r}\frac{\mathrm{d}P_{pl}}{\mathrm{d}r}\,\mathrm{d}^3 r} \tag{III.5b}$$

Now, using spherical coordinates, the angular integrations can be performed to yield:

$$\langle r^2 \rangle = \frac{\displaystyle\int_0^\infty r^3 \frac{\mathrm{d}P_{pl}}{\mathrm{d}r}\,\mathrm{d}r}{\displaystyle\int_0^\infty r \frac{\mathrm{d}P_{pl}}{\mathrm{d}r}\,\mathrm{d}r} \tag{III.5c}$$

After integration by parts, this becomes:

$$\langle r^2 \rangle = \frac{3\displaystyle\int_0^\infty r^2 P_{pl}(r)\,\mathrm{d}r}{\displaystyle\int_0^\infty P_{pl}(r)\,\mathrm{d}r} \tag{III.5d}$$

since $P_{pl}(r)$ may be expected to go to zero exponentially at infinity. Hence:

$$\langle r^2 \rangle_{pt} = 3\langle z^2 \rangle_{pl} \tag{III.5e}$$

Thus, in general, the age is one sixth the mean square slowing down length, the generalization of the diffusion area, L^2, to slowing down.

B. The Spherical Harmonics Method†

4. Reduction of the Boltzmann Equation

In attempting to solve the Boltzmann equation one soon discovers that most of the difficulty is introduced by the degradation erm (again $U = u - u'$):‡

$$J(k, u, \mu) = \sum_i N_i \int \mathrm{d}\mathbf{\Omega}' \int \mathrm{d}u'\, \sigma_s(u', U, \mu_0)\, \bar{\Phi}(k, u', \mu') = \sum_i N_i J_i$$
$$= \int S(\mathbf{r}')\, P(\mathbf{r}' \to \mathbf{r}, E, \mu)\, \mathrm{d}^3 r' \tag{III.6}$$

† This method is equally applicable in transform or configuration space.
‡ The bars (denoting Fourier transforms) will be omitted in the remainder of this chapter.

where N_i is the number density of nuclei of species i. One would like a simple approximation with which to deal with this term. One such approximation might be the estimation of this integral by a quadrature formula and indeed such procedures are frequently used and will be applied later. Another scheme that comes to mind is a series expansion of the flux (or, possibly, the entire integrand) in terms of a set of orthogonal functions, followed by a truncation of the series, yielding, it is hoped, a useful low-order approximation.

If the flux and scattering cross section (or scattering frequency) are to be expanded separately, it is clear that it is very important to choose a set of functions for which one is able to express functions of the angle between two vectors $(\mathbf{\Omega} \cdot \mathbf{\Omega}')$ in terms of functions of the individual vectors. Because of their relationship to the Laplace equation, the spherical harmonics have this property (the addition theorem) and for this reason are the most convenient set of functions for this purpose.

For expanding the flux, one need only use the Legendre polynomials since azimuthal symmetry has already been assumed:

$$\bar{\Phi}(k, u, \mu) = \sum_{L=0}^{\infty} \frac{2L+1}{2} \phi_L(k, u) P_L(\mu) \qquad \text{(III.7a)}$$

where the expansion coefficients are:

$$\phi_L(k, u) = -\int_{-1}^{1} d\mu \Phi(k, u, \mu) P_L(\mu) \qquad \text{(III.7b)}$$

The same expansions are, of course, possible in ordinary configuration $(\mathbf{r})$ space; the expansion coefficients are designated by $\phi_L(z, u)$ and are simply the inverse Fourier transforms of the $\phi_L(k, u)$. For the (elastic) scattering cross section we already have the expansion (I.49a): †

$$\sigma_s(\mu_0, u', U) = \sum_{L=0}^{\infty} \frac{2L+1}{4\pi} B_L^0(u', U) P_L(\mu_0) \qquad \text{(III.7c)}$$

† In Chapter I $\sigma_s(\mu_0, u', U)\, d\mu_0$ was defined as the cross section for scattering into solid angle $2\pi\, d\mu_0$. Here $\sigma_s(\mu_0, u', U)\, d\mathbf{\Omega}$ is the cross section for scattering into $d\mathbf{\Omega}$; hence the factor of 2π in Equation (III.7c).

Using these expansions, the $J_i(k, u, \mu)$ of Equation (III.6) can be written:

$$J_i(k, u, \mu) = \int d\mathbf{\Omega}' \int_{u-\ln 1/\alpha_i}^{u} du' \sum_{L,L'=0}^{\infty} \frac{2L'+1}{4\pi} \frac{2L+1}{2} \phi_L(k, u')\, B_{L'}^{0}(u', U) \cdot$$
$$\cdot\, P_L(\mu')\, P_{L'}(\mu_0) \qquad \text{(III.7d)}$$

It is here that the choice of Legendre polynomials as the orthogonal set bears fruit. $P_L(\mu_0)$ may be expanded by means of the addition theorem for Legendre polynomials and the integration over $d\mathbf{\Omega}'$ becomes rather trivial. The theorem may be stated as:

$$P_{L'}(\mu_0) = P_{L'}(\mu)\, P_{L'}(\mu') + 2 \sum_{m=1}^{L'} (-)^m \frac{(L'-m)!}{(L'+m)!} P_L^m(\mu) P_L^m(\mu') \times$$
$$\times \cos m(\varphi - \varphi') = \sum_{m=-L'}^{L'} Y_L^m(\mathbf{\Omega})\, Y_L^{m*}(\mathbf{\Omega}') \qquad \text{(III.7e)}$$

where the $Y_L^m(\mu)$ are the associated Legendre polynomials and the $Y_L^m(\mathbf{\Omega})$ are the spherical harmonics [1]. The angles φ and φ' are the azimuthal angles associated with $\mathbf{\Omega}$ and $\mathbf{\Omega}'$ respectively. Upon inserting this expansion into Equation (III.7d), one observes that the only φ dependence appearing is that arising from the sum in the addition theorem (due to the assumed azimuthal symmetry of the flux) and that each of these terms integrates to zero because of the periodicity properties of the cosine functions. Then we are left with

$$J_i(k, u, \mu) = \int_{u-\ln 1/\alpha_i}^{u} du' \int_{-1}^{1} d\mu' \sum_{L,L'=0}^{\infty} \frac{2L+1}{2} \frac{2L'+1}{2} \phi_L(k, u') \times$$
$$\times B_{L'}(u', U)\, P_L(\mu')\, P_{L'}(\mu')\, P_{L'}(\mu) \qquad \text{(III.8a)}$$

Application of the orthogonality property of the Legendre polynomials (I.46e) reduces Equation (III.8a) to

$$J_i(k, u, \mu) = \sum_{L=0}^{\infty} \frac{2L+1}{2} J_{iL}(k, u)\, P_L(\mu) \qquad \text{(III.8b)}$$

where the $J_{iL}(k, u)$ may be interpreted as the Legendre expansion coefficients of the degradation integral $J_i(k, u, \mu)$ and are related to

[1] For definitions of these functions as well as a proof of this theorem see E. T. WHITTAKER and G. N. WATSON, *Modern Analysis*, Oxford, 1953.

the flux expansion coefficients by

$$J_{iL}(k, u) = \int_{u-\ln 1/\alpha_i}^{u} du' \, \phi_L(k, u') \, B_L^{0^i}(u', U)$$

$$= \int_{u-\ln 1/\alpha_i}^{u} du' \, \phi_L(k, u') \sum_{L'=0}^{\infty} B_{L'}^{ci}(u') \, T_{LL'}(U) \quad \text{(III.8c)}$$

where the $T_{LL'}(U)$ are defined by Equation (I.50b).

From the definition of the Fourier transform it may be seen that $\phi_0(0, u)$, hereafter abbreviated to $\varphi(u)$, corresponds to the value of $\Phi(\mathbf{r}, \mu, u)$ integrated over all space and angle and is the total neutron flux in the reactor. Setting $k = 0$ in the Fourier transformed Boltzmann equation (III.3b) (this amounts to dropping the leakage term and hence corresponds to an infinite medium), integrating over angle, and again using the orthogonality properties of the Legendre polynomials, we find

$$\Sigma_t(u) \, \varphi(u) = \sum_i N_i J_{i0}(k, u) + S(u) \quad \text{(III.8d)}$$

with $J_{i0}(u)$ given by

$$J_{i0}(u) = \int_{u-\ln 1/\alpha_i}^{u} du' \, \varphi(u') \sum_{L'=0}^{\infty} B_{L'}^{c}(u') \, T_{LL'}(U) \quad \text{(III.8e)}$$

which are the basic equations of Chapter II.

Using the above expansions and a Legendre expansion of the source, the Fourier transformed Boltzmann equation may be written:

$$[\Sigma_t(u) - ik\mu] \sum_{L=0}^{\infty} \phi_L(k, u) \, P_L(\mu)$$

$$= \sum_i N_i \sum_{L=0}^{\infty} J_{iL}(k, u) \, P_L(\mu) + \sum_{L=0}^{\infty} S_L(k, u) \, P_L(\mu) \quad \text{(III.9)}$$

5. The P_L Approximation

The most straightforward procedure for solving Equation (III.9) is to multiply the entire equation by $P_j(\mu)$ and integrate over all μ, to obtain the coupled set of integro-differential equations

$$\Sigma_t(u) \, \phi_j(k, u) - \frac{ik}{2j + 1} [i\phi_{j-1}(k, u) + (j + 1) \, \phi_{j+1}(k, u)]$$

$$= N_H J_j^H(k, u) + N_D J_j^D(k, u) + \sum_{i \neq H, D} N_i J_{ij}(k, u) + S_j(u) \quad \text{(III.10a)}$$

where $\phi_{-1}(k, u)$ is zero by convention and the recursion relation for Legendre polynomials [2]

$$\mu P_L(\mu) = \frac{1}{2L+1}\left[(L+1)\,P_{L+1}(\mu) + LP_{L-1}(\mu)\right] \quad \text{(III.10b)}$$

has been employed in the evaluation of the second term. For an isotropic source, $S_j(u) = 0$ for $j \neq 0$ as a result of the orthogonality of the Legendre polynomials. The first two terms of the right-hand side of Equation (III.10a) represent neutron degradation (slowing down) due to hydrogen and deuterium, respectively, while the third term represents the degradation due to heavier elements, which may frequently be treated as a group.

For hydrogen, which scatters isotropically at neutron energies relevant to reactor calculations, $B^c_{L'}(u) = 0$ for $L' > 0$, so that $J^H_j(k, u)$ reduces to a single term:

$$J^H_j(k, u) = \int_0^u \phi_j(k, u')\,\sigma^H_s(u')\,\mathrm{e}^{-(u-u')}\,P_j(\mathrm{e}^{-(u-u')/2})\,\mathrm{d}u' \quad \text{(III.11)}$$

where Equation (I.50 b) has been used for $T_{L0}(U)$. If deuterium is present, the $J^D_j(k, u)$ term will also appear in Equation (III.10), but scattering from deuterium is not isotropic so that the degradation integral does not reduce to a single term. Again, deuterium moderation will be difficult to handle.

The technique employed here is often called the "Spherical Harmonics Method" [3], because the orthogonal set of expanding functions used are the spherical harmonics (although in the case of azimuthal symmetry only the ordinary Legendre polynomials are necessary).†

This technique succeeds in converting the integro-differential Boltzmann equation into an infinite set of simpler, but coupled, integro-differential equations, essentially by a variable transformation from the continuous variable μ to the discrete variable j. This, in and of itself, represents no real simplification and possibly an increase in complexity; the advantages of the method are realized only when it is noted that the neutron flux in a large reactor is not

† This procedure was used Chapter I to derive the Diffusion equation.

[2] *Ibid.*

[3] P. M. Morse and H. Feshbach, *Methods of Theoretical Physics*, McGraw–Hill (1950), Vol. I.

very anisotropic, and hence a few Legendre components might suffice to describe its angular variation accurately. This, plus the fact that the jth of Equations (III.10) is coupled only to those immediately above and below it $(j - 1, j + 1)$, suggests setting $\phi_j(k, u) = 0$ for some j, say $j = L + 1$, thus uncoupling the first $L + 1$ equations from all others and leaving a set involving only $\phi_0, \ldots, \phi_L$. The truncated set so obtained can then be solved for an approximate flux. The accuracy of the method may be tested by using higher-order approximations (increasing L) until the fluxes (or multiplication constants) predicted are the same, within a predetermined limit, for two successive approximations. This is called the P_L approximation.

The simplest of the P_L approximations is, of course, the P_0. However, it is based on the assumption of isotropic flux and hence can predict no neutron transport; Equations (III.10) then reduce to Equation (II.2a). The next simplest is the P_1 approximation, which is probably the most frequently used reactor calculational technique. Methods of solving the P_1 equations, under various assumptions, will be discussed in the next few sections. The P_2 and other even-L P_L approximations are of little value for reasons to be discovered later. Higher P_L approximations are applied mainly to special problems, especially where great accuracy is required or where the flux is highly anisotropic, as in regions near heavily absorbing rods.

6. The Spherical Harmonics Method in Other Geometries [4]

In plane geometry, we have just seen that the spherical harmonics method leads to an extremely simple set of equations for an approximate flux. This, unfortunately, is due in great measure to the unusual amount of symmetry found in this problem. Since the flux is a function of only one angular variable, an expansion of the flux in ordinary Legendre polynomials is possible. In the more general case, not only will the flux be a function of two angular variables but it will depend on more than a single spatial coordinate as well. In this situation, the flux must be expanded in terms of the full set of spherical harmonics.

[4] J. C. MARK, *The Spherical Harmonics Method*, Reports MT92, MT97, National Research Council of Canada. See also B. DAVISON, *op. cit.*, Chapter 12.

The equivalent of the P_1 approximation for general geometry was given in Chapter I when an expansion of the type

$$\Phi(\mathbf{\Omega}) = \phi_0 + \sum_m \phi_{1m} Y_1^m(\mathbf{\Omega}) \tag{III.11a}$$

was used. Here, $Y_1^m(\mathbf{\Omega})$ is the spherical harmonic discussed earlier.

Methods for handling this problem in general geometries are given in the references cited in footnote [4]. Mark uses tensor notation to simplify the problem. In general the Lth order equations can always be cast into the form

$$\sum_{n=0}^{L} a_n \nabla^{2n} \phi_0 = \sum_{n=0}^{L-1} b_n \nabla^{2n} S \tag{III.11b}$$

where the a_n and b_n are in general energy dependent operators. Since the a_n and b_n are the same for all geometries it suffices to make this calculation in plane geometry. The only difficulty thus arises in treating the boundary conditions; this is discussed in the following section.

7. Boundary Conditions for the Spherical Harmonics Method

Some of the conditions that must be imposed on the components of the flux (the $\phi_l(\mathbf{r}, u)$) are probably quite clear; each of them must be finite everywhere within the reactor volume. Furthermore, the fact that $\Phi(\mathbf{r}, u, \mathbf{\Omega})$ must be continuous everywhere within the reactor volume requires that each of its components be continuous at any internal point of the system. The conditions to be imposed on the spherical harmonics equations are then completely specified except for those which must be applied at the outer boundary of the system, i.e., at a material–vacuum interface.

First we recollect that the rigorous condition to be imposed at a "free" interface is that no neutrons enter the system from outside; that is,

$$\Phi(\mathbf{r}, u, \mathbf{\Omega}) = 0 \qquad \mathbf{\Omega} \cdot \mathbf{n} < 0 \tag{III.12a}$$

where $\mathbf{n}$ is an outward normal, for $\mathbf{r}$ on a surface. It is, however, impossible to make any finite series of the form†

$$\Sigma \phi_L P_L(\mu) \tag{III.12b}$$

fit a function which is zero for $\mu < 0$ except in the trivial (and un-

† This discussion is limited to the one-dimensional case for convenience but is readily generalized.

physical) case for which $\Phi(z, u, \mu)$ is also zero for $\mu > 0$. This then implies (1) that a set of approximate conditions will have to be found and (2) that this approximate set of conditions is not unambiguously defined. Furthermore, in an odd L approximation one can in general apply only $(L + 1)/2$ boundary conditions at an interface, or in an even approximation $L/2$. The reason for this lies in the fact that in plane geometry there will always be two free surfaces in a finite system and that in any other geometry half of the solutions will diverge at some point in the system [5]. Since the boundary conditions are ambiguous, we are free to choose any conditions we like, provided that they are consistent with the rigorous condition (this implies, for one thing, that they approach the correct condition in the limit $L \to \infty$). The most frequently applied boundary conditions are those which were proposed by Mark [6] and Marshak [7].

Mark pointed out that the problem of a slab imbedded in an absorbing, but non-scattering, medium is completely equivalent to that of the same medium in a vacuum. Then one can use the continuity boundary conditions at the interface together with finiteness conditions in the surrounding medium at infinity to solve the problem. Mark then showed that the conditions were equivalent to the requirement that

$$\Phi(z, u, \mu_j) = 0 \qquad j = 1, 2 \cdots \frac{L+1}{2}, \quad L \text{ odd} \qquad \text{(III.12c)}$$

where the μ_j are the positive roots of

$$P_{L+1}(\mu) = 0 \qquad \text{(III.12d)}$$

This provides just the correct number of conditions. Alternatively, one could apply condition (III.12c) at any $(L + 1)/2$ points, μ_j, provided that they cover the interval $0 > \mu \geqq -1$ as $L \to \infty$. The set given by Mark (III.12d) does however give the best results of any conditions of this type.

Marshak, on the other hand, pointed out that it follows from (III.12a) that

$$\int_{-1}^{0} f(\mu)\, \Phi(z, u, \mu)\, d\mu = 0 \qquad \text{(III.12e)}$$

[5] See DAVISON, *op. cit.*, Chapter 10, for a more complete discussion. This follows from Equation (III.11b) for which half the solutions are known to blow up at the origin of any three-dimensional coordinate system.

[6] J. C. MARK, *op. cit.*

[7] R. E. MARSHAK, *Revs. Mod. Phys.* **19**, 185 (1947).

for any regular function $f(\mu)$ in the interval $-1 < \mu < 0$. Then by choosing $(L + 1)/2$ such functions (for L odd, again), a set of applicable conditions will result. Clearly, if the $P_j(\mu)$ are chosen Equation (III.12e) will result in a set of simple relations among the $\phi_j(z)$. Hence, one should choose $(L + 1)/2$ of the $P_j(\mu), j < L$. But, on the interval $-1 < \mu < 0$ the set $\{P_{2k}(\mu)\}$ is complete as is the set $\{P_{2k+1}(\mu)\}$, $k = 0, 1, 2, \ldots$ Condition (III.12a) is thus met in the limit $L \to \infty$ if either of these sets of boundary conditions is chosen. The first conditions of these sets are

$$\int_{-1}^{0} \Phi(z, u, \mu)\, d\mu = 0 \tag{III.12f}$$

and

$$\int_{-1}^{0} \mu\Phi(z, u, \mu)\, d\mu = 0 \tag{III.12g}$$

The second set is generally chosen because it includes Equation (III.12g) which states that no neutrons enter the system. In Chapter I the Milne problem boundary condition was seen to be

$$\frac{1}{\phi}\,\frac{\partial\phi}{\partial n} = -\,\frac{\Sigma_{tr}}{z_0} \tag{III.12h}$$

with $z_0 = 0.7104$. Equation (III.12f) gives a similar result with $z_0 = 0.500$ while Equation (III.12g) gives $z_0 = 0.6667$ and is therefore preferable.

Calculations have shown that the Marshak conditions are more accurate in the P_1 approximation while the Mark conditions are superior for higher order approximations. However, the Marshak conditions are generally somewhat simpler and are therefore more frequently used.

8. Heavy Elements Approximations

The degradation integral $J_j(k, u)$ is still relatively difficult to handle. First, if more than one moderator is present, η_j consists of a sum of integrals over different lethargy ranges. Secondly, if the scattering is grossly anisotropic a large number of $B_L^c(u)$ are required to describe J_j accurately. An approximation valid for the general case of mixtures would be useful, but as we have seen in in Chapter II the flux may vary rapidly in energy if resonances

[8] B. Davison, *op. cit.*

in the cross sections exist. From this standpoint the slowing-down density is therefore a better quantity with which to work since, in general, it is more slowly varying than the flux. From Equation (II.14a) we have, for the infinite medium without capture, $\varphi(u) = S/\xi\Sigma_t(u)$ at lethargies far from the sources, S being the integrated source density. Physically, the slowing-down density in this case must be equal to S (all neutrons slow down all the way to thermal) so that

$$q(u) \approx \xi\Sigma_t(u)\,\varphi(u) \tag{III.13a}$$

In a finite medium this may not follow, but in a large reactor it should be a good approximation. In many cases ξ is nearly constant and resonance capture is fairly small so that $\Sigma_t(u)\ \phi(k,u) \approx B_0^c\phi$ should be slowly varying. It may be that $B_{L'}^c(u)\,\phi_L(k,u)$ for $L' > 0$ is not slowly varying, but one hopes that it is slowly varying enough in one collision interval to permit the following Taylor series expansion of $B_{L'}^c(u)\ \phi_L(k, u)$ about $u = u'$ (a Maclaurin series in U):

$$\begin{aligned} B_{L'}^c(u')\ \phi_L(k,u') &= \sum_\nu \frac{(u'-u)^\nu}{\nu!}\ \frac{\partial^\nu}{\partial u^\nu}\,[B_{L'}^c\,(u)\ \phi_L(k,u)] \\ &= \sum_\nu (-)^\nu \frac{U^\nu}{\nu!}\ \frac{\partial^\nu}{\partial u^\nu}\,[B_{L'}^c\,(u)\,\phi_L(k,u)] \end{aligned} \tag{III.13b}$$

This series, of course, does not converge rapidly for hydrogen for which U may be infinite. The series is best for heavy moderators for which U takes on only small values. Deuterium moderation is a case for which the series converges slowly and a reasonable approximation may require more terms than could easily be handled. This problem will receive special attention later. Fortunately, scattering from hydrogen is nearly isotropic in the center-of-mass system at reactor energies and the integrals run from 0 to ∞ on U so that this element is again easy to handle in a straightforward manner.

Upon inserting Equation (III.13b) into Equation (III.8c), we have

$$\begin{aligned} J_L(k,u) &= \sum_{L',\nu=0}^{\infty} \sum_i N_i \frac{\partial^\nu}{\partial u^\nu}\,[B_{L'}^{ci}\,(u)\ \phi_L(k,u)] \int_0^{\ln 1/\alpha_i} \frac{(-U)^\nu}{\nu!}\,T_{LL'}^i\,(U)\,\mathrm{d}U \\ &= \sum_i N_i \sum_{L',\nu=0}^{\infty} T_{LL'}^{\nu i} \frac{\partial^\nu}{\partial u^\nu}\,[B_{L'}^{ci}\,(u)\ \phi_L(k,u)] \end{aligned} \tag{III.14a}$$

where the $T_{LL}^{\nu i}$ have been previously defined in Section I.24.

Using this expansion for all elements except hydrogen and deuterium, we have:

$$\Sigma_t(u)\,\phi_j(k,u) - \frac{ik}{2j+1}\,[j\phi_{j-1}(k,u) + (j+1)\,\phi_{j+1}(k,u)]$$

$$= N_H J_j^H(k,u) + N_D J_j^D(k,u) + \sum_{i \neq H,D} N_i \sum_{\nu,L'=0}^{\infty} T_{jL'}^{\nu i} \times$$

$$\times \frac{\partial^\nu}{\partial u^\nu}\,[B_{L'}^{ci}(u)\,\phi_j(k,u)] + S_j(u) \qquad \text{(III.14b)}$$

Quite often, for heavy elements, one is satisfied to retain all terms of order γ, neglecting terms of higher order in γ. Selection may be made using Amster's theorem, quoted in Section I.24. For $j = 0$ and $j = 1$ only terms with $\nu = 0, 1$ need be kept, while for $j \geqq 2$ all terms may be ignored in this approximation. Frequently, the $\nu = 1$ term for $j = 1$ is also neglected, an approximation which is somewhat inconsistent† but which leads to some valuable results. In this approximation, $J_L(k, u)$ becomes

$$J_0(k,u) = \sum_i N_i \sum_{L'=0}^{\infty} \Big\{ T_{0L'}^{0i}\, B_{L'}^{ci}(u)\,\phi_L(k,u) + T_{0L'}^{1i} \frac{\partial}{\partial u} \times$$

$$\times\,[B_{L'}^{ci}(u)\,\phi_L(k,u)]\Big\} \qquad \text{(III.15a)}$$

$$J_1(k,u) = \sum_i N_i \sum_{L'=0}^{\infty} T_{1L'}^{0i}\, B_{L'}^{ci}(u)\,\phi_1(k,u) \qquad \text{(III.15b)}$$

The sums $\sum_{L'} T_{LL'}^{\nu i}\, B_{L'}^{c}(u)$ may be evaluated using Equations (I.62a) and (I.62b). We have:

$$\xi(u)\,\Sigma_s(u) = -\sum_i N_i \sum_{L'} T_{0L'}^{1i} B_{L'}^{ci}(u)$$

$$\langle\mu(u)\rangle\,\Sigma_s(u) = \sum_i N_i \sum_{L'} T_{1L'}^{0i}\, B_{L'}^{ci}(u)$$

so that

$$J_0(k,u) = \Sigma_s(u)\,\phi_0(k,u) + -\frac{\partial}{\partial u}\,[\xi\Sigma_s(u)\,\phi_0(k,u)] \qquad \text{(III.15c)}$$

$$J_1(k,u) = \Sigma_s(u)\,\langle\mu\rangle\,\phi_1(k,u) \qquad \text{(III.15d)}$$

† It should be noted that T_{10}^{1}, the leading term being dropped here, is approximately $2\gamma/3$ for small γ. More will be said about this point later.

C. The P_1 Approximation

9. The Neutron Age

It is clear that the most accurate P_1 approach should employ the full P_1 equations, but this set has been solved in closed form for only a few problems. Many solutions to the P_1 equations have been obtained in various approximations given later, but the advent of large high-speed computers has made it possible to solve the complete P_1 equations numerically. However, numerical methods cannot completely supplant analytical methods as the latter give insight into the problems. Thus the methods supplement each other. Hence we shall endeavor to give analytical solutions whenever they are obtainable, for problems which are as realistic as possible.

The P_1 equations follow from (III.10a) by setting $\phi_2(k, u) = 0$:

$$\Sigma_t(u)\,\phi_0(k, u) - ik\phi_1(k, u) = \sum_i N_i J_0^i(k, u) + S(k, u) \qquad \text{(III.16a)}$$

$$\Sigma_t(u)\,\phi_1(k, u) - \frac{ik}{3}\,\phi_0(k, u) = \sum_i N_i J_1^i(k, u) \qquad \text{(III.16b)}$$

where the $J_j(k, u)$ are written in the rigorous form of Equation (III.8c). In most work heavy element scattering is approximated in some manner. If the Taylor series approximation is made for all elements (including hydrogen and deuterium), we have the so-called Age-Diffusion approximation; if hydrogen is excluded from this approximation and is treated rigorously, the consistent P_1 Selengut–Goertzel (SG) approximation is obtained; if both hydrogen and deuterium are excluded and the T_{LL}^1 terms are retained a "consistent" P_1 approximation is obtained.

In many cases still another approximation is applied. This method, called the P_1 Greuling–Goertzel (GG) method, reduces the problem of solving an equation with integrals over different energy ranges to a much simpler one in which all integrals extend to infinity by replacing the scattering frequencies of all elements except hydrogen by approximate kernels. The procedure is, of course, not necessary for hydrogen as the rigorous kernel is already of the type required. This method is particularly useful in problems involving deuterium moderation.

It is important to note that some sort of approximation is almost always made to heavy element scattering, but the difficulty in obtaining an accurate approximation for hydrogen and the simplicity of the hydrogen degradation integral compels us to treat hydrogen slowing down rigorously. This last point is to be emphasized—*hydrogen slowing down should always be treated rigorously*.†

An important property of the P_1 approximation is that with it *one can calculate the neutron age rigorously*; that is, angular Legendre moments of the flux higher than the first do not contribute to the second spatial moment of the flux. Accurate calculation of the age is thus a criterion by means of which the validity of approximations to the P_1 equations may be examined. Later in the chapter the results of some of the approximate calculations will be compared with each other and with experiment and it will be seen that the methods mentioned above do give the age accurately in the problems for which they are designed.

The fact that the age is given rigorously in the P_1 approximation will now be proved. First, each term in the P_1 equations is expanded in a power series in k, the Fourier transform variable:

$$\varphi_j(k, u) = \sum_{l=0}^{\infty} \varphi_{jl}(u) \frac{(ik)^l}{l!} \qquad \text{(III.17a)}$$

for the flux, and

$$J_j^i(k, u) = \sum_{l=0}^{\infty} J_{jl}^i(u) \frac{(ik)^l}{l!} \qquad \text{(III.17b)}$$

for the degradation integrals. Now, from Taylor's theorem the expansion coefficients of the flux are given by

$$\varphi_{jl}(u) = \left[\frac{\partial^l}{\partial (ik)^l} \varphi_j(k, u) \right]_{k=0} = \int_{-\infty}^{\infty} (z)^l \phi_j(z, u) \, dz \qquad \text{(III.17c)}$$

where the $\phi_j(z, u)$ are the coefficients of the Legendre expansion of the flux in configuration (z) space.

Substitution of the flux expansion (III.17a) into the definition of $J_j(k, u)$, Equation (III.8c), shows that the expansion coefficients of the degradation integral are given by:

$$J_{jl}^i(u) = \int_{u - \ln 1/\alpha}^{u} \varphi_{jl}(u') \, B_l^{0i}(u', U) \, du' \qquad \text{(III.17d)}$$

† Approximations are sometimes made to $J_1^H(k, u)$ but $J_0^H(k, u)$ is always treated rigorously.

Also from the definition of the age we have

$$\tau(u) = \frac{1}{2} \frac{\int\limits_{-\infty}^{\infty} z^2 \, dz \int\limits_{-1}^{1} d\mu \, \Phi(z, u, \mu)}{\int\limits_{-\infty}^{\infty} dz \int\limits_{-1}^{1} d\mu \, \Phi(z, u, \mu)} = \frac{1}{2} \frac{\varphi_{02}(u)}{\varphi_{00}(u)} \tag{III.17e}$$

One further result is required before the statement made above can be proven, namely

$$\varphi_{jl}(u) = 0 \quad \text{for} \quad j > l \quad \text{or} \quad j + l \text{ odd} \tag{III.18a}$$

To show this, it will be most convenient to use the spherical harmonics equations without resorting to the P_1 approximation although a group-theoretical proof is also possible. Substituting the expansions (III.17a, b) into the spherical harmonics Equations (III.10a), one obtains:

$$\Sigma_t(u) \sum_{l=0}^{\infty} \frac{(ik)^l}{l!} \varphi_{0l}(u) - ik \sum_{l=0}^{\infty} \frac{(ik)^l}{l!} \varphi_{1l}(u) = \sum_{L=0}^{\infty} N_i \frac{(ik)^l}{l!} J_{0l}^i(u) + S(u) \tag{III.18b}$$

from the first equation. Here it has been assumed that the source is planar. In the expansion of the source in a power series in k only the first term is non-zero since S does not depend upon k.

From the spherical harmonics equations with $j > 0$, one obtains:

$$\Sigma_t(u) \sum_{l=0}^{\infty} \frac{(ik)^l}{l!} \varphi_{jl}(u) - \frac{ik}{2j+1} \left[j \sum_{l=0}^{\infty} \frac{(ik)^l}{l!} \varphi_{j-1,l}^{(u)} + (j+1) \sum_{l=0}^{\infty} \frac{(ik)^l}{l!} \varphi_{j+1,l}^{(u)} \right] = \sum_{i,l=0}^{\infty} N_i \frac{(ik)^l}{l!} J_{jl}^i(u) \tag{III.18c}$$

Now one can collect coefficients of the various powers of k in these equations and equate them. For example, from Equations (III.18c) if one collects powers of k^0 one obtains for any $j \neq 0$:

$$\Sigma_t(u) \, \varphi_{j0}(u) = \sum N_i J_{j0}^i(u)$$

which is a homogeneous Volterra integral equation for $\varphi_{j0}(u)$ which has only the trivial solution:

$$\varphi_{j0}(u) = 0$$

Using this result, one quickly finds by collecting the coefficients of k^1 in (III.18b) that

$$\varphi_{01}(u) = 0$$

Continuing, collecting coefficients of k^2 in the $l = 1$ equation, one has

$$\varphi_{21}(u) = 0$$

By iterating in this manner, always choosing to collect odd powers of k in the even j equations and even powers of k in the odd j equations, one arrives at the result (III.18a).

Now, collecting coefficients of k^0 and k^2 in the $j = 0$ equation and coefficients of k^1 in the $j = 1$ equation leads to a set of three coupled equations in the three lowest index non-zero $\varphi_{jl}(u)$:

$$\Sigma_t(u)\,\varphi_{00}(u) = S(u) + \sum_i N_i J^i_{00}(u) \qquad \text{(III.19a)}$$

$$\tfrac{1}{2}\Sigma_t(u)\,\varphi_{02}(u) - \varphi_{11}(u) = \tfrac{1}{2}\sum_i N_i J^i_{02}(u) \qquad \text{(III.19b)}$$

$$\Sigma_t(u)\,\varphi_{11}(u) - \tfrac{1}{3}\varphi_{00}(u) = \sum_i N_i J^i_{11}(u) \qquad \text{(III.19c)}$$

This set of equations, when taken together with the defining relations for the $J^i_{jl}(u)$, Equations (III.17d), three for each moderator present, is just sufficient to determine the quantities $\varphi_{00}(u)$ and $\varphi_{02}(u)$, and hence the age from Equation (III.17e). None of the $\varphi_{jl}(u)$ with $j > 1$ enter Equations (III.19) since $\phi_2(k, u)$ contributes nothing to the coefficient of k^1 in the second of the equations, hence the neutron age can be calculated rigorously using the P_1 approximation.

10. Age-Diffusion Theory

The simplest special case of the P_1 approximation is that for which the hydrogen and deuterium concentrations vanish; two terms on the right side of each of Equations (III.16) are eliminated. We are then dealing only with the case of "heavy" element (e.g., carbon, beryllium) moderation, and the Taylor series expansion (III.13b) can be expected to converge rapidly. It has been seen that the second and higher derivative terms in these equations have coefficients of order at least γ^2, i.e., neglect of them can be expected to introduce an error of approximately 1/81 for beryllium, but would leave an error of magnitude 1/4 for deuterium. Hence, as a first approximation they may be neglected for heavy elements, and we arrive at a consistent P_1 approximation for heavy elements:

$$\Sigma_t(u)\,\phi_0(k,u) - ik\phi_1(k,u)$$
$$= \Sigma_s(u)\,\phi_0(k,u) - \frac{\partial}{\partial u}\sum_i \xi_i\Sigma_{si}(u)\,\phi_0(k,u) + S_0(k,u) \quad \text{(III.20a)}$$

$$\Sigma_t(u)\,\phi_1(k,u) - \frac{ik}{3}\,\phi_0(k,u)$$
$$= \sum_i \Sigma_{si}(u)\langle\mu\rangle_i\,\phi_1(k,u) + \frac{\partial}{\partial u}\sum_i \langle\mu U\rangle_i\,\phi_1(k,u) \quad \text{(III.20b)}$$

Although this set can be reduced to a single second order differential equation for $\phi_0(k, u)$, in practice this is rarely done because of the difficulty involved in solving the resulting equation; the set is sometimes solved numerically however.

Dropping the last term in the second equation (a procedure which is inconsistent) leads to a simple linear relationship between $\phi_0(k, u)$ and $\phi_1(k, u)$, i.e., between the Fourier transforms of the flux and the current

$$\phi_1(k,u) = \frac{ik\,\phi_0(k,u)}{3[\Sigma_t(u) - \sum_i \Sigma_{si}\langle\mu\rangle_i]} = ikD(u)\,\phi_0(k,u) \quad \text{(III.21a)}$$

The significance of

$$D(u) = [3(\Sigma_t(u) - \sum_i \Sigma_{si}(u)\langle\mu\rangle_i)]^{-1}$$

may be seen more easily if we take the inverse Fourier transform:

$$\phi_1(z,u) = -D(u)\frac{\partial}{\partial z}\phi_0(z,u) \quad \text{(III.21b)}$$

which is Fick's law,† which was used in Chapter I. Using Equation (III.21b) to eliminate $\phi_1(k, u)$ from Equation (III.20a) one obtains the extension of the usual "Fermi Age Equation" [9] to anisotropic scattering:

$$[D(u)\,k^2 + \Sigma_a(u)]\,\phi_0(k,u) = -\frac{\partial}{\partial u}[\xi(u)\,\Sigma_s(u)\,\phi_0(k,u)] + S_0(k,u) \quad \text{(III.22)}$$

† This simple and desirable result is, of course, the reason why the last term in Equation (III.20b) is dropped. In the last analysis the only rationale for this procedure is its success in predicting the age.

[9] GLASSTONE and EDLUND, *op. cit.*, p. 172 *et seq.*

where use has been made of the result of Chapter II:

$$\xi(u) = \frac{\sum_i \Sigma_{si}(u)\, \xi_i}{\Sigma_s(u)}$$

Equation (III.22) is the basic equation of "Age–Diffusion" or "Fermi Age" theory. It may be put into a more conventional form by means of some variable transformations. First, we change the dependent variable to:

$$q(k, u) = \xi(u)\, \Sigma_s(u)\, \phi_0(k, u) \qquad \text{(III.23a)}$$

which is shown below to be an approximation to the slowing-down density (see also Equation (III.13a)). Substituting Equation (III.23a) into Equation (III.22), we have:

$$\left[\frac{D(u)\, k^2 + \Sigma_a(u)}{\xi\, \Sigma_s(u)}\right] q(k, u) = -\frac{\partial q(k, u)}{\partial u} + S_0(k, u) \qquad \text{(III.23b)}$$

Expressions of the form (III.23a) were derived for the slowing-down density in Chapter II. The simplest derivation of such a formula is to substitute the Taylor series expansion of $B_L^c(u')\, \phi_0(u')$, Equation (III.13b), into the expression for the slowing-down density, Equation (I.32b). We omit the details here as they are given in Section 12; the result is

$$q(\mathbf{r}, u) = \sum_{L', n=0}^{\infty} T_{0L'}^{n+1} \frac{\partial^n}{\partial u^n} [B_{L'}^c(u)\, \phi_0(\mathbf{r}, u)] \qquad \text{(III.23c)}$$

Keeping only the $n = 0$ terms we have Equation (III.23a). This procedure is consistent since ξ is $O(1/A)$ while the $n \geqq 1$ terms are all $O(1/A^2)$ or smaller.

Since Equation (III.23b) describes slowing down with absorption one might expect that it can be reduced to a problem without absorption by introducing the resonance escape probability appropriate to the case under consideration. This is indeed the case, for if we define $q' = q/p$, i.e., set

$$q(k, u) = q'(k, u)\, e^{-\int_{u'}^{u} \frac{\Sigma_a(u'')}{\xi \Sigma_s(u'')} du''} = q'(k, u)\, P(u' \to u) \qquad \text{(III.24a)}$$

and

$$S(k, u) = S'(k, u)\, e^{-\int_{u'}^{u} \frac{\Sigma_a(u'')}{\xi \Sigma_s(u'')} du''} = S'(k, u)\, P(u' \to u) \qquad \text{(III.24b]}$$

the absorption term disappears and we have

$$\frac{D(u)\,k^2}{\xi\Sigma_s(u)}\,q'(k, u) = -\frac{\partial q'(k, u)}{\partial u} + S'(k, u) \qquad \text{(III.25)}$$

From the results of Chapter II

$$\exp\left\{-\int_{u'}^{u} \frac{\Sigma_a(u'')}{\xi\Sigma_s(u'')}\,\mathrm{d}u''\right\}$$

may be recognized as the resonance escape probability for small capture; thus, q' may be interpreted as a fictitious slowing-down density that would result in the absence capture.

One further transformation, namely a change of independent variable from u to:

$$\tau_F(u) = \int_0^u \frac{D(u')}{\xi\Sigma_s(u')}\,\mathrm{d}u' \qquad \text{(III.26a)}$$

with

$$S'(\tau_F)\,\mathrm{d}\tau_F = S'(u)\,\mathrm{d}u \qquad \text{(III.26b)}$$

reduces (III.25) to the usual Fermi Age equation [10]

$$\frac{\partial q'(k, \tau_F)}{\partial \tau_F} + k^2 q'(k, \tau_F) = S'(k, \tau_F) \qquad \text{(III.26c)}$$

τ_F is as yet a purely mathematical quantity, but as the symbol suggests, it is indeed an approximation to the age.

Equation (III.26c) is readily solved by the methods employed in Chapter II to yield:

$$q'(k, \tau_F) = \int_0^{\tau_F} S(\tau'_F)\,\mathrm{e}^{-k^2(\tau_F - \tau'_F)}\,\mathrm{d}\tau'_F \qquad \text{(III.27a)}$$

The slowing-down density is thus:

$$q(k, u) = \int_0^u S(u')\,\mathrm{e}^{-k^2[\tau_F(u) - \tau_F(u')]}\,\mathrm{e}^{-\int_{u'}^{u} \frac{\xi\Sigma_s}{\Sigma_a}\mathrm{d}u''}\,\mathrm{d}u' \qquad \text{(III.27b)}$$

[10] As a historical note it may be noted that the name "age" arose from the similarity of Equation (III.26c) [or Eqn. (III.30a)] to the ordinary time-dependent diffusion equation (for heat, material, etc.), where τ would ordinarily be the time variable. The term seems to be due to E. Fermi, *Ricerca Sci.* 7, 13 (1936).

For a monoenergetic source $S(u) = \delta(u)$ this becomes

$$q(k, u) = e^{-k^2 \tau_F(u)} P(0 \to u) \tag{III.27c}$$

The discussion of Chapter I allows us to identify $q(|\mathbf{r} - \mathbf{r}'|, E)$ with the slowing-down density kernel $P^*(|\mathbf{r} - \mathbf{r}'|, E)$. Hence, in light of the discussion of Chapter I, we can write

$$\overline{P_T^*}(B_g^2) = \mathrm{e}^{-B_g^2 \tau_F(u_T)}\, \varepsilon\, P(0 \to u_T) \tag{III.28}$$

where fast fission is now accounted for. Hence, in the Age–Diffusion approximation the criticality condition for a purely thermal bare asymptotic reactor may be written as

$$\frac{\eta f\, \varepsilon\, p(0 \to u_T)\, \mathrm{e}^{-B_g^2 \tau_F(u_T)}}{1 + L^2 B_g^2} = 1 \tag{III.29}$$

provided τ_F does not vary greatly within the fission spectrum. The fast fission factor, ε, has been arbitrarily re-introduced because fast fission was not provided for in the slowing down equations. Equation (III.29) is the usual Age–Diffusion (Fermi Age) criticality condition. In this model the fast effect, ε, the resonance escape probability, $p(0 \to u_T)$, and the fast non-leakage probability are all independent, i.e., the total escape probability is the product of the individual escape probabilities. In Chapters I and II it was mentioned that this is not true in the actual physical case. Neither is it true in more sophisticated models unless the individual factors are all quite close to unity, but the result does obtain in this simple model.

The inverse Fourier transforms of some of the above expressions are familiar. For example, the inverse of (III.26c) is the well known Age–Diffusion equation: †

$$\nabla^2 q'(\mathbf{r}, \tau_F) = \frac{\partial q'(\mathbf{r}, \tau_F)}{\partial \tau_F} + S(\mathbf{r}, \tau_F) \tag{III.30a}$$

while the inverse of Equation (III.27b) is its solution for a point source:

$$q(\mathbf{r}, u) = \int_0^u S(u') \frac{\exp\{[-r^2/4[\tau_F(u) - \tau_F(u')]\}}{[4\pi\{\tau_F(u) - \tau_F(u')\}]^{3/2}} \times$$

$$\times \exp\left\{-\int_{u'}^{u} \frac{\Sigma_a}{\xi \Sigma_s}\, \mathrm{d}u''\right\} \mathrm{d}u' \tag{III.30b}$$

† Here we take the three-dimensional inversion, which is permitted because of the identity of the one- and three-dimensional Fourier transforms.

which for a monoenergetic source and no absorption reduces to the well known Gaussian age kernel:

$$q(\mathbf{r}, u) = \frac{\exp\{-r^2/4\tau_F(u)\}}{[4\pi\tau_F(u)]^{3/2}} \tag{III.30c}$$

In one dimension (taking a one-dimensional inversion) we have for a monoenergetic source and no absorption:

$$q'(z, \tau_F) = \frac{\exp[-z^2/4(\tau_F - \tau_0)]}{[4\pi(\tau_F - \tau_0)]^{1/2}} \tag{III.30d}$$

The mean square slowing down length will now be calculated to show that τ_F is indeed an approximation to the age. The most convenient means of calculating the age is provided by combining Equations (III.17c) and (III.17e). (We work here with the slowing-down density rather than the flux merely as a convenience. The two are proportional and clearly give the same result):

$$\langle z^2 \rangle = \left. \frac{-\dfrac{\partial^2 q(k, u)}{\partial k^2}}{q(k, u)} \right|_{k=0} = -2 \left. \frac{\dfrac{\partial q(k^2, u)}{\partial (k^2)}}{q(k^2, u)} \right|_{k=0} \tag{III.31a}$$

For a monoenergetic source we have using Equation (III.27c):

$$\langle z^2 \rangle = 2\tau_F(u) = \int_0^u \frac{D(u')}{\xi \Sigma_s(u')} \mathrm{d}u' \tag{III.31b}$$

so that the quantity $\tau_F(u)$ defined by Equation (III.26a) is identical to the neutron age, defined previously, in the Age–Diffusion approximation. It will therefore be denoted simply by τ henceforth.

If the source it not monoenergetic, the non-leakage probability must be written:†

$$\overline{P_T^*}(B_g^2) = \frac{q(B^2, \tau)}{q(0, \tau)} = \frac{\int_0^\tau S(\tau')\, \mathrm{e}^{-B^2(\tau - \tau')} p(\tau' \to \tau)\, \mathrm{d}\tau'}{\int_0^\tau S(\tau')\, p(\tau' \to \tau)\, \mathrm{d}\tau'} \tag{III.32a}$$

† τ here is used to represent $\tau(u_T)$.

and the age becomes

$$\tau(u) = \frac{1}{2}\langle z^2(u)\rangle = \frac{\int_0^\tau (\tau - \tau')\, S(\tau')\, p(\tau' \to \tau)\, d\tau'}{\int_0^\tau S(\tau')\, p(\tau' \to \tau)\, d\tau'} \tag{III.32b}$$

and is found by applying the formula (III.31 a) to Equation (III.27 b).

A connection between these formulae and a more commonly quoted result is obtained in the limit of small B^2.† In this limit

$$\overline{P_T^*}(B^2) \approx \frac{\int_0^\tau S(\tau')\,[1 - B^2(\tau - \tau')]\, p(\tau' \to \tau)\, d\tau'}{\int_0^\tau S(\tau')\, p(\tau' \to \tau)\, d\tau'}$$

$$= 1 - B^2 \frac{\int_0^\tau (\tau - \tau')\, S(\tau')\, p(\tau' \to \tau)\, d\tau'}{\int_0^\tau S(\tau')\, p(\tau' \to \tau)\, d\tau'}$$

If resonance capture is very small this reduces to

$$\overline{P_T}(B^2) = 1 - B^2 \int_0^\tau [\tau - \tau']\, S(\tau')\, d\tau' \approx e^{-B^2 \tau_{av}} \tag{III.33a}$$

where the source average value of τ, τ_{av}, is

$$\tau_{av} = \int_0^\tau S(\tau')\,[\tau - \tau']\, d\tau' \tag{III.33b}$$

This manipulation has been carried out in order to show the connection between the correct formula for the fast leakage probability (III.32a) and Equation (III.33a) which is frequently used in the literature, but is only approximate.

In Age–Diffusion theory, absorption has an effect on the spatial shape of the slowing-down density only through the slight dependence of D on Σ_a. This is essentially the statement that the processes of resonance escape and fast leakage are independent. The effect

† This was pointed out by M. Levine, unpublished.

of absorption is to reduce the number of neutrons available at any particular lethargy, but their spatial distribution is almost independent of the number which have been absorbed. This is not a true description of the physical processes occurring in a reactor, and it is a difficulty inherent in the method. This difficulty is indicated by the fact that the resonance escape probability calculated by this method is that for the small absorption limit. Other methods predict a stronger dependence of the age on Σ_a as we shall see in the next section.

In comparing the results of Age–Diffusion theory with experiments and with the results of more nearly rigorous calculations (as we shall do in Section 26), it is found that the Age–Diffusion approximation is surprisingly accurate for graphite and beryllium, but is, of course, not at all accurate for light or heavy water. A large part of the discrepancy may be overcome in a rather simple fashion. In the above calculations a point source, $\delta(\mathbf{r})$, was assumed. However, as neutrons do not start to slow down until they have made their first collisions, it is more reasonable to assume that the source has the distribution of the first flight kernel, $e^{-\Sigma_t r}/4\pi r^2$. Carrying through the calculations for this source, we find that there is an additional contribution, $2/\Sigma_t^2$, to the age. This correction may be quite large, particularly in water. A more complete discussion of this effect is given in the next section.

11. The Selengut–Goertzel Method [11]

The approach employed in the previous section is called *Age–Diffusion* theory because it assumes the linear relation between the neutron current, $\phi_1(r, u)$, and the gradient of the flux, $\phi_0(r, u)$, which is basic to diffusion theory. We were unable, however, to treat hydrogen or deuterium slowing down by that model as the truncation of the Taylor series expansion is a poor approximation for these moderators.

It will later be shown how Age–Diffusion theory can be derived in a manner which yields more physical insight into the nature of the approximations. In essence, Age–Diffusion theory implies that neutrons make an infinite number of infinitesimal flights in travelling any finite distance; alternatively, it assumes that a neutron

[11] D. SELENGUT and G. GOERTZEL, unpublished.

gains only an infinitesimal lethargy in a given collision. This is the reason that the method employed in the previous section is sometimes referred to as the *continuous slowing-down model.* More will be said about the implications of these considerations after the differences between the Selengut–Goertzel and Age–Diffusion results have been derived and discussed.

To begin with, instead of discarding the $J_0^H(k, u)$ term from the first of the P_1 equations, we will include it in its rigorous form; deuterium slowing down, which is best treated by the method of the succeeding section, is ignored for the present. In the second equation, in order to obtain Fick's law, one is forced to make the Taylor series expansion for all elements, thus yielding Equation (III.21 a) modified to include hydrogen, i.e., the diffusion coefficient, D, contains terms arising from hydrogen scattering. Treating heavy elements as in the previous section, we arrive at the basic equation:

$$[D(u)\,k^2 + \Sigma_a^{(u)} + \Sigma_S^H(u)]\,\phi_0(k, u) = J_0^H\,(k, u) - \frac{\partial}{\partial u}\,(\xi\Sigma_S'(u)\,\phi_0(k, u))\;S_0(k, u) \qquad \text{(III.34)}$$

Here

$$\Sigma_S'(u) = \Sigma_s(u) - \Sigma_S^H(u)$$

$\xi(u)$ retains its previous definition and does not include the effects of hydrogen. Of course, a new unknown, $J_0^H(k, u)$, has been added to the equation, so that another equation relating it to the flux, $\phi_0(k, u)$, is required. This is conveniently supplied by Equation (III.11) with $j = 0$, but for numerical calculations it is sometimes easier to use the equation obtained by differentiating the $j = 0$ Equation (III.11):

$$J_0^H\,(k, u) + \frac{\partial J_0^H\,(k, u)}{\partial u} = \Sigma_S^H\,(u)\,\phi_0(k, u) \qquad \text{(III.35)}$$

This method is sometimes called the diffusion Selengut–Goertzel or inconsistent P_1 Selengut–Goertzel method for reasons that are probably already plain. A consistent P_1 method similar to the Selengut–Goertzel scheme is sometimes used and employs the following set of equations:

$$[\Sigma_a(u) + \Sigma_s^H\,(u)]\,\phi_0(k, u) - ik\phi_1(k, u) = J_0^H\,(k,u) - \frac{\partial}{\partial u}\,(\xi(u)\,\Sigma_s(u)\,\phi_0(u)) + S_0(k, u) \qquad \text{(III.36 a)}$$

and

$$[\Sigma_t(u) - \Sigma_s'(u) \langle\mu\rangle + \Sigma_s^H \langle\mu\rangle_H] \phi_1(k, u) - \frac{ik}{3} \phi_0(k, u)$$
$$= J_1^H (k, u) - \frac{\partial}{\partial u} (\langle\mu U\rangle \Sigma_s'(u) \phi_1(k, u)) \tag{III.36b}$$

in which hydrogen slowing down is treated rigorously throughout, and all terms of order γ have been retained. Introduction of the fourth dependent variable $J_1^H(k, u)$ requires use of Equation (III.11) with $j = 1$. The quantity $J_1^H(k, u)$ has the same physical interpretation as in Chapter II. Since the introduction of the spatial (or Fourier transform) variable does nothing to alter the arguments made with respect to the lethargy variable, $J_0^H(k, u)$ is the Fourier transform of the slowing-down density of neutrons whose last collision was with a hydrogen nucleus.

An analytical solution of the diffusion Selengut–Goertzel equations is possible if $\bar{\xi} = 0$, i.e., if hydrogen slowing down dominates. Collisions with other elements affect the diffusion of neutrons but these elements are assumed so heavy that their contribution to the slowing-down density may be neglected compared to that of hydrogen.

First, the age is calculated; the power series expansion method is again the simplest [12]. However, in the SG method, one must distinguish two different "ages," the "flux age" and the "slowing-down density age" defined by

$$\tau_N = - \lim_{k \to 0} \frac{\partial\phi(k, u)/\partial(k^2)}{\phi(k, u)} = \frac{1}{2} \frac{\varphi_{02}(u)}{\varphi_{00}(u)} = \frac{1}{2} \frac{\int_{-\infty}^{\infty} \phi z^2(z, u)\, \mathrm{d}z}{\int_{-\infty}^{\infty} \phi(z, u)\, \mathrm{d}z} \tag{III.37a}$$

and

$$\tau_q = - \lim_{k \to 0} \frac{\partial q(k, u)/\partial(k^2)}{q(k, u)} = \frac{1}{2} \frac{J_{02}^H(u)}{J_{00}(u)} = \frac{1}{2} \frac{\int_{-\infty}^{\infty} z^2 q(z, u)\, dz}{\int_{-\infty}^{\infty} q(z, u)\, \mathrm{d}z} \tag{III.37b}$$

[12] This method was used by H. Hurwitz, Jr., and P. F. Zweifel, *J. Appl. Phys.* **26**, 8, 923 (1955) and originally by R. Marshak, *op. cit.*

These two definitions are not equivalent in the SG approximation, although they are in the Age–Diffusion approximation due to the proportionality of q and ϕ. The difference between them leads, however, to some interesting results.

Under the approximations made, the equations required for the age calculation are obtained by making the Taylor series expansions (III.17a) and (III.17b) and collecting coefficients of k^0 and k^2 in Equations (III.34) and (III.35):

$$[\Sigma_s^H(u) + \Sigma_a(u)]\,\varphi_{00}(u) = J_{00}(u) + S(u) \qquad \text{(III.38a)}$$

$$J_{00}(u) + \frac{\partial J_{00}(u)}{\partial u} = \Sigma_s^H(u)\,\varphi_{00}(u) \qquad \text{(III.38b)}$$

$$-2D(u)\,\varphi_{00}(u) + [\Sigma_a(u) + \Sigma_s^H(u)]\,\varphi_{02}(u) = J_{02}(u) \qquad \text{(III.38c)}$$

$$J_{02}(u) + \frac{\partial J_{02}(u)}{\partial u} = \Sigma_s^H(u)\,\varphi_{02}(u) \qquad \text{(III.38d)}$$

These four differential equations for the four required quantities may readily be solved. Simultaneous solution of the first two yields $J_{00}(u)$ and $\varphi_{00}(u)$:

$$J_{00}(u) = \int_0^u \frac{\Sigma_s^H(u')}{\Sigma_s^H(u') + \Sigma_a(u')}\,S(u')\exp\left\{-\int_{u'}^{u} \frac{\Sigma_a(u'')}{\Sigma_a(u'') + \Sigma_s(u'')}\,du''\right\} \qquad \text{(III.39a)}$$

and

$$\varphi_{00}(u) = \frac{1}{\Sigma_a(u) + \Sigma_s^H(u)}\,[J_{00}(u) + S(u)] \qquad \text{(III.39b)}$$

which may be used to solve the last two of Equations (III.38). Before proceeding to the calculation of φ_{02} and J_{02} it will be instructive to determine the difference between the two ages defined above. From Equation (III.38c), we have

$$\tau_N = \frac{1}{2}\,\frac{\varphi_{02}(u)}{\varphi_{00}(u)} = \frac{D(u)}{\Sigma_a(u) + \Sigma_s^H(u)} + \frac{1}{2}\,\frac{J_{02}(u)}{[\Sigma_a(u) + \Sigma_s^H(u)]\,\varphi_{00}(u)} \qquad \text{(III.39c)}$$

if there are no sources at lethargy u. But, from Equation (III.38c) we immediately find:

$$\tau_q = \frac{1}{2}\,\frac{J_{02}(u)}{J_{00}(u)} = \frac{1}{2}\,\frac{J_{02}(u)}{[\Sigma_a(u) + \Sigma_s^H(u)]\,\varphi_{00}(u)} \qquad \text{(III.39d)}$$

It is thus apparent that the difference between the flux age and the slowing-down age is just $D/(\Sigma_a + \Sigma_s^H)$ evaluated at the lethargy of measurement; this term arises from the fact that a neutron travels a finite distance at lethargy u before experiencing the collision which will either absorb it or slow it down further, and is therefore called the "last flight correction".

The physical significance of this correction is that a neutron contributes to the slowing down density instantaneously upon *reaching or crossing* lethargy u, but continues to contribute to the flux until it is *degraded to lethargies above u.* From one-speed diffusion theory it is known that the mean square distance travelled by a diffusing neutron is $6D/\Sigma_a = 6L^2$; the effective "absorption" cross section here is $[\Sigma_a(u) + \Sigma_s^H(u)]$ since either a true absorption or a slowing-down collision will remove the neutron from the lethargy of interest. The added contribution to the flux age is just one-sixth of this mean square distance.

Let us now calculate τ_q. Eliminating $\varphi_{02}(u)$ from the last two of Equations (III.38) and then using Equation (III.39b) to eliminate $\varphi_{00}(u)$, one finds by straightforward integration:

$$J_{02}(u) = 2\int_0^u \left[\frac{D\Sigma_s^H}{(\Sigma_a + \Sigma_s^H)^2}\right]_{u'} \times$$

$$\times\ [J_{00}(u') + S(u')] \exp\left\{-\int_{u'}^{u}\left(\frac{\Sigma_a}{\Sigma_a + \Sigma_s^H}\right)_{u''} du''\right\} \qquad \text{(III.40a)}$$

Then τ_q can be found by dividing $J_{02}(u)$, Equation (III.40a), by $J_{00}(u)$, Equation (III.39a). As the result is complicated, only the monoenergetic source case will be investigated. The Selengut–Goertzel slowing-down density age becomes, under this assumption:

$$\tau_q = \int_0^u \left[\frac{D\Sigma_s^H}{(\Sigma_a + \Sigma_s^H)^2}\right]_{u'} du' + \left[\frac{D}{\Sigma_a + \Sigma_s^H}\right]_{u=0} \qquad \text{(III.40b)}$$

Two points of difference between this expression and the Age–Diffusion expression are noted. First, an additional term $D/(\Sigma_a + \Sigma_s^H)$ evaluated at the lethargy is included. It is the contribution to the age due to the distance that neutrons travel at the source energy before starting to slow down and hence is designated the "first flight correction". That the first flight correction should be $D/(\Sigma_a + \Sigma_s^H)$ can be verified by the method used in considering the last flight correction. Secondly, the integrand of the first term on the right-hand side of Equation (III.40b) contains a term $(\Sigma_s^H/\Sigma_a + \Sigma_s^H)^2$ that was not found in the Fermi Age expression. As a result, the resonance escape probability p and the non-leakage probability λ are now coupled.

Flight corrections arise from a failing in the Age–Diffusion approximation that was mentioned previously. Age–Diffusion theory assumes that neutrons travel an infinitesimal distance between consecutive collisions. This may be a reasonable approximation for neutrons which have started to slow down and thus are not all at one energy. However, at the source energy, neutrons scatter with the one flight collision distribution

$$\frac{\exp\left[-\Sigma_t(E_0)\,r\right]}{4\pi r^2} \tag{III.41}$$

before starting to slow down if all elements present are assumed to slow down. If the first kernel represents only a short flight of the neutrons from the source, little error is introduced by neglecting this effect. Hydrogen, however, is once again peculiar; its scattering cross section falls off rather rapidly above 1 MeV (see Fig. III.3) so that the first flight correction is considerable at fission energies. This is further aggravated by the fact that slowing down in hydrogen requires relatively few collisions so that the age tends to be relatively small in hydrogenous systems. In other moderators this effect is not nearly so pronounced; in fact, the last flight correction is small even for hydrogen.

Even the diffusion Selengut–Goertzel method does not treat the first andlast flight corrections properly. It has been seen that the SG theory predicts these corrections as $[D/(\Sigma_a + \Sigma_s^H)]_{u=0}$ and $[D(\Sigma_a + \Sigma_s^H)]_u$ respectively, i.e., it gives the diffusion theory result; the rigorous correction will be given later. The nature of the flight correction suggests that the flux in the SG approximation, as given above, is the convolution of an age kernel and two one-speed diffusion kernels:

$$P_0(r' \to r, u) = \int d^3r''' \int d^3r'' \, \frac{\exp\left[-|r'' - r'|/L\,(0)\right]}{4\pi\, D(0)\,|r'' - r'|} \times$$

$$\times \frac{\exp\left[-|r''' - r''|/4\tau\right]}{(4\pi\tau)^{3/2}} \, \frac{\exp\left[-|r''' - r|/L(u)\right]}{4\pi D(u)\,|r''' - r|} \tag{III.42}$$

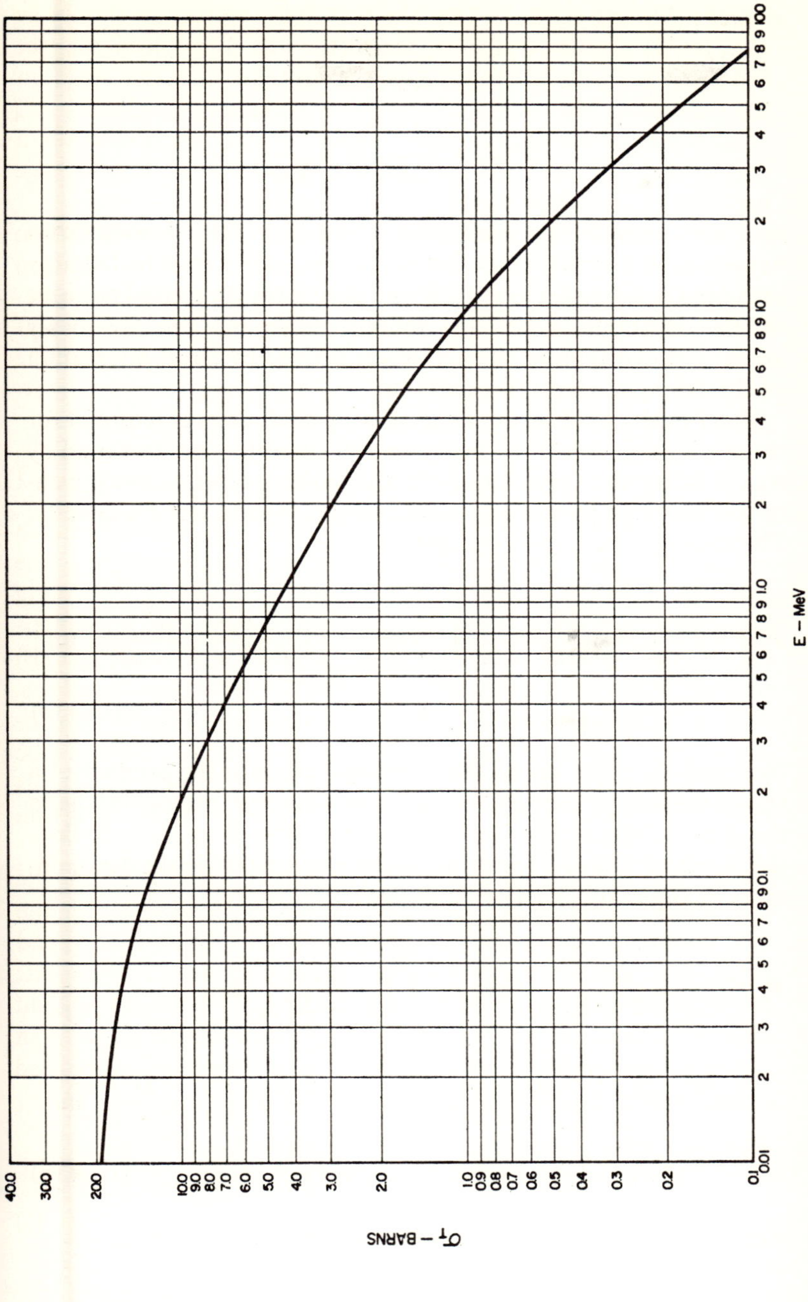

FIG. III.3. The total cross section of hydrogen. From D. J. Hughes and R. B. Schwartz, *Neutron Cross Sections*, BNL-325, 2nd ed. (1958).

where

$$L^2(0) = \frac{D(0)}{\Sigma_a^{(0)} + \Sigma_s^H(0)}$$

$$L^2(u) = \frac{D(u)}{\Sigma_a(u) + \Sigma_s^H(u)}$$

and one expects the analogous result for the slowing-down density to be a similar convolution of a one flight kernel (at lethargy zero) and a Gaussian kernel. If one calculates the second moment of this distribution he does indeed obtain the age previously given. The distribution (III.42) can be derived by direct integration of the Selengut–Goertzel equations.

The Fourier transform of the slowing-down kernel required for the Asymptotic Reactor Theory critical condition can be found by transforming the convolution for the slowing-down density mentioned above or by solving Equations (III.34) and (III.35) directly. The result is, for a monoenergetic source:

$$\bar{P}(B^2) = \left(\frac{\Sigma_s^H}{\Sigma_s^H + DB^2 + \Sigma_a}\right)_{u=0} \exp\left\{-\int_0^{u_{th}} \left(\frac{DB^2 + \Sigma_a}{DB^2 + \Sigma_a + \Sigma_s^H}\right)_{u'} du'\right\} \tag{III.43a}$$

Note that here the kernel does not separate into the product of a resonance escape probability and a non-leakage probability as in the Age–Diffusion case. Only in the limit $B^2 \to 0$ (or $\Sigma_a \to 0$) does this separation occur, for then one can neglect DB^2 in the denominators in Equation (III.43a) to obtain:

$$\bar{P}(B^2) = \left(\frac{\Sigma_s^H}{\Sigma_s^H + \Sigma_a}\right)_{u=0} \exp\left[-\int_0^{u_{th}} \left(\frac{\Sigma_a}{\Sigma_a + \Sigma_s^H}\right)_{u'} du'\right] \times$$

$$\times \exp\left[-B^2 \int_0^{u_{th}} \left(\frac{D}{\Sigma_a + \Sigma_s^H}\right) du'\right] = p \exp\left[-B^2 \int_0^{u_{th}} \left(\frac{D}{\Sigma_s^H + \Sigma_a}\right) du'\right]$$

$$\approx p\, e^{-B^2 \tau_{th}} \tag{III.43b}$$

where τ_{th} is now the Selengut–Goertzel age evaluated at u_{th}.

12. The Greuling–Goertzel Procedure [13, 14]

The simplicity of the equations that arise in hydrogen moderation problems and the subsequent ease of solution of the equations for such systems suggests seeking similar approximate equations for applications to systems containing other moderators. In particular, one would like to have an equation analogous to Equation (III.35) of the type

$$q(k, u) + \lambda(u) \frac{\partial q(k, u)}{\partial u} = \xi(u) \Sigma_s(u) \phi_0(k, u) \qquad \text{(III.44)}$$

for all moderators. One hopes the parameters λ and ξ can be chosen such that the error inherent in the approximation will be minimized, e.g., proportional to γ^2 instead of γ as in the Age–Diffusion method. The symbol ξ will be shown to retain its previous meaning. This approximation may be applied to any element other than hydrogen but the success of the Age–Diffusion model for heavy element moderation has limited its use almost exclusively to deuterium,† which is in the anomalous position of being too light to permit application of age theory and yet does not enjoy the simple scattering kernel of hydrogen.

The derivation given here is rather formal but is in some respects simpler than the original method of derivation. In a rather arbitrary fashion, a new function $\eta_L(k, u)$ is defined by a Taylor series expansion‡

$$\eta_L(k, u) = -\sum_{n=0}^{\infty} \frac{\partial^n}{\partial u^n} \sum_{L'=0}^{\infty} T_{LL'}^{n+1} B_{L'}^c(u) \phi_L(k, u) \qquad \text{(III.45a)}$$

Then, by differentiation:

$$\frac{\partial \eta_L(k, u)}{\partial u} = -\sum_{n=0}^{\infty} \frac{\partial^{n+1}}{\partial u^{n+1}} \sum_{L'=0}^{\infty} T_{LL'}^{n+1} B_{L'}^c(u) \phi_L(k, u) \qquad \text{(III.45b)}$$

† E. Gelbard (private communication) has pointed out that the GG method is also more accurate than the AD method for heavy elements, the difference being in many cases, surprisingly large.

‡ For simplicity the development will be made for a single moderator but the extension to the general case will be obvious. It will also be convenient to momentarily consider the $B_L^c(u)$ to include a factor of atom density, i. e., to be regarded as the expansion coefficients for the macroscopic cross section.

[13] E. Greuling and G. Goertzel, *Nucl. Sci. Eng.* 7, 69 (1960).
[14] The development here closely parallels that of H. Amster, *J. Appl. Phys.* 29, 4, 623 (1958).

For notational purposes, indices referring to particular nuclear types are suppressed. In general, however, a sum over all nuclear types weighted by the atom density of each nuclear type is to be implied. Multiplying Equation (III.45b) by a parameter, λ_L (which may be a function of u), and adding it to Equation (III.45a) we obtain, after some rearrangement,

$$\eta_L(k, u) + \lambda_L \frac{\partial \eta_L(k, u)}{\partial u} = - \sum_{L'=0}^{\infty} T^1_{LL'} B^c_{L'}(u)\, \phi_L(k, u) \left(1 - \frac{d\lambda_L}{du}\right)$$

$$- \frac{\partial}{\partial u} \sum_{L'=0}^{\infty} B^c_{L'}(u)\, \phi_L(k, u)\, [T^2_{LL'} + \lambda_L T^1_{LL'}]$$

$$- \sum_{n=2}^{\infty} \sum_{L'=0}^{\infty} \left[\frac{\partial^n}{\partial u^n} (B^c_{L'}(u)\, T^{n+1}_{LL'}) + \lambda_L \frac{\partial^n}{\partial u^n} (B^c_{L'}(u)\, \phi_L(k, u)\, T^n_{LL'})\right] \tag{III.45c}$$

The parameters λ_L may now be chosen such that the second term on the right-hand side of Equation (III.45c) vanishes, i.e., all first derivatives of the flux drop out:

$$\lambda_L(u) = - \frac{\sum_{L'=0}^{\infty} T^2_{LL'} B^c_{L'}(u)}{\sum_{L'=0}^{\infty} T^1_{LL'} B^c_{L'}(u)} = \frac{\langle U^2 P_L(\mu)\rangle}{2\langle U P_L(\mu)\rangle} \tag{III.45d}$$

Now, by Amster's theorem (see p. 59), since all of the components of the third term on the right-hand side of Equation (III.45c) are of order γ^2 or smaller, this term may be neglected to the required degree of approximation. Thus we have:

$$\eta_L(k, u) + \lambda_L \frac{\partial \eta_L k, u)}{\partial u} = - \left(1 - \frac{d\lambda_L}{du}\right) \phi_L(k, u) \sum_{L'=0}^{\infty} T^1_{LL'} B^c_{L'}(u) \tag{III.45e}$$

λ_0 will be denoted simply as λ; for isotropic scattering in the center-of-mass system it is a constant, so that the $L = 0$ Equation (III.45e) will be the desired result (III.45a) if it can be demonstrated that $\eta_0(k, u) = q(k, u)$, the slowing-down density.

At this point, we shall digress momentarily to show the equivalence of η_0 and q and also display the physical significance of the

$\eta_L(k, u)$ [15]. First, the concept of an angular dependent slowing-down density is required. This quantity, designated by $q(\mathbf{r}, u, \mu)$, is defined as the number of neutrons slowing down past lethargy u per unit volume at $\mathbf{r}$ per second which are travelling in direction μ (per unit μ) after the scattering collision. By the arguments given in the derivation of Equation (I.32b) for $q(\mathbf{r}, u)$ (which is now seen to be the zeroth angular moment of $q(\mathbf{r}, u, \mu)$) we can write

$$q(\mathbf{r}, u, \mu) = \int_{u-\ln 1/\alpha}^{u} \mathrm{d}u' \int_{u}^{u'+\ln 1/\alpha} \mathrm{d}u'' \int \mathrm{d}\mu' \, \Sigma_s(u' \to u'', \mu' \to \mu) \, \Phi(\mathbf{r}, u', \mu') \tag{III.46a}$$

Now if the flux and scattering cross section are expanded in Legendre polynomials according to Equations (III.7a,c) and the addition theorem (III.7e) is applied, it is possible to reduce the above expression for $q(\mathbf{r}, u, \mu)$ much as the expression for the degradation integral $J(\mathbf{r}, u, \mu)$ was reduced in the development of the spherical harmonics method (cf. pp. 135-7). The calculations, which are very similar to those used in the spherical harmonics development, are not shown here but the result is:

$$q(\mathbf{r}, u, \mu) = \sum_{L=0}^{\infty} \frac{2L+1}{2} P_L(\mu) \sum_{L'=0}^{\infty} \int_{u-\ln 1/\alpha}^{u} \mathrm{d}u' \, \phi_L(\mathbf{r}, u') \, B_{L'}^{c}(u') \times \\ \times \int_{u}^{u'+\ln 1/\alpha} T_{LL'}(u'' - u') \, \mathrm{d}u'' \tag{III.46b}$$

where all of the quantities involved are as previously defined. Changing the integration variable from u'' to $U = u'' - u'$ in the second integral and from u' to $U' = u - u'$ in the first integral we have

$$q(\mathbf{r}, u, \mu) = \sum_{L=0}^{\infty} \frac{2L+1}{2} P_L(\mu) \sum_{L'=0}^{\infty} \int_{0}^{\ln 1/\alpha} \mathrm{d}U' \, \phi_L(\mathbf{r}, u - U') \, B_{L'}^{c}(u - U') \times \\ \times \int_{U'}^{\ln 1/\alpha} T_{LL'}(U) \mathrm{d}U \tag{III.46c}$$

[15] The following is due to H. Amster, private communication.

Now expanding $\phi_L(u - U')\, B^c_{L'}(u - U')$ in a Taylor series as in Equation (III.13), this becomes

$$q(\mathbf{r}, u, \mu) = \sum_{L=0}^{\infty} \frac{2L+1}{2} P_L(\mu) \sum_{n,L'=0}^{\infty} \frac{\partial^n}{\partial u^n} [B^c_{L'}(u)\, \phi_L(\mathbf{r}, u)] \int_0^{\ln 1/\alpha} \frac{(-U')^n}{n!} \mathrm{d}U' \int_{U'}^{\ln 1/\alpha} T_{LL'}(U)\, \mathrm{d}U \qquad \text{(III.46d)}$$

Finally, one does the U' integral by parts, viz.,

$$\int_0^{\ln 1/\alpha} \mathrm{d}U' \frac{(-U')^n}{n!} \int_{U'}^{\ln 1/\alpha} \mathrm{d}U\, T_{LL'}(U) = \frac{(-)^n}{(n+1)!} \left[U'^{n+1} \int_{U'}^{\ln 1/\alpha} T_{LL'}(U)\, \mathrm{d}U \right]_0^{\ln 1/\alpha} - \int_0^{\ln 1/\alpha} \mathrm{d}U' \frac{(-U')^{n+1}}{(n+1)!} T_{LL'}(U') \qquad \text{(III.46e)}$$

The first term is clearly zero while the second is just the definition of $-T^{n+1}_{LL'}$, Equation (I.56), so that we have

$$q(\mathbf{r}, u, \mu) = \sum_{L=0}^{\infty} \frac{2L+1}{2} P_L(\mu) \sum_{n=0}^{\infty} \sum_{L'=0}^{\infty} T^{n+1}_{LL'} \frac{\partial^n}{\partial u^n} [B^c_{L'}(u)\, \phi_L(\mathbf{r}, u)] \qquad \text{(III.46f)}$$

A comparison of the Fourier transform of Equation (III.46f) with the definition of $\eta_L(k, u)$, Equation (III.45a), reveals that

$$q(k, u, \mu) = \sum_{L=0}^{\infty} \frac{2L+1}{2} \eta_L(k, u)\, P_L(\mu) \qquad \text{(III.46g)}$$

Thus the $\eta_L(k, u)$ defined earlier are seen to be the Legendre coefficients of the angular slowing-down density. In particular,

$$\eta_0(k, u) = q(k, u) \qquad \text{(III.46h)}$$

Another important relationship can be derived by comparing Equations (III.14a) and (III.45b):

$$J_L(k, u) = \frac{-\partial \eta_L(k, u)}{\partial u} + \phi_L(k, u) \sum_{L=0}^{\infty} T^0_{LL'} B^c_{L'}(u) \qquad \text{(III.47a)}$$

We are now in a position to write the desired equations. An equation for the slowing-down density of the desired form (III.44) has been obtained and its relation to the degradation integral has been found. All one now need do is to find $\eta_L(k, u)$ in terms of $\phi_L(k, u)$, construct $J_L(k, u)$ from Equation (III.47a) and substitute, it into the P_L equations. The mechanics of this operation follow. Solving Equation (III.45e) for $\eta_L(k, u)$ we have

$$\eta_L(k, u) = -\sum_{L'=0}^{\infty} T^1_{LL'} \int_0^u du' \, \phi_L(k, u') \left(1 - \frac{d\lambda_L}{du}\right) \frac{B^c_{L'}(u')}{\lambda_L(u')} \times$$

$$\times \exp\left[-\int_{u'}^{u} \frac{du''}{\lambda_L(u'')}\right] \quad \text{(III.47b)}$$

And on differentiating one obtains:

$$\frac{\partial \eta_L(k, u)}{\partial u} = -\sum_{L'} T^1_{LL'} \phi_L(k, u) \left(1 - \frac{d\lambda_L}{du}\right) \frac{B^c_{L'}(u)}{\lambda_L(u)}$$

$$+ \sum_{L'=0}^{\infty} \frac{T'_{LL'}}{\lambda_L(u)} \int_0^u \phi_L(k, u') \left(1 - \frac{d\lambda_L}{du}\right) \frac{B^c_{L'}(u')}{\lambda_L(u')} \exp\left[-\int_{u'}^{u} \frac{du''}{\lambda_L(u'')}\right] du' \quad \text{(III.47c)}$$

which may be substituted into Equation (III.47a) to yield $J_L(k, u)$, which, in turn, may be substituted into the P_L Equations (III.10) to yield the desired modified set:

$$\Sigma_{mj}(u) \, \phi_j(k, u) - \frac{ik}{2j + 1} [j\phi_j(k, u) + (j + 1) \, \phi_{j+1}(k, u)]$$

$$= S_0 \, \delta_{j0} - \sum_{L'=0}^{\infty} \frac{T^1_{jL'}}{\lambda_j(u)} \int_0^u \phi_j(k, u') \frac{\left(1 - \frac{d\lambda_j}{du}\right)}{\lambda_j(u')} B^c_{L'}(u') \times$$

$$\times \exp\left[-\int_{u'}^{u} \frac{du''}{\lambda_j(u'')}\right] du' \quad \text{(III.47d)}$$

A weighted sum over the various moderators present is again understood and a new cross section has been introduced:

$$\Sigma_{mj} = \Sigma_t - \sum_{L'=0}^{\infty} \left[T^0_{jL'} - \frac{T^1_{jL'}}{\lambda_j(u)} \left(1 - \frac{d\lambda_j}{du}\right)\right] B^c_{L'}(u) \quad \text{(III.47e)}$$

These are the basic equations of the Greuling–Goertzel method. The motivation behind this method and the explanation of some of the recent interest in it is now becomes clear. The Equations (III 47d) strongly resemble the equations for slowing down in hydrogen in that the limits of the degradation integral are now $(0, u)$ instead of $(u - \ln 1/\alpha, u)$. The actual and modified scattering frequencies are compared in Fig. III.4. The advantage is that we are now dealing with integral equations of a type which are readily solved. Analytical solutions are possible for some simplified problems, and, perhaps more importantly, numerical solution is considerably simplified. In fact, computer codes developed for treating slowing down in hydrogen are easily modified for solution of the Greuling–Goertzel (GG) equations. Thus far, application has been made only to the P_1 case, for which a set of two coupled integral

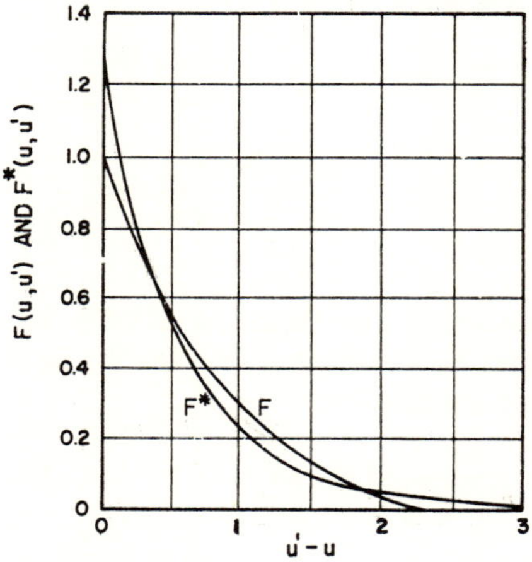

FIG. III.4. Comparison of exact and Greuling–Goertzel kernels. $F(u', u)$ is the exact kernel derived in Chapter I; $F^*(u', u)$ is the approximate kernel that replaces $F(u', u)$ in the GG method (not written explicitly in the text). From E. Greuling and G. Goertzel, *Nucl. Sci. Eng.* **7**, 69 (1960).

equations is obtained. An alternate procedure is to use the P_1 equations with $\eta_L(k, u)$ given by Equation (III.47a) together with the differential Equations (III.45e) for $\eta_L(k, u)$. Clearly, these schemes are completely equivalent and the choice is generally made on the basis of ease of programming for numerical solution. The set is written down in order to permit comparison of this method with those that have been investigated previously.

Substituting Equation (III.47a) into the P_1 equations gives:

$$\Sigma_a(u)\,\phi_0(k, u) - ik\phi_1(k, u) = S_0(k, u) - \frac{\partial q(k, u)}{\partial u} \qquad \text{(III.48a)}$$

and

$$\Sigma_t(u)\,\phi_1(k,u) - \frac{ik}{3}\phi_0(k,u) = J_1(k,u)$$

$$= \frac{\partial\eta_1(k,u)}{\partial u} + \sum_{L'=0}^{\infty} T^0_{1L'}\,B^0_{L'}(u)\,\phi_1(k,u) = \frac{\partial\eta_1}{\partial u} + \Sigma_s\langle\mu\rangle\phi_1 \quad \text{(III.48b)}$$

or, combining terms:

$$\Sigma_{t+r}(u)\,\phi_1(k,u) - \frac{ik}{3}\phi_0(k,u) = \frac{\partial\eta_1(k,u)}{\partial u} \quad \text{(III.48c)}$$

Now, remembering that

$$\sum_{L'=0}^{\infty} T^1_{0L'}\,B^c_{L'}(u) = \xi(u)\,\Sigma_s(u) \quad \text{(III.48d)}$$

and defining

$$\sum_{L'=0}^{\infty} T^1_{1L'}\,B^c_{L'}(u) = \Sigma_s\langle\mu U\rangle = \Sigma_s\,\zeta \quad \text{(III.48e)}$$

we can write Equations (III.45e) for $L = 0, 1$ as:

$$q(k,u) + \lambda(u)\frac{\partial q(k,u)}{\partial u} = \phi_0(k,u)\left(1 - \frac{d\lambda_0}{du}\right)\xi(u)\,\Sigma_s(u) \quad \text{(III.48f)}$$

$$\eta_1(k,u) + \lambda_1(u)\frac{\partial\eta_1(k,u)}{\partial u} = \phi_1(k,u)\left(1 - \frac{d\lambda_1}{du}\right)\zeta(u)\,\Sigma_s(u) \quad \text{(III.48g)}$$

For isotropic scattering, the λ_L are constants

$$\lambda_L = -\frac{T^2_{L0}}{T^1_{L0}} \quad \text{(III.49a)}$$

and in particular

$$\lambda_0 = -\frac{T^2_{00}}{T^1_{00}} = \frac{1 - \alpha\left[1 + \ln\frac{1}{\alpha} + \frac{1}{2}\ln^2\frac{1}{\alpha}\right]}{1 - \alpha\left(1 + \ln\frac{1}{\alpha}\right)} \to \frac{4\gamma}{3} \quad \text{(III.49b)}$$

$$\lambda_1 = -\frac{T^2_{10}}{T^1_{10}} \to \frac{4\gamma}{5} \quad \text{(III.49c)}$$

the arrows indicating the limits for small γ. These values will allow us to compare the GG method with others.

Both consistent and inconsistent GG calculations may be utilized. The former is the set of Equations (III.48 a, c, f, g) just written, while the latter may be derived from (III.48) by setting $\partial\eta_1/\partial u = 0$ in Equation (III.48b). Inspection reveals the following features of these equations:

(a) If $\lambda_0 = 1, \lambda_1 = 3/2$, the values which obtain for hydrogen, the consistent GG equations are identical to the consistent P_1 equations for hydrogen, i.e., hydrogen is treated rigorously by the GG method;

(b) If $\lambda_0 = 1, \lambda_1 = 0$, the inconsistent GG set is identical to the Selengut–Goertzel equations and

(c) If $\lambda_0 = \lambda_1 = 0$, the inconsistent GG set reduces to the Age–Diffusion equation.

An analytical solution is possible for the inconsistent P_1 GG set of equations when λ is constant, i.e., scattering is isotropic. It is solved here to give an expression for the age, which will facilitate comparison, at least qualitatively, with other methods. In this approximation Equation (III.48g) has no significance, while (III.48c) simply becomes Fick's law. When Fick's law is applied to Equation (III.48a) and Equation (III.47c) is used to evaluate $\partial q/\partial u$, a single diffusion equation results:

$$\left[\Sigma_a(u) + D(u)\,k^2 + \frac{\xi}{\lambda_0}\,\Sigma_s(u)\right]\phi_0(k, u)$$

$$= S(k, u) + \frac{\xi}{\lambda_0^2}\int_0^u \Sigma_s(u')\,\phi_0(u')\,\mathrm{e}^{-(u-u')/\lambda_0}\,\mathrm{d}u' \qquad \text{(III.50a)}$$

which is identical to Equation (II.29a) if $\Sigma_a(u)$ is replaced by $[\Sigma_a(u) + D(u)\,k^2]$. Hence, the solution can be written down practically by inspection. Once again, only monoenergetic sources are considered for simplicity. The solution for $\phi_0(k, u)$ is:

$$\phi_0(k, u) = \left[\frac{1}{\lambda_0(\Sigma_a + Dk^2) + \xi\Sigma_s}\right]_u \left[\frac{\xi\Sigma_s}{\lambda_0(\Sigma_a + Dk^2) + \xi\Sigma_s}\right]_{u=0} \times$$

$$\times \exp\left\{-\int_0^u \frac{\Sigma_a(u') + D(u')\,k^2}{\lambda_0(\Sigma_a(u') + D(u')\,k^2) + \xi\Sigma_s(u')}\,\mathrm{d}u'\right\} \qquad \text{(III.50b)}$$

The observation that spatial effects are obtained by replacing $\Sigma_a(u)$ in the infinite medium equations by $[\Sigma_a(u) + D(u)\,k^2]$ is a result which applies to the diffusion approximations discussed earlier. Since this term arises from neutron leakage and in a bare reactor $k^2 = B^2$, this often leads to the interpretation of $D(u)\,B^2$ as a "leakage cross section" for bare reactors. For reflected systems, one must, of course, be more careful about making this interpretation.

The age may be obtained by either of the methods previously employed. Having already obtained an explicit expression for $\phi_0(k, u)$, it is more convenient to differentiate (III.50b) directly rather than to employ the power series method. The result for τ_N is

$$\tau_N(u) = \left.\frac{D\lambda_0}{\lambda_0\Sigma_a + \xi\Sigma_s}\right|_u + \left.\frac{D\lambda_0}{\lambda_0\Sigma_a + \xi\Sigma_s}\right|_{u=0} + \int_0^u \frac{\xi\Sigma_s D}{(\lambda_0\Sigma_a + \xi\Sigma_s)^2}\,\mathrm{d}u' \tag{III.50c}$$

Again, the terms may be interpreted as a last flight correction, a first flight correction and a modified Fermi Age formula. $\tau_q(u)$, as might be inferred from the previous section, is given by $\tau_N(u)$ minus the first term on the right-hand side of (III.50c). The interesting point here is the introduction of a new fictitious absorption cross section given by λ_0 times the true absorption cross section. This result is exactly the analytical result obtained by the Selengut–Goertzel method if $\lambda_0 = \xi = 1$, the values for hydrogen. Thus, in a way, the GG procedure is a generalization of the Selengut–Goertzel method to all moderators. Values of λ_0 were given in Table 1, Chapter II. As will be seen later in this chapter, this method in its more rigorous form yields better results than the previously described methods, particularly for deuterium (heavy water) moderated reactors.

13. The Age in Hydrogen

Exact expressions for the age may be obtained analytically in several special cases. The first of these is the rather important case in which hydrogen is the only moderator. (This case was discussed in relation to the SG method earlier; again it is assumed that all slowing down is due to hydrogen but other elements may affect neutron transport.) For such a case one replaces $N^iJ_0^i(k, u)$ by

$\Sigma_s^i(u)\ \phi_0(k, u)$ and $N^i J_1^i(k, u)$ by $\langle\mu\rangle_i \Sigma_s^i \phi_1(k, u)$ for all elements except hydrogen. Making these substitutions, Equations (III.19) become the following simplified equations:

$$[\Sigma_a(u) + \Sigma_s^H(u)]\,\varphi_{00}(u) = S_0 + N_H\,J_{00}^H(u) \qquad \text{(III.51 a)}$$

$$\tfrac{1}{2}[\Sigma_a(u) + \Sigma_s^H(u)]\,\varphi_{02}(u) - \varphi_{11}(u) = \tfrac{1}{2}N_H\,J_{02}^H(u) \qquad \text{(III.51 b)}$$

$$[\Sigma_{tr}(u) + \Sigma_s^H\langle u\rangle_H]\,\varphi_{11}(u) - \tfrac{1}{3}\varphi_{00}(u) = N_H J_{11}^H(u) \qquad \text{(III.51 c)}$$

where $\langle\mu\rangle_H = 2/3$ for the range of energies of interest. The solution of the set of Equations (II.51) and the defining equations for the $J_{jl}^H(u)$ is straightforward, if rather tedious, and yields the well-known "Marshak formula" for the age [16, 17]. Only the $\varphi_{jl}(u)$ are given here as the $J_{jl}^H(u)$, are simply linear combinations of the $\varphi_{jl}(u)$, and may easily be obtained from Equations (III.51) using the results quoted below.

$$\tfrac{1}{2}\varphi_{02}(u) = \frac{1}{\Sigma_a(u) + \Sigma_s^H(u)}\left[\varphi_{11}(u) + \int_0^u (1 - \chi(u'))\,\varphi_{11}(u')\,e^{-\int_{u'}^{u} \chi(u'')\,du''}\right] \qquad \text{(III.52 a)}$$

$$\varphi_{11}(u) = \frac{1}{3(\Sigma_{tr}(u) + \frac{2}{3}\Sigma_s^H(u))}\left[\varphi_{00}(u) + \tfrac{3}{2}\int_0^u (1 - \lambda(u'))\,\varphi_{00}(u')\,e^{-\frac{3}{2}\int_{u'}^{u} \lambda(u'')\,du''}\right] \qquad \text{(III.52 b)}$$

$$\varphi_{00}(u) = \frac{1}{\Sigma_a(u) + \Sigma_s^H(u)}\left[S(u) + \int_0^u (1 - \chi(u'))\,S(u')\,e^{-\int_{u'}^{u} \chi(u'')\,du''}\right] \qquad \text{(III.52 c)}$$

The following symbols have been employed:

$$s(u) = \frac{\Sigma_a(u)}{\Sigma_a(u) + \Sigma_s^H(u)} \qquad \text{(III.52 d)}$$

$$\lambda(u) = \frac{\Sigma_{tr}(u)}{\Sigma_{tr}(u) + \frac{2}{3}\Sigma_s^H(u)} \qquad \text{(III.52 e)}$$

[16] R. E. MARSHAK, *op. cit.*
[17] HURWITZ and ZWEIFEL, *op. cit.*

The values of τ_N (or τ_q, if desired) obtained from these formulae are rigorous for the case treated here. While it may not seem to have much application to an experimental situation, this case actually approximates slowing down in water rather well.† Here, the formula obtained for the age is not easily interpretable unless one invokes assumptions so drastic as to render the expression obtained almost meaningless. Some discussion of this result is to be found in the literature [18].

The previously derived inconsistent P_1 SG results are obtained by using Fick's law

$$\varphi_{11}(u) = \frac{1}{3}\frac{\varphi_{00}(u)}{\Sigma_{tr}(u)}$$

instead of Equation (III.5c) or Equation (III.52b). Formally, the same result is obtained if in Equation (III.52) one sets:

$$e^{-\frac{3}{2}\int_{u'}^{u} \lambda(u'')\,du''} \rightarrow \frac{2}{3}\frac{\delta(u-u')}{\lambda(u')}$$

It may also be shown that for the special case of no absorption the first flight correction is given by

$$\left[\frac{\lambda(u)}{3\Sigma_s(u)\,\Sigma_{tr}(u)}\right]_{u=0} = \left[\frac{1}{3[\Sigma_{tr} + \frac{2}{3}\Sigma_s^H]\,\Sigma_s}\right]_{u=0} \qquad \text{(III.53)}$$

which is smaller than the SG result because $2/3\Sigma_s^H(u)$ $(= \Sigma_s^H\langle\mu\rangle_H)$ is added to the transport cross section. This phenomenon is explained by the fact that hydrogen scattering changes the lethargy of source neutrons and therefore effectively "absorbs" them. Equation (III.53) is the rigorous first flight correction under the assumption that heavy elements do not slow down.

14. The Age in Pure Heavy Elements

Marshak [19] has shown that it is possible to calculate the age rigorously in a single heavy non-absorbing element. Here, it is more convenient to adopt an approach different from that used above as the limits on integrals defining the $J_j^i(k, u)$ do not permit

† This is, in part, due to some of the effects of oxygen scattering cancelling each other out.

[18] *Ibid.*

[19] R. E. Marshak, *op. cit.*

the easy conversion of the integral equation into a differential equation that made the hydrogen case simple.

First, the P_1 equations are written in terms of the collision density $\chi_j(k, u) = \Sigma_t(u)\, \phi_j(k, u)$. One has, for a monoenergetic plane source, and no absorption:

$$\chi_0(k, u) - ik\phi_1(k, u) = \int_{u - \ln 1/\alpha}^{u} \chi_0(k, u')\, F_0(u - u')\, du' + \delta(u) \qquad \text{(III.54a)}$$

$$\chi_1(k, u) - \frac{ik}{3}\varphi_0(k, u) = \int_{u - \ln 1/\alpha}^{u} \chi_1(k, u')\, F_1(u - u')\, du' \qquad \text{(III.54b)}$$

where F_0 and F_1 are the first two Legendre moments for the scattering frequency. In order to facilitate use of Laplace transforms, the scattering frequencies F_0 and F_1 are defined as zero for $u > u' + \ln 1/\alpha$ and the integrals in Equations (III.54) are extended to zero. Then the integrals in Equations (III.54) are convolutions and Laplace transforms of the equations may be taken:

$$\bar{\chi}_0(k, s) - ik\,\bar{\phi}_1(k, s) = \bar{\chi}_0(k, s)\,\bar{F}_0(S) + 1 \qquad \text{(III.55a)}$$

$$\bar{\chi}_1(k, s) - \frac{ik}{3}\bar{\phi}_0(k, s) = \bar{\chi}_1(k, s)\,\bar{F}_1(s) \qquad \text{(III.55b)}$$

where s is again the transform variable and the barred quantities are the Laplace transforms of the unbarred quantities appearing in Equations (III.54). Now the expansions of the $\phi_j(k, u)$ in power series in k as given by Equation (III.17a) may be transformed to yield an analogous series expansion for $\bar{\phi}_j(k, s)$

$$\bar{\phi}_j(k, s) = \sum_{l = j, j+2 \cdots} \bar{\varphi}_{jl}(s)\, \frac{(ik)^l}{l!} \qquad \text{(III.56a)}$$

A similar series may be obtained for $\chi_j(k, s)$ by multiplying Equation (III.17a) by $\Sigma_s(u)$ before transforming:

$$\bar{\chi}_j(k, s) = \sum_{l = j, j+2 \cdots} \bar{\chi}_{jl}(s)\, \frac{(ik)^l}{l!} \qquad \text{(III.56b)}$$

Again, use is made of Equation (III.17e) to determine the age, noting that either the $\varphi_{jl}(u)$ or $\chi_{jl}(u)$ may be used; the $\chi_{jl}(u)$ are of course simply the inverse transforms of the $\chi_{jl}(s)$. These latter quantities are determined by equating coefficients of the various owers of k after the power series expansions have been inserted into Equations (III.55). They are:

$$\bar{\chi}_{00}(s) = \frac{1}{1 - \bar{F}_0(s)} \qquad \text{(III.57a)}$$

$$\bar{\chi}_{11}(s) = \frac{\bar{\Phi}_{00}(s)}{3(1 - \bar{F}_1(s))} \qquad \text{(III.57b)}$$

$$\bar{\chi}_{02}(s) = \frac{2\bar{\Phi}_{11}(s)}{1 - \bar{F}_0(s)} \qquad \text{(III.57c)}$$

A formal solution is possible if $\chi_{00}(u)$ is written as the inverse Laplace transform of $(1 - \bar{F}_0[s]^{-1})$ and is denoted by $\varkappa_0(u)$. Note that $\varkappa_0(u)$ is just the infinite medium collision density (Placzek function) derived in Sections 5 and 6 of Chapter II. Now $\chi_{11}(u)$ and $\chi_{02}(u)$ may be written as convolutions involving $\varkappa_0(u)$:

$$\chi_{11}(u) = \frac{1}{3} \int_0^u \varkappa_1(u - u') \frac{\varkappa_0(u')}{\Sigma_s(u')} \, du' \qquad \text{(III.58a)}$$

with

$$\varkappa_1(u) = \mathscr{L}^{-1} \left\{ \frac{1}{1 - \bar{F}_1(s)} \right\} \qquad \text{(III.58b)}$$

and

$$\chi_{02}(u) = 2 \int_0^u \varkappa_0(u - u') \frac{\chi_{11}(u')}{\Sigma_s(u')} \, du' \qquad \text{(III.58c)}$$

These may be combined to yield the rigorous age for slowing down in a single heavy element:

$$\tau(u) = \frac{1}{3\varkappa_0(u)} \int_0^u \frac{\varkappa_0(u - u')}{\Sigma_s(u')} \, du' \int_0^{u'} \frac{\varkappa_0(u'') \, \varkappa_1(u' - u'')}{\Sigma_s(u'')} \, du'' \qquad \text{(III.59)}$$

This formula, while rigorous, is quite formal and rather difficult to apply, but it does reduce to the formula of the previous section for hydrogen. Even in the case of isotropic elastic scattering, the forms of $\varkappa_0(u)$ and $\varkappa_1(u)$ are too complicated to allow simple application of Equation (III.59) to practical situations. Again, $\varkappa_0(u)$ is just the solution for the flux as a function of lethargy that was obtained in Chapter II; except for hydrogen it is an infinite series and is clearly difficult to apply here.

D. Methods Related to the P_L Approximations

15. The Gauss Quadrature Procedure [20, 21]

Probably the most used methods in reactor calculations are the P_L approximations, particularly the P_1 approximation and its modifications. Several other methods, which at first sight appear to be entirely different, are in reality closely related to the P_L approximation and are also frequently employed. In few cases are they of greater value than the P_L approximation, but they do point out some of the difficulties inherent in the P_L methods; at times there is advantage in using these approximations themselves.

An alternative to expansion of the angular neutron flux $\Phi(\mathbf{r}, E, \mathbf{\Omega})$ in spherical harmonics is the approximation of the angular portion of the degradation integral by a quadrature formula. Many such schemes exist and have been applied to this problem. Of these, only two will be discussed here: the simplest scheme, which is based on the trapezoid rule, and is discussed later as the S_N method, and the scheme which gives the best approximation to the integral in a least mean square sense, the Gauss quadrature method. The latter method will now be discussed and its relation to the P_L method will be demonstrated.

In any quadrature method one approximates an integral on the range $(-1, 1)$ by a sum of the form:

$$\int_{-1}^{1} f(x)\,\mathrm{d}x = \sum_{i=0}^{N} \omega_i f(x_i)$$

(an integral on any other finite interval is easily converted to one on $(-1, 1)$). It is shown by Chandrasekhar [22] that for a given N the best result is obtained if the x_i are taken to be the zeros of the Legendre polynomial $P_{N+1}(\mu)$ and the ω_i are the Chris-

[20] G. C. Wick, *Z. Phys.* **121**, 702 (1943).

[21] The Gauss quadrature method is discussed in many places, but the best discussion is probably S. Chandrasekhar, *Radiative Transfer*, Dover, New York, 1960, pp. 54–69.

[22] *Ibid.*

toffel numbers:

$$\omega_i = \frac{1}{P'_{N+1}(x_i)} \int_{-1}^{1} \frac{P_{N+1}(x')}{x' - x_i} \mathrm{d}x$$

In particular, this method rigorously evaluates any integral of a polynomial of degree up to $2N + 1$. We now apply this formula to the Boltzmann equation.

In applying this procedure, however, one must be careful, since the Gauss quadrature formula using an odd number of points ("discrete ordinates") employs $\mu = 0$ as one of the points at which the integrand is evaluated. However, the angular flux $\Phi(z, E, \mu)$ need not be continuous in the $\mu = 0$ direction [23]. For example, consider a plane moderator–vacuum interface. At the surface there will be no neutrons in the vacuum travelling parallel to the interface, but there will be some in the moderator. For example, in the Milne problem $\Phi(0,0)/\phi_0(0) = 0.707$. Then $\Phi(z, E, \mu)$ is not a continuous function of μ at $\mu = 0$ at the boundary. Therefore, this procedure may be successfully employed only in an approximation using an even number of points.

Applying the $L + 1$ point (L odd) Gauss formula to the Fourier transformed one-dimensional Boltzmann equation, we obtain the equation

$$\Sigma(u)\,\Phi(k, u, \mu) - ik\mu\Phi(k, u, \mu)$$

$$= S_0(u) + \int_{u-\ln 1/\alpha}^{u} \sum_{j=0}^{L} F(u' \to u, \mu_j \to \mu)\Sigma_s(u')\,\Phi(k, u', \mu_j)\mathrm{d}u' \quad \text{(III.60)}$$

$F(u' \to u, \mu_j \to \mu)$, as before, is the probability for transition from μ_j to μ while scattering from u' to u, per unit μ per unit lethargy.

If, in particular, μ is chosen to be each of the μ_j consecutively, this procedure results in a set of $L + 1$ simultaneous equations in the $(L + 1)$ functions $\Phi(k, u, \mu_j)$. The angular flux $\Phi(k, u, \mu)$ for all μ is obtained by substituting into Equation (III.60). As presented here, this method is strictly applicable only to the one-dimensional problem. Extensions to other geometries are possible, however.

[23] This argument is for the one-dimensional case and is related to the fact that the one-dimensional Boltzmann equation has a singular point at $\mu = 0$. For three dimensions the arguments are similar but more difficult to make. The authors are indebted to B. I. SPINRAD, private communication, for pointing out this argument.

16. Relation to the P_L Method

The Gauss quadrature method is equivalent to the P_L method in some approximation [24]. Noting that $F(u' \to u, \mu_j \to \mu_i)$ is a function only of the angle between the vectors $\mathbf{\Omega}$ and $\mathbf{\Omega}'$ we can write, using the addition theorem (III.7e) (the terms depending on the azimuthal angle are suppressed as they drop out when the expansion is inserted into the degradation integral and the azimuthal integration is carried out):

$$F(u' \to u, \mu_j \to \mu_i) = \sum_{n=0}^{\infty} \frac{2n+1}{2} F_n(u' \to u)\, P_n(\mu_j)\, P_n(\mu_i) \tag{III.61 a}$$

Furthermore, it is always possible to write

$$\Phi(k, u, \mu_i) = \sum_{l=0}^{L} \frac{2l+1}{2} \phi_l(k, u)\, P_l(\mu_i) \tag{III.61 b}$$

since the right-hand side is a polynomial of degree L in μ_i and can always be made to pass through the $L+1$ points, $[\mu_i, \Phi(k, \mu_i, u)]$ by suitable choice of the $(L+1)$ $\phi_l(k, u)$. Using these expansions, which show some similarity to the expansions used in the P_L method, we can now rewrite Equation (III.60) as

$$\Sigma_t(u)\, \Phi(k, u, \mu_i) - ik\mu_i \Phi(k, u, \mu_i) = S_0(u)$$

$$+ \int_{u-\ln 1/\alpha}^{u} \mathrm{d}u'\, \Sigma_s(u') \sum_{n=0}^{\infty} \frac{2n+1}{2} F_n(u' \to u)\, P_n(\mu_i) \left[\sum_{l=0}^{L} \frac{2l+1}{2} \phi_l(k, u') \times \right.$$

$$\left. \times \sum_{j=0}^{L} \omega_j P_n(\mu_j)\, P_l(\mu_j) \right] \tag{III.62 a}$$

The last summation (over the index j) is recognized as the $L+1$ point Gauss quadrature of the integral

$$\int_{-1}^{1} P_n(\mu)\, P_l(\mu)\, \mathrm{d}\mu \tag{III.62 b}$$

Now we have noted that the $L+1$ point Gauss formula rigorously evaluates integrals of polynomials of degree $2L+1$ or less. Then if n is no greater than $L+1$, the integrand of (III.62 b) will

[24] R. Gast, *On the Equivalence of the Gauss Quadrature and P_L Methods* (WAPD-TM-118).

be a polynomial of degree no greater than $2L + 1$ (since $l < L$) and the sum over j in Equation (III.62a) will be rigorously equal to the integral (III.62b). But this integral is just the orthogonality integral for Legendre polynomials, so the entire quantity in the bracket in Equation (III.62a) is just $\phi_n(k, u')$. Using this along with the recursion relation (III.10b) for Legendre polynomials and the definition of $\phi_n(k, u')$, we obtain from Equation (III.60)

$$- \sum_{l=0}^{L} \{ik [(l + 1) \phi_{l+1}(k, u) P_l(\mu_i) (1 - \delta_{lL}) + l\phi_{l-1}(k, u) P(\mu_i)]$$

$$+ \Sigma_t(u) (2l + 1) \phi_l(k, u) P_l(\mu_i) - (2l + 1) \int_0^u B_l^0(u', U) \phi_l(k, u') \times$$

$$\times F_l(\mu_i) \, du' - S_l(u) \delta_{l0} P_l(\mu_i)\} = 0 \tag{III.63}$$

Finally, each term on the left- hand sides of these equations is a polynomial of degree $L + 1$; hence Equations (III.63) can be satisfied only if the coefficients of each Legendre polynomial are equated to zero. The P_L equations then result.

Thus we have shown that the P_L approximation and the $(L + 1)$ point quadrature method are equivalent provided the scattering kernel (and the source, if it is anisotropic, although this was not shown) can be expressed in terms of the first $L + 1$ Legendre polynomials. Put another way, the Gauss quadrature method is equivalent to the P_L method if the actual scattering frequency is replaced by

$$\sum_{n=0}^{L} \frac{2n + 1}{2} F_n(u' \to u) P_n(\mu_0)$$

In light of the above it is apparent that the even LP_L approximations, which are equivalent to odd-point quadrature schemes, are not as useful as the even P_L methods. They suffer from the fact that a discontinuity of the flux at $\mu = 0$ cannot be tolerated [25].

17 The Double P_L and Double Gauss Quadrature Methods [26]

The possibility of a flux discontinuity at $\mu = 0$ leads to possible difficulties in the P_L approximation. Although a series of Legendre

[25] Other reasons for not using even L approximations, based on arguments concerning boundary conditions, are given by DAVISON, *op. cit.*

[26] J. YVON, *J. Nucl. Ener.* **4**, 305 (1957).

polynomials gives the best fit to a function for given L in a least mean squares sense, it need not give the best fit at any particular point, i.e., the P_L approximation may be very poor near $\mu = 0$. Similar difficulties arise in the Gauss quadrature method.

Probably the most obvious means of overcoming this difficulty is to use different expansions for the flux on either side of $\mu = 0$ or, what is the same thing, to apply the quadrature formula separately to each region. A discontinuity of the flux at $\mu = 0$ is then of no consequence; even the lowest order approximation will be capable of dealing with a step discontinuity. In this way, one arrives at the Double P_L (DP_L) and Double Gauss quadrature (DGQ) approximations. As it is readily shown [27] that these schemes are equivalent only the DP_L procedure is discussed. As yet, the methods have not seen much application in slowing-down theory *per se*, but as multigroup methods are essentially schemes for adapting one-velocity transport theory to multi-energy problems, the methods see some application to slowing down in this sense.

First, the μ space $[-1, 1]$ is divided into two subspaces $[-1, 0]$ and $[0, 1]$. In each subspace, the angular flux and scattering functions are expanded in Legendre polynomials, but, in order to retain the orthogonality properties, the expansion is made in terms of polynomials of $2\mu - 1$ for $\mu > 0$ and of $2\mu + 1$ for $\mu < 0$. These expansions will be then substituted into the Boltzmann equation written in the following form:

$$
\begin{aligned}
(\Sigma_t(u) - ik\mu)\,\Phi(k, u, \mu) &= \int_{u-\ln 1/\alpha}^{u} \mathrm{d}u' \Big[\int_{-1}^{0} \mathrm{d}\mu'\, \Phi(k, u', \mu')\, \Sigma_s \times (u' \to u, \mu' \to \mu) \\
&+ \int_{0}^{1} \mathrm{d}\mu'\, \Phi(k, u', \mu')\, \Sigma_s(u' \to u, \mu' \to \mu) \Big] + S_0(u) \qquad \text{(III.64)}
\end{aligned}
$$

As before, the expansion of the scattering frequency may be simplified by means of the addition theorem (III.7e), yielding:

$$
\Sigma_s(u' \to u, \mu_0) = \sum_{j=0}^{\infty} \frac{2j+1}{2} \Sigma_{sj}(u' \to u)\, P_j(\mu')\, P_j(\mu) \qquad \text{(III.65a)}
$$

where the terms involving the azimuthal angle φ have been suppressed as they will again integrate to zero when inserted into the degradation integral. To obtain the desired form, however, it is necessary to express the $P_j(\mu)$ as functions of the Legendre polynomials of $2\mu + 1$. Two expansions are required:

$$\begin{aligned} P_j(\mu) &= \sum_{l=0}^{j} a_j^l\, P_l(2\mu + 1) \quad \mu < 0 \\ &= \sum_{l=0}^{j} b_j^l\, P_l(2\mu - 1) \quad \mu > 0 \end{aligned} \tag{III.65b}$$

Only j terms are needed in this expansion since $P_j(\mu)$ is a polynomial of degree j. The constants a_j^l and b_j^l appearing in these equations can be determined by using the orthogonality properties. That is, we have

$$\begin{aligned} \int_{-1}^{0} P_l(2\mu + 1)\, P_j(2\mu + 1)\, \mathrm{d}\mu &= \int_{0}^{1} P_l(2\mu - 1)\, P_j(2\mu - 1)\, \mathrm{d}\mu \\ &= \delta_{jl}/(2l + 1) \end{aligned}$$

and

$$(2l + 1)^{-1}\, a_j^l = \int_{-1}^{0} P_j(\mu)\, P_l(2\mu + 1)\, \mathrm{d}\mu$$

$$(2l + 1)^{-1}\, b_j^l = \int_{0}^{1} P_j(\mu)\, P_l(2\mu - 1)\, \mathrm{d}\mu$$

In this manner, one obtains four expansions for the scattering frequency:

$$\begin{aligned} &\Sigma_s(u' \to u, \mu' \to \mu) \\ &= \sum_{j=0}^{\infty} \frac{2j + 1}{2} \Sigma_{sj}(u' \to u) \sum_{l=0}^{j} b_j^l P_l(2\mu' + 1) \sum_{q=0}^{j} b_j^q P_q(2\mu + 1) \qquad \begin{matrix} \mu' < 0 \\ \mu < 0 \end{matrix} \\ &= \sum_{j=0}^{\infty} \frac{2j + 1}{2} \Sigma_{sj}(u' \to u) \sum_{l=0}^{j} b_j^l P_l(2\mu' + 1) \sum_{q=0}^{j} a_j^q P_q(2\mu - 1) \qquad \begin{matrix} \mu' < 0 \\ \mu > 0 \end{matrix} \\ &= \sum_{j=0}^{\infty} \frac{2j + 1}{2} \Sigma_{sj}(u' \to u) \sum_{l=0}^{j} a_j^l P_l(2\mu' - 1) \sum_{q=0}^{j} b_j^q P_q(2\mu + 1) \qquad \begin{matrix} \mu' > 0 \\ \mu < 0 \end{matrix} \\ &= \sum_{j=0}^{\infty} \frac{2j + 1}{2} \Sigma_{sj}(u' \to u) \sum_{l=0}^{j} a_j^l P_l(2\mu' - 1) \sum_{q=0}^{j} a_l^q P_q(2\mu - 1) \qquad \begin{matrix} \mu' > 0 \\ \mu > 0 \end{matrix} \end{aligned} \tag{III.66}$$

In a similar manner, the flux expansion may be written

$$\Phi(k, u, \mu) = \sum_{l=0}^{\infty} (2l + 1)\, \psi_l(k, u)\, P_l(2\mu - 1) \quad \mu > 0$$

$$= \sum_{l=0}^{\infty} (2l + 1)\, \chi_l(k, u)\, P_l(2\mu + 1) \quad \mu < 0$$

where the expansion coefficients are now expressible as integrals on the intervals $[-1, 0]$ and $[0, 1]$:

$$\psi_l(k, u) = \int_0^1 \Phi(k, u, \mu)\, P_l(2\mu - 1)\, \mathrm{d}\mu$$

and

$$\chi_l(k, u) = \int_{-1}^{0} \Phi(k, u, \mu)\, P_l(2\mu + 1)\, \mathrm{d}\mu$$

These expansions may be inserted into the Boltzmann equation, which then separates into two sets of coupled equations in the $\chi_l(k, u)$ and $\psi_l(k, u)$; sources are again assumed to be isotropic:

$$- ik\left[\frac{j}{2j + 1}\psi_{j-1}(k, u) - \psi_j(k, u) + \frac{j + 1}{2j + 1}\psi_{j+1}(k, u)\right] + 2\Sigma_t(u)\, \psi_j(k, u)$$

$$= \int_{u - \ln 1/\alpha}^{u} \mathrm{d}u' \sum_{l=0}^{\infty} \left[\chi_l(k, u') \sum_{k=0}^{\infty} \frac{2k + 1}{2j + 1} \Sigma_{sk}(u' \to u)\, a_k^j b_k^l \right.$$

$$\left. + \psi_l(k, u') \sum_{k=0}^{\infty} \frac{2k + 1}{2j + 1} \Sigma_{sk}(u' \to u)\, a_k^j b_k^l\right] + S_0(k, u) \quad \text{(III.67a)}$$

and

$$- ik\left[\frac{j}{2j + 1}\chi_{j-1}(k, u) - \chi_j(k, u) + \frac{j + 1}{2j + 1}\chi_{q+1}(k, u)\right] + 2\Sigma_t(u)\, \chi_j(k, u)$$

$$= \int_{u - \ln 1/\alpha}^{u} \mathrm{d}u' \sum_{l=0}^{\infty} \left[\chi_l(k, u') \sum_{k=0}^{\infty} \frac{2k + 1}{2j + 1} \Sigma_{sk}(u' \to u)\, b_k^j b_k^l \right.$$

$$\left. + \psi_l(k, u') \sum_{k=0}^{\infty} \frac{2k + 1}{2j + 1} \Sigma_{sk}(u' \to u)\, a_k^l b_k^j\right] \quad \text{(III.67b)}$$

Again, an infinite set of integro-differential equations has been obtained. In this case, the coupling assumes a slightly more complicated form; all of the $\psi_j(k, u)$ and $\chi_j(k, u)$ are coupled to all of the others. This set cannot be truncated in the same manner as were the spherical harmonics equations. Here, one must set $\psi_j(k, u) = \chi_j(k, u) = 0$ for all $j > L$ to obtain the Double P_L (DP_L) equations. Alternatively, one could set $\Sigma_{sk}(u' \to u) = 0$ for $k > L$ and $\psi_{L+1}(k, u) = \chi_{L+1}(k, u) = 0$ since, by definition, the a_k^l and b_k^l are zero if $l > k$. It is evident that a combination of these methods is also possible.

In contrast to the properties of the P_L approximation, the DP_0 equations may be applied to finite medium problems. The DP_0 equations merely assume a step function for the angular variation of the flux, and, in fact, for some problems involving plane interfaces the DP_0 approximations have been found to give results which are as accurate as the P_1 results [28]. Similarly, the DP_1 equations give results of accuracy comparable to that of the P_3 equations in such problems. Another important point is that the even L–DP_L equations do not possess the inherent difficulties of the even L–P_L equations and may be successfully employed.

The equivalence of the DP_L and DGQ methods may be demonstrated without resort to the kernel and source truncations that were necessary for the analogous proof with the ordinary GQ method [28]. Thus, it will be advantageous to retain extra terms in the expansion of the scattering kernel. A reasonable estimate is that $2L + 1$ terms should be retained: The DP_L scheme employs (in DGQ notation) $2L + 2$ discrete ordinates; comparable accuracy in the expansion of the scattering frequency is obtained only if $2L + 1$ terms are retained.

The DP_L equations may be set up in geometries other than plane following the method used here; in multigroup calculations of reflected reactors such developments are highly necessary, particularly if non-planar boundaries are present. For these developments, the reader is referred to the work of Schiff and Ziering [29].

It has been suggested by Schiff and Ziering [30] that higher order multiple P_L equations, in which the angular region is broken into

[28] Gast, *op. cit.*
[29] S. Schiff and S. Ziering, *Nucl. Sci. Eng.* **3**, 635 (1958).
[30] S. Schiff and S. Ziering, *Nucl. Sci. Eng.* **1**, 172 (1960).

more than two parts, could be developed, but there is some doubt that the advantages gained are worth the increase in complexity. For example, in a quadruple P_L approximation, sixteen separate expansions of the scattering frequency are required.

18. The B_L Approximation

Another useful, and more rapidly converging, approximation may be obtained from the spherical harmonics expansions of the Boltzmann equation [Eqn. (III.9)]. To obtain it, we divide Equation (III.9) by $(1 - ik\mu/\Sigma_t(u))$ before multiplying by $P_j(\mu)$ and integrating over μ. This results in a set of coupled integral equations different from the P_L equations:

$$\Sigma_t(u)\,\phi_j(k,u) = \sum_{l=0}^{\infty}(2l+1)\,A_{jl}(\eta)\,[N_H J_l^H(k,u) + N_D J_l^D(k,u) + \sum_{i\neq H,D} N_i J_l^i(k,u) + S_0(k,u)\,\delta_{l0}] \qquad \text{(III.68a)}$$

where the $A_{jl}(\eta)$ are defined by

$$A_{jl}(\eta) = A_{lj}(\eta) = \tfrac{1}{2}\int_{-1}^{1}\frac{P_j(\mu)\,P_l(\mu)}{1-\eta\mu}\,d\mu \qquad \text{(III.68b)}$$

and the parameter η is defined by

$$\eta = iy = \frac{ik}{\Sigma_t(u)} \qquad \text{(III.68c)}$$

By dividing Equation (III.10b), the recursion relation for Legendre polynomials, by $(1 - ik\mu/\Sigma_t)/P_j(\mu)$ and integrating over μ, we obtain recursion relations for the A_{jl}. For the relation relative to the second index we have:

$$\frac{2l+1}{\eta}A_{jl}(\eta) - (l+1)\,A_{j,\,l+1}(\eta) - lA_{j,\,l-1}(\eta) = \frac{\delta_{jl}}{\eta} \qquad \text{(III.69a)}$$

Since the $A_{jl}(\eta)$ are symmetric in j and l, the relation on the first index is quite similar

$$\frac{2j+1}{\eta}A_{jl}(\eta) - (j+1)\,A_{j+1,\,l}(\eta) - jA_{j-1,\,l} = \frac{\delta_{jl}}{\eta} \qquad \text{(III.69b)}$$

By successive use of these two recursion relations, any of the $A_{jl}(\eta)$ may be expressed in terms of $A_{00}(\eta)$ which can be found by direct integration:

$$A_{00}(\eta) = \eta^{-1} \tanh^{-1} \eta = \frac{Q_0(1/\eta)}{\eta} \qquad \text{(III.69c)}$$

where $Q_l(x)$ is a Legendre function of the second kind [31]. In general, it may be shown that [32]

$$A_{jl}(\eta) = \frac{Q_j(1/\eta)\, P_l(1/\eta)}{\eta} \qquad l \leqq j \qquad \text{(III.69d)}$$

Numerical values of the first few $A_{jl}(\eta)$, up to $A_{22}(\eta)$ have been tabulated by Bethe, Tonks, and Hurwitz [33], who also wrote them out explicitly:

$$A_{01} = \frac{A_{00} - 1}{\eta}$$

$$A_{11} = \frac{A_{00} - 1}{\eta^2}$$

$$A_{02} = \frac{3}{2\eta^2}\left[\left(1 - \frac{\eta^2}{3}\right) A_{00} - 1\right]$$

$$A_{12} = \frac{A_{02}}{\eta} \qquad A_{22} = \left[\frac{3}{2\eta^2} - \frac{1}{2}\right] A_{02}$$

Before proceeding, let us look into the nature of this scheme. The operation of dividing by $(1 - ik\mu/\Sigma_t)$ in Fourier transform space corresponds, in ordinary configuration space, to operating with $\left(1 + \frac{\mu\partial/\partial z}{\Sigma_t}\right)^{-1}$, the inverse of the "streaming" operator in the Boltzmann equation. Stated less abstractly, this operation amounts to the conversion of the integro-differential Boltzmann equation to a pure integral equation. Hence the above analysis is really equivalent to making a Legendre analysis of the integral form of the Boltzmann equation. In configuration space this analysis is rather difficult to make; the results would clearly be the same as those obtained by taking the inverse Fourier transforms of Equations (III.68a) and are convolutions of the Legendre components of the flux, $\varphi_j(\mathbf{r}, u)$ with the inverse transforms of the $A_{jl}(\eta)$, which are

[31] E. T. Whittaker and G. N. Watson, *A Course in Modern Analysis*, Oxford, 1950.

[32] I. Waller, *Arkiv. Math. Astron. Fysik* **34**, 3, 1 (1947).

[33] H. A. Bethe, L. Tonks, and H. Hurwitz, Jr., *Phys. Rev.* **80**, 11 (1950).

exponential integral functions. These are the integral Boltzmann equations. For isotropic scattering, only the first equation is required since the J_l $(l > 0)$ are zero; for anisotropic scattering, a number of coupled integral equations results. Since integral equations are generally difficult to solve numerically, the approximations obtained by truncating Equations (III.68 a), called the B_L approximations [34], are applied mainly to special problems, as for example the calculation of moments of the flux (e.g., the age) or the computation of group constants for multigroup calculations (see Chapter IV). A modified B_L approach has been used for cell calculations at thermal energies [35] with considerable success.

A low order approximation to the infinite set of coupled equations may be obtained by setting

$$J_l(k, u) = 0 \text{ for all} \qquad l > L$$

This is the B_L approximation mentioned above. Again, a set of $L + 1$ coupled integral equations is obtained. In contrast to the method of uncoupling the equations in the P_L case, it is here essential to set all the $J_l(k, u)$ above the Lth equal to zero. In practice, this makes little difference since the physically interesting quantities are just the flux $\varphi_0(k, u)$ and the current $\varphi_1(k, u)$. The superiority of the B_L approximation over the P_L approximation arises from the fact that J_l is generally smaller than φ_l (see the following section).

19. Relation to the P_L Approximation †

A relation between the P_L and B_L approximation is readily obtained. From Equations (III.68 a) and (III.69 d) it can be seen that if $J_j(k, u) = 0$ for all $j > L$, then

$$\frac{\phi_l(k, u)}{\phi_{l+1}(k, u)} = \frac{Q_l(1/\eta)}{Q_{l+1}(1/\eta)} \qquad l \geqq L \tag{III.70 a}$$

The B_L equations may, in fact, be obtained by setting

$$\phi_{L+1}(k, u) = \frac{Q_{L+1}(1/\eta)}{Q_L(1/\eta)} \phi_L(k, u) \tag{III.70 b}$$

† The bulk of the results quoted in this section were originally given by Hurwitz and Zweifel, *ibid.*

[34] See H. Hurwitz, Jr., and P. F. Zweifel, *J. Appl. Phys.* **26**, 923 (1956).

[35] H. C. Honeck, *Nucl. Sci. Eng.* **8**, 193 (1960); H. C. Honeck and I. Kaplan, *Nucl. Sci. Eng.* **8**, 203 (1960).

in the spherical harmonics expansion of the Boltzmann equation, instead of setting $\phi_{L+1} = 0$ as in the P_L approach.

It is also possible to cast the P_L equations in a form similar to that of the B_L equations by replacing the $A_{jl}(\eta)$ by $A'_{jl}(\eta)$, defined by †

$$A'_{jl} = \frac{1}{2} \int\limits_{G_{L+1}} \frac{P_j(\mu)\, P_l(\mu)}{1 - \eta\mu}\, d\mu \tag{III.71a}$$

The subscript G_{L+1} on the integral indicates that the Gauss quadrature formula with $L + 1$ discrete ordinates is to be used in evaluation of the integral. In other words, the P_L equations can be rearranged to read:

$$\Sigma_t(u)\, \phi_j(k, u) = \sum_{j=0}^{L} A'_{jl}(\eta)\, (2l + 1)\, [J_l(k, u) + S_0\, \delta_{0l}] \tag{III.71b}$$

This procedure gives another, and perhaps clearer, means of viewing the inconsistencies in even $L - P_L$ approximations. First, it is to be noted that for $j \neq L$, $l \neq L$, the $A_{jl}(\eta)$ satisfy the same recursion relations as the A_{jl} so, in effect, the P_L equations may be obtained by replacing A_{00} by A'_{00}. Now, a property of Fourier transforms is that the transform of a bounded function must go to zero for large k. But, in the $L + 1$ point Gauss integration scheme where $L + 1$ is odd $\mu = 0$ is always one of the discrete ordinates chosen; the denominator of the integrand of (III.71a) is always unity at $\mu = 0$ so A'_{00} does not go to zero for large k. This, in turn prevents $\phi_0(k, u)$ from going to zero in this limit, but instead makes it to go to a constant, and introduces delta functions in the spatial variation of the flux. This behavior does not obtain when $L + 1$ is even.

It is interesting to note, however, that in the B_L method there is special merit in dealing with even L approximations. The argument is made for a particular case, but is not limited to this case. Rewrite the power series expansion (III.13) for the hydrogen degradation integral, $J_j^H(k, u)$ as

$$J_j^H(k, u) = \sum_{n=0}^{\infty} T_{j0H}^{n} \frac{\partial^n}{\partial u^n} (\Sigma_s^H(u)\, \phi_j(k, u)) \tag{III.72}$$

† For a proof see Appendix F.

Noting that the most important terms are those with $n = 0$, we have, from Equation (I.54a):

$$T_{00}^0 = 1 \qquad T_{30}^0 = T_{50}^0 = 0$$

$$T_{10}^0 = \frac{2}{3} \qquad T_{40}^0 = -\frac{1}{24}$$

$$T_{20}^0 = \frac{1}{4}$$

and, in general,

$$T_{2l+1,0}^0 = 0 \quad l \geqq 1$$

and

$$T_{2l,0}^0 = \frac{(-)^{l+1}(2l-3)!}{2^{2l-2}(l-2)!\,(l+1)!} \to \left(\frac{e}{l}\right)^{3/2} \frac{(-)^{l+1}}{2\sqrt{\pi}}$$

in the limit of large l. For hydrogen T_{jl}^0 with l different from zero need not be considered since scattering is isotropic in C. In general, higher l must be considered but they are usually small compared with the $l = 0$ terms. Thus, in going from the B_{L-1} approximation to the B_L approximation for L odd, the extra contributions come only from the derivative terms ($n > 1$) in Equation (III.72) and the improvement is relatively small since the T_{jl}^1 are generally smaller than the T_{jl}^0. In the P_L case, on the other hand, there is an additional improvement in going from the P_{L-1} approximation to the P_L approximation due to the fact that the A_{jl} are evaluated by the $L + 1$ point rather than the n-point Gaussian formula.

The smallness of the T_{L0}^0 contributes to the superiority of the B_L method over the P_L method since in the P_L case one ignores ϕ_{L+1} whereas in the B_L scheme the most important neglected term is $T_{L+1,0}^0\,\phi_{L+1}(k, u)$. This is just what one expects on intuitive grounds, i.e., the P_L approximation neglects ϕ_{L+1} whereas the largest term neglected in the B_L approximation is J_{L+1}, which is generally much smaller than ϕ_{L+1}.

Considering specifically the P_1 and B_1 approximations, one notes that the two point Gauss scheme gives

$$A_{00}' = \left(1 + \frac{y^2}{3}\right)^{-1} = 1 - \frac{y^2}{3} + \frac{y^4}{9} + \cdots \qquad \text{(III.73a)}$$

where A_{00} is, as before, given by

$$A_{00} = y^{-1} \tanh^{-1} y = 1 - \frac{y^2}{3} + \frac{y^4}{5} + \cdots \qquad \text{(III.73b)}$$

so that for small y (small k, and hence large reactors) the two methods are equivalent. In general, the expansions of A_{00} and A'_{00} agree to terms of order y^{2L}, which is a consequence of the theorem that integration using the $L + 1$ point Gaussian quadrature is rigorous for any polynomial of degree up to and including $2L + 1$. In particular, both the B_1 and P_1 methods give the age rigorously.

The Age-Diffusion, Selengut-Goertzel, and Greuling-Goertzel inconsistent P_1 approximations have analogs in the B_1 approximation. In the absence of sources, Fick's law may again be obtained by replacing $J_1(k, u)$ by $\Sigma_s(u)\bar{\mu}\varphi_1(k, u)$, i.e., by retaining only one term of the Taylor series expansion of $J_1\,(k, u)$. However, in this case the diffusion coefficient is found to be:

$$D \to 'D' = \frac{1}{3'\Sigma_{tr}'} = \frac{1 - A_{00}(y)}{y^2[A_{00}(y)\,\Sigma_t(u) - 3A_{11}(y)\,\langle\mu\rangle\Sigma_s(u)]} \tag{III.74a}$$

Note that as $y \to 0$, $'\Sigma_{tr}' \to \Sigma_{tr}$. Expanding $'\Sigma_{tr}'$ in powers of y, one finds

$$'\Sigma_{tr}' = \Sigma_{tr} + \Sigma_s\left(\frac{4}{15}y^2 + \frac{9}{25}y^4 + \cdots\right) \tag{III.74b}$$

In a similar manner B_1 Selengut-Goertzel and Greuling-Goertzel calculations may be set up.

The B_1 adaptations of the inconsistent approximations may be used provided the lethargy expansions that lead to them are valid. On the other hand, the validity of the P_1 versions of these approximations depends not only on the validity of these expansions, but also on the validity of replacing A_{00} by A'_{00}. This last mentioned error is large when $(4/15)\,\Sigma_s(u)\,y^2$ is large compared to $\Sigma_{tr}(u)$. Thus, for small reactors, the B_1 approximations are considerably superior to their P_1 analogs.

20. The Moments Method [36]

The moments method, originally devised by Fermi for his early work on slowing down in paraffin [37], is equivalent to the P_L method already discussed although its outward appearance is quite different. The starting point here is the Legendre-analyzed,

[36] J. E. WILKINS, Jr., R. HELLENS and P. F. ZWEIFEL, *PICG* **5**, 62 (1955).
[37] E. FERMI, *Ricerca Scientifica* **7** (2), 13 (1936).

one-dimensional Boltzmann equation, which may be obtained by inverting the Fourier transformed equations (III.10a):

$$\Sigma_t(u)\,\phi_j(z,u) - \frac{j}{2i+1}\,\frac{\partial\phi_{j-1}(z,u)}{\partial z} - \frac{j+1}{2j+1}\,\frac{\partial\phi_{j+1}(z,u)}{\partial z} = \sum_i{}' N_i J_j^i(z,u) + S(u)\delta_{0j} \qquad \text{(III.75)}$$

To obtain the moments equations, Equation (III.75) is multiplied by z^n and integrated over all z. The *moments* of the flux $\varphi_{jn}(u)$ are now defined as

$$\phi_{jn}(u) = \frac{1}{n!}\int_{-\infty}^{\infty} z^n\phi_j(z,u)\,\mathrm{d}z = \frac{1}{n!}\int_{-\infty}^{\infty} z^n\,\mathrm{d}z\int_{-1}^{1}\mathrm{d}\mu\,\Phi(z,u,\mu)\,P_j(\mu) \qquad \text{(III.76a)}$$

and are the jth Legendre and nth spatial moments of the angular neutron flux. For the case of a plane unit isotropic source, we obtain, upon taking the spatial moments of Equations (III.75),

$$\Sigma_t^{(u)}\varphi_{jn}(u) = \frac{(j+1)\,\varphi_{j+1,n-1}(u) + j\varphi_{j-1,n-1}(u)}{2j+1} + S(u)\,\delta_{no}\,\delta_{jo} + N_i\,J_{jl}^i(u) \qquad \text{(III.76b)}$$

where the moments of the degradation integral $J_{jn}(u)$ are defined by:

$$J_{jn}^i(u) = \int_{u-\ln 1/x_i}^{u}\mathrm{d}u\,\varphi_{jn}(u')\,B_j^{oi}(u',U) \qquad \text{(III.76c)}$$

The set of coupled integral equations arising out of this method is identical to the set previously obtained from the Fourier-transformed Boltzmann equation by expansion of the $\varphi_j(k,u)$ in power series in k. When coefficients of the various powers of k are equated in those equations, the set of Equations (III.76b) results; the first few of these are Equations (III.19). It was, in fact, shown [see Eqn. (III.17c)] that the Taylor series expansion coefficients are identically the moments defined by (III.76a). Hence, the two methods are really only different derivations of the same set of equations.

Clearly, many of the results previously obtained apply to this method. For example, Equation (III.18a) reads:

$$\varphi_{jn} = 0 \qquad \begin{array}{l}\text{if } j+n \text{ is odd}\\ \text{or}\quad \text{if } n > j\end{array}$$

The moments method is particularly useful for obtaining the flux at large distances from the source and has therefore found considerable application in shielding calculations. The reason for this is provided by the Tauberian theorem [38] which states that in order to obtain accurate values of a function for large values of its argument, it necessary to have accurate values of the Fourier transform of the function for small values of its argument. For small k, the Taylor expansion (III.17a) will converge rapidly and only a few moments are required. On the other hand, for reactor calculations larger values of k are of importance and the moments method is impractical. Of course, for comparison with measurements of the age or of higher flux moments, the moments method is ideal. For comparison with the flux distribution at short and intermediate distances, the moments method is probably about equivalent in difficulty to inversion of the Fourier transform solutions.

Naturally, the Age Diffusion, Selengut–Goertzel, and Greuling–Goertzel approximations may be adapted for use in the moments method. They are obtained by making approximations to the $J_{jn}(u)$ which parallel those made in the P_1 approach. In view of the exact equivalence of the two methods, these approximations will not be developed here.

21. Polynominal Approximations

The P_L approximation to the Boltzmann equation is the one most frequently used and is, historically, the most important, having been developed long before its application to nuclear reactors. Other approximations are, of course, possible and many have been developed. Expansions in Legendre polynomials are chosen most often because of their simple orthogonality relation on the interval of interest and the addition theorem (III.7e) which makes handling of the scattering integral particularly simple. This does not mean that Legendre polynomials necessarily give the best approximation for a given number of terms in the series. It can, in fact, be shown that Legendre polynomials give the best approximation in the sense of a least squares fit, that is, the mean square error in the angular distribution of the flux is minimal for a given number of terms when Legendre polynomials are used. However, the error at a particular

[38] N. WEINER, *The Fourier Integral and Certain of its Applications*, Dover, New York, 1958.

point may be very great. We have seen that transport theory does not require continuity of the neutron flux at $\mu = 0$ and that in some cases discontinuities are observed. Yvon's method, covered in Section 16, was specially devised to counteract such difficulties.

It can be shown that if the aim of a polynomial approximation is to minimize the maximum error at any μ, the expansion should be carried out in terms of Tschebycheff Polynomials [39].

However, in practice, this method suffers so much from the lack of a simple addition theorem that its use must be restricted to problems involving isotropic scattering in the laboratory system. For this reason, Tschebycheff polynomial approximations are not likely to see any great use in the future.

E. Other Approximations to the Boltzmann Equation

22. The S_N Method [40]

We now come to a method which is designed specifically for use on high-speed computers, the so-called S_N procedure. In this method no attempt is made at obtaining analytical results. Rather, the techniques used are selected so as to minimize the difficulty in solving the Boltzmann equation numerically, that is, for ease in programming for computer solution. Since analytical results are not sought by this method, calculations utilizing it always employ multigroup methods. Thus, it will be sufficient to consider the one-velocity Boltzmann equation as the starting point; it is to be understood that multigroup methods will generally be used to supply the physical constants required as well as the source term.

The one-velocity, one-dimensional Boltzmann equation can be written:

$$\Sigma_t(u)\ \phi(z, \mu) + \mu \frac{\partial \phi}{\partial z}(z, \mu) = \int_{-1}^{1} \phi(z, \mu')\, \Sigma(\mu' \to \mu)\, \mathrm{d}\mu' + S(z, \mu) \tag{III.77}$$

This equation can be set up for numerical solution once a procedure for evaluating the integral term has been specified. The Gaussian scheme already used is one such method and has advantages that were pointed out earlier. However, as nearly all reactor calculations

[39] W. Conkie, *Nucl. Sci. Eng.* **6**, 260 (1959). An important application of a related property of these functions is given in Chapter IV.

[40] B. G. Carlson and G. I. Bell, PICG P/2386 (1958).

are currently done by computers, there is some advantage in using a method which sacrifices accuracy for simplicity. That is to say, it may be more economical to use a scheme which requires a greater number of terms for convergence but in which the equations are simpler and easier to program.

As a start, the interval $[-1, 1]$ in μ space is divided into N intervals, $[\mu_{i-1}, \mu_i]$, which are generally, but need not be, equal in size. Most often, the number of intervals is even; S_2, S_4, S_8, and S_{16} are the most frequently used approximations. Within each sub-interval the variation of the flux is assumed to be linear:†

$$\phi(z\ \mu) = \left[\frac{\mu_i - \mu}{\mu_i - \mu_{i-1}}\,\phi(z,\mu_{i-1}) + \frac{\mu - \mu_{i-1}}{\mu_i - \mu_{i-1}}\,\phi(z,\mu_i)\right] \qquad \text{(III.78 a)}$$

$$\mu_{i-1} \leqq \mu \leqq \mu_i \quad i = 1,2 \cdots N$$

where μ_i is the end of the ith interval and $\mu_0 = -1$. A similar expression may be written for the source if it is anisotropic:

$$S(z, \mu) = \frac{\mu_i - \mu}{\mu_i - \mu_{i-1}}\,S(z,\mu_{i-1}) + \frac{\mu - \mu_{i-1}}{\mu_i - \mu_{i-1}}\,S(z, \mu_i) \qquad \text{(III.78 b)}$$

When these expressions are substituted into Equation (III.77) and the equation integrated from μ_{i-1} to μ_i a set of N coupled equations results:

$$\frac{[\mu_i - \mu_{i-1}]}{2}\,\Sigma_t\,[\phi(z, \mu_i) + \phi(z, \mu_{i-1})] + \frac{[\mu_i - \mu_{i-1}]}{6} \times$$

$$\times \left[(\mu_i + 2\mu_{i-1})\frac{\partial}{\partial z}\,\phi(z, \mu_i) + (2\mu_i + \mu_{i-1})\frac{\partial}{\partial z}\,\phi(z, \mu_{i-1})\right]$$

$$= \sum_{j=1}^{N} \int_{\mu_{j-1}}^{\mu_j} d\mu'\ \phi(z, \mu') \int_{\mu_{i-1}}^{\mu_i} d\mu\ \Sigma(\mu' \to \mu) + \frac{(\mu_i - \mu_{i-1})}{2}\,[S(z, \mu_i)$$

$$+ S(z, \mu_{i-1})] \qquad \text{(III.79 a)}$$

† An alternate approach is to estimate the scattering integral by the trapezoid rule. The mechanics of this approach are similar to those used in the Gauss quadrature method and are not given here. The approach given here is the one most frequently used in the literature.

Finally, $\Sigma(\mu' \to \mu)$ can be approximated by a linear expression of the type (III.78a) in both μ and μ'. If this expression and Equation (III.78a) are substituted into the integral in Equation (III.79a), there results a set of algebraic equations for the $\phi(z, \mu_i)$ which involve the quantities $F(\mu_j \to \mu_i)$. For simplicity, we consider the case of isotropic scattering in the laboratory system.† Then

$$\Sigma(\mu' \to \mu) = (1/2)\Sigma_s$$

and Equation (III.79a) becomes

$$\Sigma_t[\phi(z, \mu_i) + \phi(z, \mu_{i-1})] + \frac{(\mu_i + 2\mu_{i-1})}{3} \frac{\partial\phi(z, \mu_i)}{\partial z} + \frac{(2\mu_i + \mu_{i-1})}{3} \frac{\partial\phi(z, \mu_{i-1})}{\partial z} = \frac{\Sigma_s}{2} \sum_{j=1}^{N} (\mu_j - \mu_{j-1}) \times$$
$$\times [\phi(z, \mu_j) + \phi(z, \mu_{j-1})] + [S(z, \mu_j) + S(z, \mu_{j-1})] \quad \text{(III.79b)}$$

Equations (III.79b) are a set of N coupled ordinary differential equations; there are, however, $(N + 1)\ \mu_i$; hence, one more equation involving the $\varphi(z, \mu_i)$ is required. Such an equation is obtained by setting $\mu = \mu_0 = -1$ in Equation (III.77)

$$\Sigma_t\phi(z, \mu_0) - \frac{\partial}{\partial z}\phi(z, \mu_0)$$
$$= \frac{\Sigma_s}{4} \sum_{j=i}^{N} (\mu_j - \mu_{j-1}) [\phi(z, \mu_{j-1}) + \phi(z, \mu_j)] + S(z, \mu_0) \quad \text{(III.79c)}$$

This completes the set of S_N equations. Evidently the solution is not completely determined until boundary conditions are specified. From the correct boundary condition at a free surface

$$\phi(R_S, \mu) = 0 \quad \text{(III.80a)}$$

for R_S on the boundary and $\mu = \mathbf{\Omega} \cdot \hat{\boldsymbol{n}} < 0$, where $\hat{\boldsymbol{n}}$ is an outward normal, it is clear that a satisfactory proper boundary condition is

$$\phi(R_S, \mu_i) = 0 \quad \text{(III.80b)}$$

† Most currently used S_N codes do not allow for anisotropic scattering. However, in many calculations Σ_t is replaced by Σ_{tr}. This is frequently called the "transport approximation" and accounts for anisotropic scattering in much the same way as diffusion theory does.

for all $\mu_i < 0$. Also, of course, one must insist on $\phi(z, \mu_i)$ being continuous (except possibly for $\mu_i = 0$) everywhere inside the reactor. Schemes for adapting the S_N method to other geometries have been given and are frequently useful but space does not permit their inclusion here [41]. It is noteworthy that, although Equations (III.79b) and (III.79c) are a coupled set of differential equations for the $\phi(z, \mu_i)$, they are generally solved iteratively. One guesses

$$\phi_0(z) = \sum_j \frac{(\mu_j - \mu_{j-1})}{2} [\phi(z, \mu_{j-1}) + \phi(z, \mu_j)] \quad \text{(III.80c)}$$

which appears on the right-hand sides of the equations, solves the equations for the $\phi(z, \mu_j)$, recomputes $\phi_0(z)$ from Equation (III.80c) and iterates in this manner until convergence is obtained. This procedure is somewhat more efficient for numerical calculations than the more obvious straightforward method.

A number of one-speed neutron transport test problems were solved by Carlson [42]. Clearly, the accuracy of a given S_N calculation will depend on flux anisotropy and hence on reactor dimensions, but it was found that for systems approximately the size of existing reactors, the S_4 method gives the criticality to within about one per cent of that obtained by nearly exact variational calculations. Since one generally does not care to attempt to calculate k_{eff} to an accuracy better than this (due to uncertainties in cross sections and in reactor composition), Carlson recommends the S_4 method for general calculational purposes. Higher order S_N procedures are generally reserved for specialized problems.

Carlson further states that the accuracy of the S_2 method is comparable to that of the P_3 method and that the S_4 method is comparable to the P_5 method for computing k_{eff} in one speed transport problems. These statements were based on test problems in which it was found that the P_L approximations usually yielded values lower than the exact results. These results may seem a little peculiar in light of the fact that for a given N the P_N method treats the angular distribution more correctly than the S_N method, but they derive from the better treatment of the boundary conditions in the S_N method.

[41] B. DUANE, *The DSN Programs*, Report XDC 59-9-118 (1959).
[42] B. G. CARLSON, *op. cit.*

F. Statistical Methods

23. Monte Carlo [43]

All of the methods discussed up to now are schemes for solving slowing-down problems by means of some approximation to the Boltzmann equation. Here we give a brief outline of methods which are frequently used but which are based on the fact that, after all, the physical processes occurring in a reactor are probabilistic in nature.

In all of the previous work in this book several implicit statistical assumptions were made. In treating the neutron density,

$$n(\mathbf{r}, E, \mathbf{\Omega}) = \frac{\Phi(\mathbf{r}, E, \mathbf{\Omega})}{v}$$

it was assumed that at criticality a steady state existed such that a time independent neutron density could be assigned to every $\mathbf{r}$, E, and $\mathbf{\Omega}$. Actually, even in a critical reactor, the neutron density fluctuates somewhat about the average value; in reality we have described only the behavior of the neutron density averaged over a time long compared to the period of the fluctuations. The first method described here, the Monte Carlo method (so called because it uses random numbers), relies on the statistical nature of the neutron density.

Instead of attempting to study the statistical behavior of large numbers of neutrons directly, as use of the Boltzmann equation implies, we shall study the behavior of individual neutrons and obtain their individual characteristics: their energies, positions, and directions of travel. These properties may then be averaged over many neutron "histories". For statistical reliability, a large number of neutron histories must be used; naturally, this is possible only on a large computer. Perhaps the method can best be illustrated by a simple example (not connected with slowing down)—the determination of the diffusion length. For further simplification, plane geometry will be chosen.

[43] An excellent description of the principles of this method as well as some of the techniques employed in shortening the computation and improving convergence are to be found in an article by G. GOERTZEL and M. KALOS in *Progress in Nuclear Energy*, Vol. I, Pergamon Press, New York, 1956. A further list of references is also found there.

Suppose a neutron emanates from a plane isotropic source at $z = 0$. The probability of its being emitted in a direction $\theta = \cos^{-1}\mu$ relative to the z-axis is

$$P(\mu)\,\mathrm{d}\mu = \frac{\mathrm{d}\mu}{2} = \frac{\sin\theta}{2}\,\mathrm{d}\theta \qquad \text{(III.81 a)}$$

To ensure that after large numbers of neutrons have been considered they will have this distribution, use may be made of random numbers. A new variable y is introduced whose distribution $Q(y)$ is related to $P(\mu)$ by

$$Q(y)\,\mathrm{d}y = P(\mu)\,\mathrm{d}\mu \qquad \text{(III.81 b)}$$

In order to facilitate recourse to random numbers, it is necessary for y to be randomly distributed. Mathematically:

$$Q(y) = 1 \qquad 0 \leqq y \leqq 1 \qquad \text{(III.81 c)}$$

Then y may be selected from a set of random numbers between zero and unity. Clearly, this is possible only if

$$\mathrm{d}y = P(\mu)\,\mathrm{d}\mu \qquad \text{(III.81 d)}$$

or, on integrating, one may write (with $P(\mu) = \frac{1}{2}$)

$$y = \int_{-1}^{\mu} P(\mu')\,\mathrm{d}\mu' = \frac{\mu + 1}{2} \qquad \text{(III.81 e)}$$

(Note that y is the probability that the cosine of the scattering angle be less than μ.) Then, having chosen a random number for y, one obtains the required value of μ by inverting Equation (III.81 e)

$$\mu = 2y - 1 \qquad \text{(III.81 f)}$$

Having the direction in which the neutron is emitted, one then wants to know how far the neutron travels before experiencing a collision; this is easily done by applying the above technique to the distribution for path lengths:

$$P(r)\,\mathrm{d}r = \mathrm{e}^{-\Sigma r}\,\Sigma \mathrm{d}r \qquad \text{(III.81 g)}$$

We then have

$$y = \int_{0}^{r} P(r)\,\mathrm{d}r = 1 - \mathrm{e}^{-\Sigma r} \qquad \text{(III.81 h)}$$

or, inverting,

$$r = \frac{1}{\Sigma} \ln \frac{1}{1 - y} \qquad \text{(III.81i)}$$

The question of whether the collision is a scattering or an absorption is answered by choosing a third random number. If this number is less than Σ_a/Σ_t the collision is assumed to be an absorption; if not, the collision is taken to be a scattering. Proceeding in this manner, one may use random numbers to decide: (1) with what type of nucleus the collision takes place, and (2) if the collision is a scattering, the new direction of motion. By repeating this process for each flight, and keeping track of its position as it travels, we may follow the neutron until it is absorbed. At this point the distance z from the source may be computed and squared. By averaging z^2 over many such histories, one can obtain a value for L^2.

This scheme is called "straight" or "analog" Monte Carlo. While it is a valid procedure, a considerable amount of computation is required in order to obtain sufficient histories for statistically reliable results. Statistical theory predicts that the relative error should be inversely proportional to the square root of the number of histories so that approximately 10,000 histories are required for one per cent accuracy. Many improvements to this basic scheme have been suggested. Generally, they achieve better accuracy for a given amount of computation by changing the relative probabilities of certain events and then compensating for this by weighting the neutron histories accordingly when averages are taken. As a discussion of these techniques would require a chapter by itself, it is not given here. Instead, the reader is referred to the review article by Goertzel and Kalos cited in footnote [43] where a complete set of references may be found.

This method is generally used to best advantage in solving problems that are difficult to solve analytically, such as those involving heterogeneous media or rapidly varying cross sections. Due to their nature, Monte Carlo problems of any difficulty require large digital computers and even then they frequently require large amounts of machine time. This is one of the major drawbacks in the method, others being the lack of generalized results that enable one to apply the results of one calculation to other problems and the inability to obtain results as functions of parameters without extremely large expenditures of computer time.

24. Stochastic Process Methods

In the previous section a description of the processes undergone by individual neutrons was given in which the process of neutron diffusion was represented as being quite similar to diffusion of other types of particles and to Brownian motion. A general theory has been developed to treat such problems which involve stochastic processes, i.e., those governed by probabilistic considerations [44]. As this method is almost entirely unrelated to those preceding it here, only a brief introduction to the method may be attempted here. Readers interested in this approach will find considerable material in the references already given [44] and in the work of Doob [45].

First of all, one notes that neutron slowing down and diffusion is a Markoff process, that is, a neutron's future may be predicted from a knowledge of its present state without referring to its past history. This allows the application of Markoff's method to the problems.

Secondly, all calculations of this nature necessarily involve time dependence; the method strictly applies only to the calculation of the space–energy coordinates of a neutron with increasing time. Thus, the attack will have to be that of solving the problem for a pulsed (delta function in time) source and then to treat this solution as a Green's function for any time dependent, or constant, source.

For a Markoff process, one can write the Smoluchowski equation:

$$P(\mathbf{R}, t) = \int P(\mathbf{R} - \Delta\mathbf{R}, t - \Delta t)\, \varrho(\mathbf{R} - \Delta\mathbf{R}, \Delta\mathbf{R}, t) + S(\mathbf{R}, t) \qquad \text{(III.82)}$$

Here $P(\mathbf{R}, t)\, d\mathbf{R}$ is the probability that a neutron will be in the phase space volume element $d\mathbf{R}\, (= d^3r\, d\mathbf{\Omega}\, du)$ at $\mathbf{R}\, (= \mathbf{r}, \mathbf{\Omega}, u)$ at time t after having been emitted from an isotropic source at the origin at $t = 0$, and $\varrho(\mathbf{R}, \Delta\mathbf{R}, \Delta t)$ is the probability that a neutron at $\mathbf{R}$ will suffer displacement $\Delta\mathbf{R}$ in time Δt. The major part of the task is that of finding the displacement function $\varrho(\mathbf{R}, \Delta\mathbf{R}, \Delta t)$.

To this attempt, Markoff's method may be applied. The essence of this method is the following (not derived here for reasons of

[44] An excellent review article by S. Chandrasekhar, *Revs. Mod. Phys.* **15**, 1 (1943), reprinted in *Noise and Stochastic Processes*, N. Wax, ed., Dover, New York, 1954, is available. Also of particular value is the work of M. C. Wang and G. E. Uhlenbeck, *Revs. Mod. Phys.* **17**, 323 (1945), also reprinted in the above. The first application of this method to neutron problems was made by C. N. Klahr, *Nucl. Sci. Eng.* **3**, 269 (1958).

[45] J. Doob, *Stochastic Processes*, Wiley, New York, 1953.

space), quoted from the work of Chandrasekhar.† If $\tau(\Delta\mathbf{R}_i)\,d\Delta\mathbf{R}_i$ is the probability that the ith flight will lie in the volume element $d\Delta\mathbf{R}_i$ about $\mathbf{R}_i$, ‡ and if all flights have this same distribution, the N flight distribution function $W_N(\mathbf{R})\,d\mathbf{R}$, i.e., the probability that a neutron will lie in volume element $d\mathbf{R}$ at $\mathbf{R}$ after N flights, may be calculated as follows:

First, take the multi-dimensional Fourier transform of $\tau(\Delta\mathbf{R})$:

$$A(\varrho) = \int_{\text{all }\mathbf{R}} e^{i\varrho\cdot\Delta\mathbf{R}}\,\tau(\Delta\mathbf{R})\,d\Delta\mathbf{R} \tag{III.83a}$$

Then take the Nth power of $A(\boldsymbol{\rho})$ and invert the Fourier transform

$$W_N(\Delta\mathbf{R}) = \frac{1}{(2\pi)N}\int_{\text{all }\varrho} e^{-\Delta i\varrho\cdot\Delta\mathbf{R}}\,[A(\varrho)]^N\,d\varrho \tag{III.83b}$$

Assuming that $\Delta\mathbf{R}$ is small, i.e. that the change in a neutron's coordinates on a given flight are small, one may find $[A(\boldsymbol{\rho})]^N$ approximately by expanding $e^{i\varrho\cdot\Delta\mathbf{R}}$ in the integral defining $A(\boldsymbol{\rho})$ and retaining only the first three terms. One then has:

$$[A(\boldsymbol{\rho})]^N = \left[1 + i\sum_{j=1}^{n}\langle\Delta x_j\rangle\varrho_j - \frac{1}{2}\sum_{j,k=1}^{n}\langle\Delta x_j\,\Delta x_k\rangle\,\varrho_j\,\varrho_k\right]^N$$

$$\approx \exp\left\{iN\sum_{j=1}^{n}\langle\Delta x_j\rangle\,\varrho_j - \frac{N}{2}\sum_{j,k=1}^{n}\langle\Delta x_j\,\Delta x_k\rangle\,\varrho_j\,\varrho_k\right\} \tag{III.83c}$$

where the averages $\langle\Delta x_j\,\Delta x_k\rangle$ are defined by

$$A_i = \langle\Delta x_i\rangle = \int \Delta x_i\,\tau(\Delta\mathbf{R})\,d\Delta\mathbf{R}$$

$$B_{jk} = \langle\Delta x_j\,\Delta x_k\rangle = \int \Delta x_j\,\Delta x_k\,\tau(\Delta\mathbf{R})\,d\Delta\mathbf{R} \tag{III.83d}$$

Now, taking the inverse transform of Equation (III.83c) one obtains a Gaussian distribution for $W_N(\mathbf{R})$

$$W_N(\mathbf{R}) = \prod_{i=1}^{n}\frac{1}{2\pi N\langle\Delta x_i^2\rangle}\exp\left[-\langle\Delta x_i\rangle^2/2N\langle\Delta x_i^2\rangle\right] \tag{III.83e}$$

provided one neglects the average of products $\langle\Delta x_j\,\Delta x_k\rangle$.†† If one now assumes that neutrons make many collisions while suffering small changes in position, lethargy, or direction, $W_N(\mathbf{R})$

† See footnote [44].

‡ Note that the definition of a "flight" includes the changes in energy and direction that occur in a collision. Thus $\Delta\mathbf{R}$ is a six component vector.

†† These terms may readily be included but are omitted to make the development somewhat neater. They will be reinserted later.

as given by Equation (III.83e) may be taken as the required function $\varrho(\mathbf{R}, \Delta\mathbf{R}, \Delta t)$.

The assumption that only small changes in a neutron's coordinates occur in a small time interval is rather severe and implies that slowing down in light elements cannot be properly treated by this method since light moderators cause large lethargy changes; also, the angular variation of the neutron density should not be given correctly by the method as a neutron almost always suffers large angular changes in a collision. Equation (III.83e) may be cast into a more convenient form if one defines m as the mean number of collisions a neutron suffers per unit time (approximately $v\Sigma_t$) so that

$$N = m\Delta t$$

and Equation (III.83e) becomes

$$\varrho(\mathbf{R}, \Delta\mathbf{R}, \Delta t) = \prod_{i=1}^{n} \frac{1}{2\pi m\,\Delta t\,\langle \Delta x_i^2 \rangle} \exp\left[-\langle \Delta x_i \rangle^2 / 2m\Delta t \langle \Delta x_i^2 \rangle\right] \tag{III.83f}$$

Now, upon substituting Equation (III.83f) into the Smoluchowski equation (III.82) and expanding $P(\mathbf{R} - \Delta\mathbf{R}, \mathbf{R}, t - \Delta t)$ in a Taylor series about $\mathbf{R}$ and t, keeping only second derivative terms, one obtains the Fokker–Planck equation:

$$\frac{\partial P}{\partial t} = \sum_i \left[-\frac{\partial}{\partial x_i}[m\langle \Delta x_i \rangle P] + \frac{1}{2}\frac{\partial^2}{\partial x_i^2}[m\langle \Delta x_i^2 \rangle P]\right] + S \tag{III.84a}$$

It may be noted here that if the cross product averages $\langle \Delta x_i\, \Delta x_j \rangle$ had not been neglected, we would have arrived at the more general equation:

$$\frac{\partial P}{\partial t} = \sum_i \frac{\partial}{\partial x_i}[m\langle \Delta x_i \rangle P] + \sum_{i,j} \frac{\partial^2}{\partial x_i \partial x_j}[m\langle \Delta x_i\, \Delta x_j \rangle P] + S \tag{III.84b}$$

Equation (III.84b) is the basic equation of the stochastic process method.

As an example, we shall neglect the angular variables. Now we define

$$\frac{1}{6} m\langle (\Delta r)^2 \rangle = \frac{1}{2} m\langle (\Delta x)^2 \rangle = \frac{1}{2} m\langle (\Delta y)^2 \rangle = \frac{1}{2} m\langle (\Delta z)^2 \rangle = D \tag{III.85a}$$

and

$$\langle(\Delta u)^2\rangle = \gamma \tag{III.85b}$$

We have previously defined

$$\langle\Delta u\rangle = \xi \tag{III.85c}$$

As defined above, D will be found to be the previously defined diffusion coefficient divided by v. Treating only the case of one space dimension, we arrive at the equation

$$\frac{\partial P}{\partial t} = -m\xi\frac{\partial P}{\partial u} + \frac{m\gamma}{2}\frac{\partial^2 P}{\partial u^2} + D\frac{\partial^2 P}{\partial x^2} + \delta(u)\,\delta(x)\,\delta(t) \tag{III.86}$$

for a pulsed monoenergetic plane source of neutrons. It is interesting to note that Equation (III.86) is just the time-dependent Age–Diffusion equation without absorption if $\gamma = 0$, since $m = \Sigma_s v$. In this manner the Age–Diffusion equation has been derived from a completely different approach which makes some of the assumptions implicit in it more comprehensible. In particular, the statement that Age–Diffusion theory implies small lethargy changes in collisions and small flights between collisions has been validated in a manner which shows the nature of the approximation more explicitly.

Differences between the Age–Diffusion equation and Equation (III.86) are readily pointed out. The results given here are not derived but are quoted from Klahr [46]. Solving Equation (III.86), one has

$$P(x, u, t) = \frac{1}{2\pi t}\left(\frac{1}{\gamma Dm}\right)^{\frac{1}{2}} \exp\left\{-\left[\frac{x^2}{4Dt} + \frac{\xi^2}{\gamma mt}\left(\frac{u}{\xi} - m\,t\right)\right]\right\} \tag{III.87a}$$

Integration of this result in time yields the solution for a steady state source

$$P(x, u) = \int_0^\infty P(x, u, t)\,\mathrm{d}t = \frac{1}{2\pi}\left(\frac{1}{\gamma Dm}\right)^{\frac{1}{2}} e^{\xi u/\gamma} K_0\left[\frac{\xi}{\gamma}\left(\frac{\gamma mx^2}{2D} + u^2\right)^{\frac{1}{2}}\right] \tag{III.87b}$$

[46] KLAHR, *op. cit.*

where $K_0(x)$ is the modified Bessel function of the second kind. This result is rather difficult to work with as it stands. It is interesting to note that the Fermi age kernel is yielded in the limit

$$\frac{m\gamma x^2}{2Du^2} = \frac{x^2/\langle(\Delta x)^2\rangle}{u^2/\langle(\Delta u)^2\rangle} \ll 1 \qquad \text{(III.87c)}$$

for Equation (III.87b) then yields, as expected,

$$m\xi P(x, u) = q(x, u) = \frac{1}{\sqrt{(4\pi\tau)}} e^{-x^2/4\tau} \qquad \text{(III.87d)}$$

$$\left(\tau = \frac{D}{u\xi\Sigma_s} u\right)$$

for the slowing-down density. This is precisely what is expected since this limit is that for which Equation (III.86) reduced to the Age–Diffusion equation. Also, we note that for heavy moderators

$$\langle(\Delta u)^2\rangle \approx \frac{2\xi}{3} = \frac{4}{3A}$$

$$\langle(\Delta x)^2\rangle = \frac{2}{\Sigma^2}$$

so that

$$\frac{\langle\Delta u^2\rangle}{\langle\Delta x^2\rangle} = \frac{2}{3}\frac{\Sigma^2}{A}$$

that is, this limit is valid for heavy elements and for large cross sections, in agreement with results given earlier.

Extensions of this method for use in reactor calculations are being made [47]. This method does not, however, appear to have any significant advantages at the present time over the methods already given.

G. Small Source Theory

25. Heterogeneous Method

The final method presented here is one designed specifically for application to heterogeneous systems. As it treats fuel assemblies as line (or point) sources and sinks, it has been named "Small Source

[47] C. N. KLAHR and K. HELD, *Stochastic Process Method Calculation of Neutron Flux Spectra*, Reports TRG 13-28, TRG 112, QPR 7-8 (1959)

Theory" by W. A. Horning [48] who developed it in the United States independently of S. M. Feinberg [49] who devised it in the U.S.S.R. This method, too, has undergone considerable development as a tool for reactor calculations in recent years; hence, only a bare outline of the method can be given here.

As stated above, in a cylindrical system fuel assemblies are assumed line sources and sinks infinite in the axial direction.† Ignoring epithermal fission, the source strength S_k, i.e., the number of neutrons emitted per second by the kth rod, is related to a_k, the number of absorptions per second in the kth rod by

$$S_k = \eta_k a_k \tag{III.88a}$$

where η_k is the number of fast neutrons emitted by the kth rod per thermal neutron absorbed. Note that η_k is not necessarily equal to η, the fuel efficiency previously defined, unless the actual rod size is negligible, i.e., if there is no fast fission.

If $P_T(|\mathbf{r} - \mathbf{r}'|)$ is the thermal slowing-down density kernel (assumed a displacement kernel) as defined in Chapter I, the slowing-down density into the thermal group at point $\mathbf{r}$ due to the kth rod will be

$$S_k P_T(|\mathbf{r} - \mathbf{r}_k|) = S_k \frac{e^{-|\mathbf{r} - \mathbf{r}_k|^2/4\tau}}{4\pi\tau} \tag{III.88b}$$

if Age–Diffusion theory may be used to describe slowing down. The kernel $P_T(|\mathbf{r} - \mathbf{r}'|)$ should be defined with respect to the neutron spectrum emerging from the rods, but the fission spectrum kernel is usually quite accurate enough.

The thermal diffusion kernel $f(\mathbf{r}, \mathbf{r}')$ is now introduced and defined as the probability per unit volume at $\mathbf{r}$ that a neutron thermalized at $\mathbf{r}'$ will diffuse to $\mathbf{r}$. For line sources:

$$f(\mathbf{r}, \mathbf{r}') = \frac{K_0\left(\dfrac{|\mathbf{r} - \mathbf{r}'|}{L}\right)}{2\pi\Sigma_a D} \tag{III.88c}$$

† Neutron leakage in the axial direction is generally treated by diffusion theory.

[48] W. A. Horning, HW 34021 (Hanford).
[49] S. M. Feinberg, PICG P/669 (1955).

is the required kernel in the diffusion approximation. Defining the convolution of the slowing-down density and diffusion kernel as

$$F(\mathbf{r}, \mathbf{r}') = \int_{\mathbf{r}''} P_T(|\mathbf{r}'' - \mathbf{r}'|) f(|\mathbf{r}'' - \mathbf{r}|)\, d^2 r'' \qquad \text{(III.88d)}$$

and noting that each rod absorbs a_k thermal neutrons per second, we can write the thermal flux

$$\begin{aligned} \phi_T(\mathbf{r}) &= \sum_k [S_k F(\mathbf{r}, \mathbf{r}_k) - a_k f(\mathbf{r}, \mathbf{r}_k)] \\ &= \sum_k a_k [\eta_k F(\mathbf{r}, \mathbf{r}_k) - f(\mathbf{r}, \mathbf{r}_k)] \end{aligned} \qquad \text{(III.89a)}$$

using relation (III.88a). It has been assumed that the effect of the rods on slowing down can be ignored. This method has difficulty in dealing with resonance absorption accurately in a simple way. However, resonance absorption is usually important and must be allowed for. The usual approximation is to treat epithermal absorption artifically by assuming that it all takes place at one energy.

At a point $\mathbf{r}_n$ on the surface of the nth rod, the distance to the kth rod may be approximated by the distance between centers, except when $n = k$. For $n = k$, one takes the "distance" to the rod as the radius of the rod ϱ_n. Thus

$$\begin{aligned} \phi_T(\mathbf{r}_n) = \sum_{n \neq k} & a_k [\eta_k F(|\mathbf{r}_n - \mathbf{r}_k|) - f(|\mathbf{r}_n - \mathbf{r}_k|)] \\ & + a_n[\eta_n F(\varrho_n) - f(\varrho_n)] \end{aligned} \qquad \text{(III.89b)}$$

The essence of the method lies in the assumption that a linear relationship exists between a_n and $\varphi(\mathbf{r}_n)$, i.e., between the absorption in the rod and the flux on the surface:

$$a_n = \frac{1}{\gamma_n} \phi_T(\mathbf{r}_n) \qquad \text{(III.89c)}$$

Then, equating $\phi_T(\mathbf{r}_n)$ in Equations (III.89b) and (III.89c) one has

$$\begin{aligned} \gamma_n a_n = \sum_{n \neq k} & a_k [\eta_k F(|\mathbf{r}_n - \mathbf{r}_k|) - f(|\mathbf{r}_n - \mathbf{r}_k|)] \\ & + a_n[\eta_n F(\varrho_n) - f(\varrho_n)] \end{aligned} \qquad \text{(III.90a)}$$

For simplicity we set

$$\begin{aligned} \alpha_{kn} &= \eta_k F(|\mathbf{r}_n - \mathbf{r}_k|) - f(|\mathbf{r}_n - \mathbf{r}_k|) \qquad k \neq n \\ &= \eta_n F(\varrho_n) - f(\varrho_n) \qquad k = n \end{aligned} \tag{III.90b}$$

to obtain

$$\gamma_n a_n = \sum_k \alpha_{kn} a_k$$

This is a set of homogeneous linear equations in the a_n, there being as many a_n and equations as there are rods in the system. A solution will exist only if the determinant

$$A = \begin{vmatrix} \alpha_{11} - \gamma_1 & \alpha_{12} \cdots & & & \\ \alpha_{21} & \alpha_{22} - \gamma_2 & & & \\ \cdot & \cdot & \cdot & & \\ \cdot & \cdot & & \cdot & \\ \cdot & & & & \cdot \\ \cdot & & & & \cdot \quad \alpha_{NN} - \gamma_N \end{vmatrix}$$

is equal to zero. This is then the critical equation for the system.

It is presumed that L and τ are well known. In a typical calculation, one wants to find a combination $\mathbf{r}_n - \mathbf{r}_k$, ϱ_n, and the number of rods N which makes the system critical. This is clearly possible only if the two parameters η_k and γ_k are known. For the case of identical rods they need be known for only one rod.

Herein lies the advantage of this method. One need only measure the parameters η and γ for one rod, say in a thermal column, to be able to calculate the critical size of a reactor. It is possible to include resonance absorption, epithermal fission, and any other effect of interest by adding more parameters of this type, although it should again be mentioned that the treatment is generally quite artificial.

Most reactors currently being built do not meet the conditions required for application of the method. Of the many types currently under consideration, heavy water- or graphite-moderated reactors can be said to definitely fall into this category.

This method aroused considerable interest at one time but now seems to have been superseded by the cell methods described elsewhere in this book.

H. Comparison of Results with Experiment

26. Comparison of Theory and Experiment

Of the results obtained by the calculational methods developed in this chapter, the two which are most readily compared with experiment are the predictions of neutron distributions from point or plane sources (in particular, the moments of distribution (e.g., the age) and predictions of criticality (or non-leakage probabilities). Generally, the age is a simple quantity to measure and age experiments tend to be rather "clean". In a typical experiment a "point" source of neutrons (e.g., a very small polonium–beryllium source or

TABLE 1. AGE OF U-235 FISSION NEUTRONS TO INDIUM RESONANCE IN WATER†

Experiment	Measured Age	Reactor	Source geometry
Anderson, Fermi, Nagle (1944)	32.3	CP-2	Point
Hill, Roberts, Fitch (1948)	30.8	ORNL X-10	
Hoover, Arnett (1950)	30.0	ORNL X-10	
Wade (1956)	31.0	SR U-235C Test	
Barkov, Mukhin (1956)	29.4 ± 1.5	MIP (USSR)	
Blosser, Trubey (1959)	26.7	ORNL-X10	
Lombard, Blanchard (1960)	27.3 ± 0.9	P.S.U.	
Pettus (1960)	27.0 ± 0.9	BW LPR	
Doerner, Armani, Zagotta, Martens (1961)	27.9 ± 0.1	ZPR IV	Plane
Jaworowski (1962)	26.7 ± 0.5	SPR	

† References:

H. L. ANDERSON, E. FERMI and P. NAGLE, *Metallurgy Lab. Prog. Rept.* CP 1531 (1944).
J. E. HILL, L. D. ROBERTS and T. E. FITCH, *J. Appl. Phys.* **26**, 1013 (1955).
J. I. HOOVER and T. J. ARNETT, *ORNL* 641 (1950).
J. W. WADE, *Nucl. Sci. Eng.* **4**, 12 (1958).
L. M. BARKOV and K. N. MUKHIN, *Atomnaya Energiya* **1**, 31 (1956); *ibid.* **1**, 33 (1956) translated in *J. Nucl. Ener.* **2**, 91 (1957); *ibid.* **2**, 94 (1957).
D. B. LOMBARD and E. H. BLANCHARD, *Nucl. Sci. Eng.* **1**, 448 (1960).
T. V. BLOSSER and D. K. TRUBEY, *ORNL* 2842 (1959).
W. G. PETTUS, BAW-146 (1960).
R. C. DOERNER, R. J. ARMANY, W. E. ZAGOTTA and F. H. MARTENS, *Nucl. Sci. Eng.* **9**, 221 (1961).
T. R. JAWOROWSKI, *Trans. Amer. Nucl. Soc.* **5**, No. 2 (1962).

TABLE 2. COMPUTED VALUES OF THE AGE

Reference	Method	Date	Value
1	Fourier transform	1954	25.3
2	P_1 and B_1	1955	25.3
			23.6†
			26.8‡
	SG	1955	30.9
			28.8†
			32.8‡
1	Mod. SG	1955	30.7
1	Moments	1955	25.7
	P_1 and B_1		24.8
2	Moments	1955	25.8
1	Monte Carlo	1956	25.6 ± 0.3
2	Moments	1957	26.0
2	Fourier transform	1957	25.9
2	Monte Carlo	1957	26.7
2	Moments	1957	26.5
2	Monte Carlo	1958	25.9

† Oxygen scattering assumed isotropic.
‡ Oxygen slowing down neglected.

[1] J. E. WILKINS, R. L. HELLENS and P. F. ZWEIFEL, *PICG* **5**, 62 (1955).
[2] H. GOLDSTEIN, P. F. ZWEIFEL and D. G. FOSTER, *PICG* **16**, 379 (1958).

a small piece of U-235 bathed in a thermal neutron flux) is placed in a large sample of the medium whose age is to be measured. (Recent experiments have utilized "plane" sources, e.g., plates of U-235 bathed in a thermal neutron flux.) Cadmium covered indium foils are then irradiated at various distances from the source and the moment of the measured distribution calculated. Indium is chosen because it has a very strong capture resonance at 1.44 eV which leads to a conveniently measured radioactive isotope. The purpose of the cadmium is to shield thermal neutrons from the indium, i.e., to insure that the activity is entirely due to neutron capture in the resonance. Details of these experiments have been discussed most recently by Doerner *et al.* [50]

Experimental values of the age of fission neutrons in water to 1.44 eV are given in Table 1; the currently accepted experimental

50] R. C. DOERNER, F. J. ARMANI, W. E. ZAGOTTA, and F. H. MARTENS, *Nucl. Sci. Eng.* **9**, 221 (1961).

value is about 27.6 cm², which is in good agreement with the theoretical values given in Table 2. It is noted that the early experiments gave results which were too high. This was due to the necessity of making hard-to-evaluate experimental corrections. Recent experiments have been able to illuminate many of the difficulties and seem to have cleared up the discrepancy between theory and experiment.

A comparison of the various theoretical models may be obtained from Tables 2 and 3. The Age–Diffusion result is clearly much too low; this is due mainly to the lack of a first flight correction. On the other hand, the Selengut–Goertzel method gives a result which is larger than the correct result due to the overestimation of the first flight correction by diffusion theory. From Table 3 it is seen that oxygen scattering has little effect on the age in water, i.e., the Marshak formula gives nearly correct results. The small differences in the values calculated by the P_1, B_1, Moments and Monte Carlo methods in Table 2 are due to slight differences in the cross sections used in the various calculations.

The higher moments of the distribution of U-235 fission neutrons in water is given in Table 4. Again the agreement between theory and experiments is rather good. This table allows us to compare the various methods used in the calculations. Although both methods give the second moment accurately, we see that for the higher moments indeed the B_1 approximation is superior to the P_1 approximation although both methods become increasingly poor with increasing n.

TABLE 3†. AGE OF FISSION NEUTRONS AT 1.44 eV IN WATER

	Rigorous	SG	Age–Diffusion
a) No approximations	25.3 cm²	30.8 cm²	–
b) Oxygen scattering assumed isotropic	23.6	28.7	–
c) Oxygen assumed not to slow down	26.8	32.8	17.5 cm²
d) Approximations b and c both made	25.3	–	–

† Taken from J. E. WILKINS, R. L. HELLENS and P. F. ZWEIFEL, *PICG* **5**, 62 (1955).

TABLE 4. MOMENTS $(\overline{r^n})$ OF THE FLUX OF 1.4 eV NEUTRONS FROM A POINT SOURCE OF U-235 FISSION IN WATER

n	2	4	6	8
Experiment †	151 cm^2	0.82×10^5 cm^4	1.169×10^8 cm^6	3.5×10^{11} cm^8
Experiment ††	185	1.22×10^5	2.27×10^8	8.6×10^{11}
B_3 approximation, moderation by oxygen neglected ††	154	0.98×10^5	1.93×10^8	8.7×10^{11}
Selengut–Goertzel method †	184	1.42×10^5	2.54×10^8	4.9×10^{11}
Three-group approximation ‡	185	0.99×10^5	1.09×10^8	2.1×10^{11}
P_1 ‡‡	149	0.74×10^5	0.92×10^8	1.5×10^{11}
B_1 ‡‡	149	0.91×10^5	1.39×10^8	2.5×10^{11}

† R. C. Doerner, R. J. Armani, W. E. Zagotta and F. H. Martens, *Nucl. Sci. Eng.* **9**, 221 (1961).

†† J. E. Hill, L. D. Roberts and T. E. Fitch, *J. Appl. Phys.* **26**, 1013 (1955).

‡ A. M. Weinberg and E. P. Wigner, *The Physical Theory of Neutron Chain Reactors*, Univ. of Chicago Press, 1958.

‡‡ J. E. Wilkins, R. L. Hellens and P. F. Zweifel, *PICG* **5**, 62 (1955).

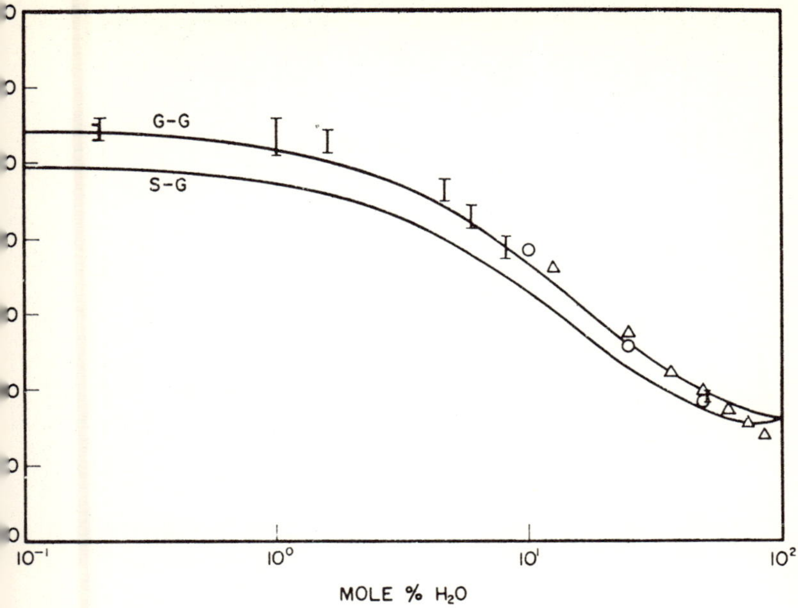

FIG. III. 5 The age in heavy water–light water mixtures. (From M. M. Levine, K. E. Roach, D. B. Wehmeyer and P. F. Zweifel, *Nucl. Sci. Eng.* **7**, 14 (1960).)

The experimental points are from:

I J. W. Wade, *Nucl. Sci. Eng.* **4**, 12 (1958).

⊐ R. N. Olcott, *Nucl. Sci. Eng.* **1**, 327 (1956).

△ R. Coveyou and L. Sullivan, cited in M. Tobias, ORNL-2575 (1958).

⊙ G. Brown, cited in M. Tobias, ORNL-2575 (1958).

Table 4 shows that the Selengut–Goertzel method overestimates all of the moments, again due to a too large first flight correction. The SG method was, at one time, quite popular because it agreed with the experimental results (compare the SG results to those on line 2 of Table 4). However, the more recent experimental results given on line 1 of Table IV are in better agreement with the rigorous P_1 and B_1 methods than with the SG method. The three group calculation on line 5 of Table 4 was done by the methods of Chapter IV with the group constants selected to agree with the experimental data on line 2.

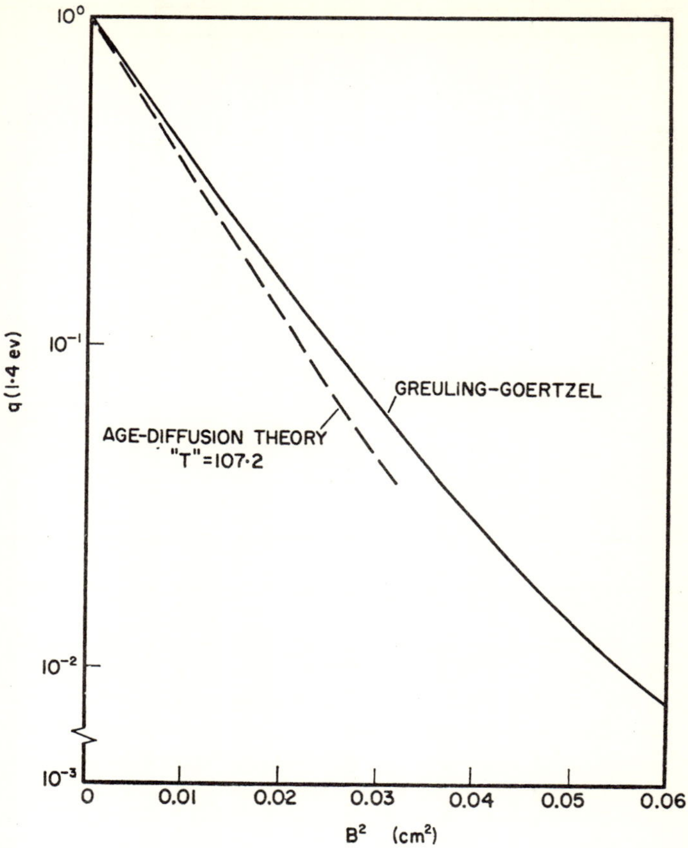

FIG. III.6 The calculated slowing-down density (non-leakage probability) in 99.8% D_2O. From M. M. Levine, K. E. Roach, D. B. Wehmeyer and P. F. Zweifel, *Nucl. Sci. Eng.* 7, 14 (1960).

The applicability of the Greuling–Goertzel approximation to the calculation of the slowing down of neutrons in heavy water is illustrated by Figs. III.5 and III.6. From Fig. III.5 we see that the age calculated by the GG method is in excellent agreement with experiment whereas the SG method (in which deuterium is treated as in age theory) gives results which are low by about 10%. Again this is due to the lack of a first flight correction. Figure III.6 shows calculations of the slowing-down density at 1.44 eV (essentially the

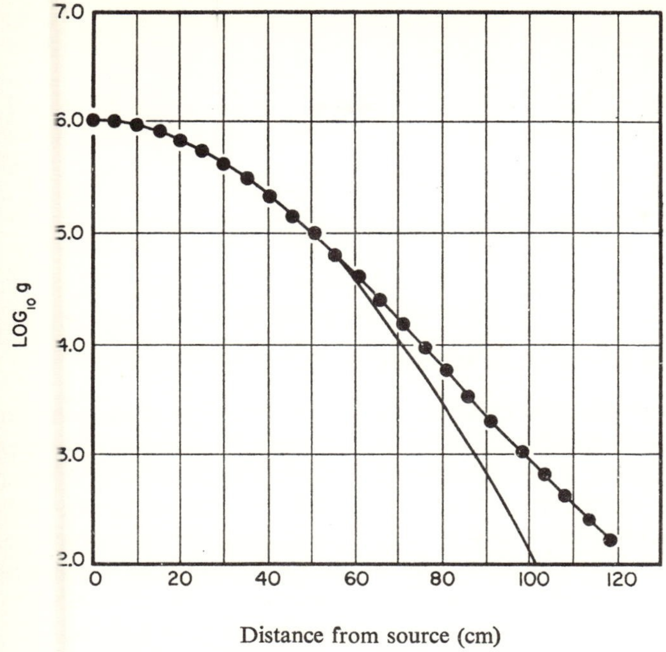

FIG. III.7 Slowing down of fission neutrons in graphite. (From J. E. Hill and L. D. Roberts, *Slowing Down of Fission Neutrons in Graphite* [ORNL-187].) The lower curve corresponds to the results of age theory.

non-leakage probability) vs. the buckling as calculated by the GG approximation and by Age Diffusion Theory ($e^{-B^2\tau}$). It is seen that the GG approximation gives results which are in much better agreement with experiment.

Figure III.7 shows the slowing-down density of U-235 fission neutrons in graphite. The agreement with age theory is seen to be quite good close to the source, but age theory predicts too small a slowing-down density at large distances from the sources. This is due to several factors, chief among which are that the neutrons far from the source result from (1) neutrons with long first flights and (2) neutrons which make many small angle collisions. Since Age–Diffusion theory treats neither of these correctly, the disagreement is to be expected.

CHAPTER IV

NUMERICAL SOLUTIONS OF THE BOLTZMANN EQUATION

A. The Multigroup Approach

1. Failure of Analytical Methods

Most of the calculational methods of the first three chapters rested on assumptions which are not generally valid for existing reactors. First, it was assumed that the reactor consists of a single region in which the composition is uniform or at least periodic with a period small compared to the overall dimensions of the system. In such systems all neutrons leaving the outer surface of the reactor never return, since the systems were assumed to be surrounded by an infinite vacuum. These assumptions led to a rather simple exposition of reactor theory in general, and of slowing down in particular, which was useful for pointing out many basic principles. However, from the reactor designer's point of view, the disappearance of the neutrons which leave the reactor leaves fewer behind for maintaining the critical chain reaction. To compensate for this the concentration of fuel must generally be increased. As fuel is generally more expensive than moderator, this is a rather uneconomical mode of operation.

If, instead of allowing the neutrons to escape, we provide some means of returning the escaped neutrons to the reactor, i.e., of reflecting them, the fuel requirements of the "core" should be smaller. It can readily be seen that a reflector should be constructed of materials with absorption cross sections as low as possible. Furthermore, since much of the neutron leakage from a reactor will be leakage of fast neutrons and since thermal neutrons are generally more valuable for causing fission, an additional advantage is gained if the reflector is also a good moderator. Thus, the reflector materials most frequently employed are the common moderators, i.e., hydrogen, deuterium, beryllium, carbon, and/or their compounds.

In fast reactors, the situation is somewhat different. Here, a moderating reflector is frequently a disadvantage; one of the greatest advantages of fast reactors is their ability to breed, which is a direct consequence of the facts that η is higher and absorption cross sections of structural materials are lower for fast neutrons. The fuel loading of fast reactors is generally much larger than that of thermal reactors so that reflection is not so important as it is in thermal systems. A better scheme is to surround the system with fertile materials (e.g., U-238 and Th-232) in which neutron capture leads to production of more fuel (here Pu-239 and U-233); thus the breeding ratio may be increased at no penalty to the core. Here, the material surrounding the reactor is generally called a "blanket" rather than a reflector.

It was pointed out in Chapter I that the methods of the first three chapters do not apply to fast reactors. This is a consequence of the fact that much of the moderation present in fast systems is due to inelastic scattering in structural materials, fuel, and coolants (usually liquid metals). Inelastic scattering reduces the energy of the scattered neutrons rather drastically on a single collision and is difficult to describe analytically. Another approach is required if one is to be able to handle such systems.

Finally, the spatial distribution of the flux (and hence the power) in the type of reactor discussed in the first three chapters is highly peaked near the center of the reactor. This means that more heat is produced there (which is in itself poor from an engineering standpoint) and that the fuel in the center will be depleted more rapidly. Consequently, the limitations of the system, either decrease in reactivity or metallurgical limits, will be reached rather quickly in terms of total energy production by a core over its lifetime. To overcome this, it is advantageous to "flatten" the flux shape. Reflection tends to aid in this attempt as it replenishes the neutrons near the boundaries, where the flux in a bare reactor is smallest. Further flattening must, however, be accomplished by making the core itself non-uniform.

For reactors in which any of the above difficulties arise (and this includes almost all reactors), one must resort to numerical solutions of the Boltzmann equation; the Boltzmann equation will have to be approximated in some way that is convenient for machine computation. This does not mean that the development of the first three chapters is invalid. It, in fact, contains many of the approximations frequently used in numerical reactor calculations. For example, numerical methods usually employ diffusion theory or the P_L approximations, although some of the other methods of Chapter III also find frequent application.

The general approach will be essentially that of using finite difference approximations in treating the space and energy (or lethargy) variations of the flux. For the lethargy, the treatment is very similar to the "one group" approximation applied to thermal neutrons in Chapter I, but is applied to the fast neutron group (or groups) as well. The major difference is that fast neutrons are not usually treated by one group theory, but are generally treated as though they possess only a finite number of discrete energies. This is the *multigroup* approach.

After making the multigroup approximation, finite differences techniques will be applied to the spatial variation of the neutrons in each group. The Boltzmann equation will then have been reduced to a form suitable for numerical computation.

2. The Adjoint Flux

The concept of an *adjoint flux* can be quite useful in reactor physics. Basically, the concept is mathematical, but the adjoint flux actually does have a direct physical interpretation due to Hurwitz. The procedure here will be to define the adjoint flux and derive the equation governing it in a somewhat mathematically abstract manner and then to state its physical significance. First we define the *scalar product* (Ψ, X) of two functions Ψ and X by

$$(\Psi, X) = \int \Psi X \, d\tau \tag{IV.1a}$$

where the integration is understood to be taken over the complete range of variables for which the functions are defined; $d\tau$ represents the product of differentials of all variables involved. For example, if Ψ is the neutron flux $\Phi(\mathbf{r}, u, \mathbf{\Omega})$, and if X is another function defined over the same range of variables, $\mathbf{r}$, u, and $\mathbf{\Omega}$, the integration would extend over the reactor (R) (or extrapolated reactor in diffusion theory) in configuration space, over all lethargies from zero to infinity and over all solid angles $\mathbf{\Omega}$.

Now, if L is a linear operator [1] which, when operating on a function defined over these variables, say X, produces another function defined over the same variables, one can then define the scalar product (Ψ, LX). The operator adjoint to L, denoted by L^+, may then be defined by the relation

$$(\Psi, LX) = (L^+ \Psi, X) \tag{IV.1b}$$

i.e., L^+ is the operator which makes (IV.1b) hold. A word of caution is necessary here. In general, in finding an operator which satisfies Equation (IV.1b), it is found that if X satisfies a boundary condition then Ψ must satisfy a different boundary condition — the adjoint boundary condition.

In order to make these concepts somewhat more concrete we choose as an example the time independent Boltzmann operator, **B**, defined by the Boltzmann equation

$$\begin{aligned} \mathbf{B}\Phi(\mathbf{r}, u, \mathbf{\Omega}) \equiv \mathbf{\Omega} \cdot \nabla\Phi(\mathbf{r}, u, \mathbf{\Omega}) + \Sigma_t(\mathbf{r}, u)\, \Phi(\mathbf{r}, u, \mathbf{\Omega}) \\ - \int \mathrm{d}u' \int \mathrm{d}\mathbf{\Omega}'\, \Phi(\mathbf{r}, u', \mathbf{\Omega}')\, \Sigma_s(u' \to u, \mathbf{\Omega}' \to \mathbf{\Omega}) \\ - f(u) \int \mathrm{d}\mathbf{\Omega}' \int \mathrm{d}u'\, \nu(u')\, \Sigma_f(u')\, \Phi(\mathbf{r}, u', \mathbf{\Omega}') = 0 \end{aligned} \tag{IV.2a}$$

where $f(u)$ is the fission spectrum. The boundary condition is, of course,

$$\Phi(\mathbf{r}, u, \mathbf{\Omega}) = 0 \qquad \mathbf{\Omega} \cdot \hat{\mathbf{n}} < 0 \tag{IV.2b}$$

for $\mathbf{r}$ on the reactor surface S and $\mathbf{n}$ the outward normal. The scalar product of Equation (IV.2a) with $\Psi(\mathbf{r}, u, \mathbf{\Omega})$ is now taken, i.e., Equation (IV.2a) is multiplied by $\Psi(\mathbf{r}, u, \mathbf{\Omega})$ and integrated over the reactor volume, all u and $\mathbf{\Omega}$. If Ψ satisfies appropriate conditions, which will be given later, it is possible to find an operator $\mathbf{B}^+$ that satisfies the following equation

$$(\mathbf{B}^+\Psi, \Phi) = (\Psi, \mathbf{B}\Phi) = 0 \tag{IV.3a}$$

where $\mathbf{B}^+$ is the adjoint Boltzmann operator.

In particular, the solution of the equation

$$B^+ \Psi = 0 \tag{IV.3b}$$

[1] That is, it has the property $L(A\Psi + BX) = AL\Psi + BLX$ for any pair of functions Ψ and X and constants A and B. For the properties of such operators see, for example, R. Courant and D. Hilbert, *Methods of Mathematical Physics*, Vol. I, Interscience, New York, 1953.

which obeys the adjoint boundary condition (IV.5b) will be denoted by $\Phi^+(\mathbf{r}, u, \mathbf{\Omega})$ and be called the *adjoint flux*.

Since each operator comprising the Boltzmann operator is itself linear and since linear operators obey the relation

$$(L_1 + L_2)^+ = L_1^+ + L_2^+ \tag{IV.4}$$

the adjoint of the Boltzmann operator is just the sum of the adjoints of the individual operators comprising the Boltzmann operator. It is easy to verify, using Equation (IV.3a), that the adjoint Boltzmann equation is:

$$\begin{aligned} \mathbf{B}^+\Phi^+(\mathbf{r}, u, \mathbf{\Omega}) = & -\mathbf{\Omega} \cdot \nabla \Phi^+(\mathbf{r}, u, \mathbf{\Omega}) + \Sigma_t(\mathbf{r}, u)\, \Phi^+(\mathbf{r}, u, \mathbf{\Omega}) \\ & - \int \mathrm{d}u' \int \mathrm{d}\mathbf{\Omega}'\, \Phi^+(\mathbf{r}, u', \mathbf{\Omega}')\, \Sigma_s(u \to u', \mathbf{\Omega} \to \mathbf{\Omega}') \\ & + \nu(u)\, \Sigma_f(\mathbf{r}, u) \int \mathrm{d}\mathbf{\Omega}' \int \mathrm{d}u'\, f(u')\, \Phi^+(\mathbf{r}, u', \mathbf{\Omega}') = 0 \end{aligned} \tag{IV.5a}$$

with the adjoint boundary condition

$$\Phi^+(\mathbf{r}, u, \mathbf{\Omega}) = 0 \qquad \mathbf{\Omega} \cdot \mathbf{n} > 0 \tag{IV.5b}$$

if $\mathbf{r}$ is on the surface S.

One may note the "upside-down" nature of the adjoint equation. The first term has its sign reversed; the "adjoint neutrons" flow with the gradient of the adjoint flux, not against it. From the third term it is seen that "adjoint neutrons" gain energy in collisions whenever they lost energy in the original Boltzmann equation (and vice versa at thermal energies). Finally, in the last term, the roles of the fission cross section and fission spectrum are reversed, that is, "adjoint neutrons" are absorbed at high energies and produced at thermal energies if the reverse is true in the original Boltzmann equation. This "upside-downness" is a general property of adjoint operators.

The direct physical meaning of the adjoint angular flux $\Phi^+(\mathbf{r}, u, \mathbf{\Omega})$ has been given by Hurwitz [2], who showed that in a critical reactor $\Phi^+(\mathbf{r}, u, \mathbf{\Omega})$ is proportional to the fission rate at a much later time, caused by one neutron now at position $\mathbf{r}$ with lethargy u

[2] H. Hurwitz, Jr., KAPL-98 (1948).

travelling in direction $\mathbf{\Omega}$. Hence, when properly normalized,† it is equal to the *Iterated Fission Probability*, denoted by $F(\mathbf{r}, u, \mathbf{\Omega})$. Also, because it is a measure of how important a particular neutron is in maintaining the critical chain reaction, it is sometimes normalized differently and called the *Importance Function* [3].

This interpretation of the adjoint flux leads directly to its use in perturbation theory; from the above definition, one can quickly derive the effect of a small disturbance on a reactor. Thus, if, for example, one were to add an absorbing foil with macroscopic absorption cross section Σ_a to a critical reactor, the foil (considered to be small) would absorb $\Sigma_a(\mathbf{r}, u)\ \Phi(\mathbf{r}, u, \mathbf{\Omega})\ d^3r\ d\mathbf{\Omega}\ du$ neutrons in d^3r, $d\mathbf{\Omega}$ and du per second. Then, if each of the neutrons absorbed would eventually have given rise to a fission rate F, the number of such fissions will be diminished by

$$\Delta N_a = \int F(\mathbf{r}, u, \mathbf{\Omega})\ \Sigma_a(\mathbf{r}, u)\ \Phi(\mathbf{r}, u, \mathbf{\Omega})\ d^3r\ du\ d\mathbf{\Omega} \qquad \text{(IV.6a)}$$

Due to the normalization of F (see footnote this page), $\Delta N_a/N_f$, where

$$N_f = \int \Sigma_f(\mathbf{r}, u)\ \Phi(\mathbf{r}, u, \mathbf{\Omega})\ d^3r\ du\ d\mathbf{\Omega} \qquad \text{(IV.6b)}$$

is the total fission rate in the reactor, is the fractional decrease in the number of fissions at a later time due to the disturbance, i.e., it is the decrease in the multiplication factor.

Similar arguments with respect to other types of perturbations lead to the usual perturbation theory formulae. Other derivations rely directly on the mathematical properties of adjoint functions, and lead, as expected, to the same results [4].

3. Properties of the Adjoint Flux

Thus far, we have derived the equation adjoint to the Boltzmann equation. However, as the earlier chapters have shown, reactor

† The normalization is

$$F(\mathbf{r}, u, \mathbf{\Omega}) = \frac{\Phi^+(\mathbf{r}, u, \mathbf{\Omega}) \int \Sigma_f(\mathbf{r}', u')\ \Phi(\mathbf{r}', u', \mathbf{\Omega}')\ d\mathbf{r}'\ du'\ d\mathbf{\Omega}'}{\nu \int d^3\mathbf{r}' \int \Sigma_f(\mathbf{r}', u')\ \Phi(\mathbf{r}', u', \mathbf{\Omega}')\ du'\ d\mathbf{\Omega}' \int \Phi^+(\mathbf{r}', u'', \mathbf{\Omega}'') f(u'')\ du''\ d\mathbf{\Omega}''}$$

The average of F over the fission spectrum, $\int F(\mathbf{r}, u, \mathbf{\Omega}) f(u)\ du\ d\mathbf{\Omega}$, is called $F_{II}(\mathbf{r})$ and will be seen to be a useful quantity in reactor calculations.

[3] The use of the adjoint flux in reactor theory is originally due to E. P. WIGNER.

[4] See, for example, WEINBERG and WIGNER, *op. cit.*, Chapter XVI.

calculations always employ some approximation to the Boltzmann equation. It is to be expected then that actual calculations of the adjoint flux will similarly employ approximations to the adjoint Boltzmann equation. A cursory examination of Equation (IV.5a) will reveal that its structure is much the same as that of the Boltzmann equation, so that procedures identical to those used for obtaining approximations to the Boltzmann equation may be employed in obtaining workable approximations. An alternate approach is, of course, to derive the adjoint equations for a given approximation directly.

This alternate approach may, however, give different results than those obtained by approximating the adjoint Boltzmann equation. For example, the P_1 equations may be written in matrix form

$$\begin{pmatrix} \Sigma_a & \partial/\partial z \\ \frac{1}{3}\frac{\partial}{\partial z} & \Sigma_{tr} \end{pmatrix} \begin{pmatrix} \varphi_0 \\ \varphi_1 \end{pmatrix} = 0 \tag{IV.7a}$$

or, more compactly,

$$||P||\,|\varphi| = 0 \tag{IV.7b}$$

where $|\varphi|$ is a vector† with elements φ_0, φ_1 and $||P||$ is the matrix appearing in Equation (IV.7a). Now, the adjoint of a matrix can be obtained by interchanging rows and columns (i.e., by transposing) and taking the adjoint of any operator appearing as a matrix element.‡ Then the adjoint P_1 equations are

$$||\Delta^+||\,|\varphi^+| = 0 \tag{IV.7c}$$

where

$$||P^+|| = \begin{pmatrix} \Sigma_a & -1/3\ \partial/\partial z \\ -\partial/\partial z & \Sigma_{tr} \end{pmatrix} \tag{IV.7d}$$

Now, on the other hand, suppose one makes the P_1 expansion of the adjoint Boltzmann equation in the usual fashion. One finds a different adjoint P_1 operator:

$$||P^+||' = \begin{pmatrix} \Sigma_a & -\ \partial/\partial z \\ -\ 1/3\ \partial/\partial z & \Sigma_{tr} \end{pmatrix} \tag{IV.7e}$$

The same phenomenon occurs if the multigroup equations (derived in the succeeding sections) are written in matrix form, and then

† Vector in the sense of an $L \times 1$ matrix.

‡ This statement can be proven from the matrix form of Equation (IV.3a). The scalar product then denotes the operation of taking the matrix product and then integrating over all independent variables.

"adjointized" in the manner described above. We will return to this point later; for now, we merely point out that instead of "adjointing" approximations, one should approximate adjoints.

Now, the operators $||P^+||$ and $||P^+||'$ lead to the same equation for the "total adjoint flux," φ_0^+, however, they predict a different relation between φ_0^+ and φ_1^+ which will, as we shall see later, lead to different boundary conditions for the two adjoint models.

Let us consider a further interesting application of the adjoint function, i.e., the determination of the multiplication constant, k, of a reactor.

To begin with, the multiplication constant, which does not appear in the Boltzmann equation *per se*, must be introduced artificially in some manner. The usual scheme for doing this is to note that k must be proportional to ν, the mean number of neutrons per fission. Although ν is a natural constant, we may treat it as an eigenvalue, i.e., a constant to be determined by the calculation. Then we may set

$$\nu' = \frac{\nu}{k} \tag{IV.7f}$$

where ν' is the eigenvalue determined by the Boltzmann equation. (It is the value which ν would have to take to make a non-critical reactor critical.) Then the Boltzmann equation may be written

$$\begin{aligned} \mathbf{\Omega}\cdot\nabla\Phi + \Sigma_t\Phi - \int \Sigma_s(u' \to u, \mathbf{\Omega}' \to \mathbf{\Omega})\, \Phi(\mathbf{r}, u', \mathbf{\Omega}')\, \mathrm{d}u'\, \mathrm{d}\mathbf{\Omega}' \\ = \nu' f(u') \int \Sigma_f \Phi \mathrm{d}u'\, \mathrm{d}\mathbf{\Omega}' \end{aligned} \tag{IV.7g}$$

Now if the flux Φ were known, one could determine ν' by dividing Equation (IV.7g) by $f \int \Sigma_f \Phi \mathrm{d}u'\, \mathrm{d}\mathbf{\Omega}'$ and inserting the flux. This, however, is certainly equivalent to solving the complete Boltzmann equation and is generally quite impractical. For the sake of brevity we shall rewrite Equation (IV.7b) in abstract operator form as

$$\mathscr{L}\Phi = \nu' \mathscr{R}\Phi \tag{IV.8a}$$

where $\mathscr{L}$ and $\mathscr{R}$ represent the operators on the left- and right-hand sides of Equation (IV.7b) respectively. The adjoint equation could similarly be written

$$\mathscr{L}^+\Phi^+ = \nu' \mathscr{R}^+\Phi^+ \tag{IV.8b}$$

where we have explicitly used the fact that operators with real eigenvalues have the same eigenvalues as their adjoints. Now the scalar product of Equation (IV.8a) with Φ^+ yields

$$\nu' = \frac{(\Phi^+, \mathscr{L}\Phi)}{(\Phi^+, \mathscr{R}\Phi)} \tag{IV.8c}$$

The advantage in using Equation (IV.8c) to calculate ν' is that it is a variationally stationary expression. That is, the error in ν' is proportional to the product of the errors in Φ and Φ^+, a result which is readily proved as follows.

Let us use "trial functions" Φ_t and Φ_t^+ which are guesses at the true values of the flux and its adjoint:

$$\begin{aligned} \Phi_t &= \Phi + \delta\Phi \\ \Phi_t^+ &= \Phi^+ + \delta\Phi^+ \end{aligned} \tag{IV.9a}$$

and insert them into Equation (IV.8c) in an attempt to calculate ν'. The resulting value of ν is denoted by

$$\nu_t = \nu' + \delta\nu' \tag{IV.9a}$$

where the δ quantities represent the deviations from the correct values of the quantities. Substituting these into Equation (IV.8c),

$$(\nu' + \delta\nu')(\Phi^+ + \delta\Phi^+, \mathscr{R}[\Phi + \delta\Phi]) = (\Phi^+ + \delta\Phi^+, \mathscr{L}[\Phi + \delta\Phi]) \tag{IV.9b}$$

or, expanding, we have:

$$\begin{aligned} &(\nu' + \delta\nu')[(\Phi^+, \mathscr{R}\Phi) + (\delta\Phi^+, \mathscr{R}\Phi) + (\Phi^+, \mathscr{R}\delta\Phi) + (\delta\Phi^+, \mathscr{R}\delta\Phi)] \\ &= (\Phi^+, \mathscr{L}\Phi) + (\delta\Phi^+, \mathscr{L}\Phi) + (\Phi^+, \mathscr{L}\delta\Phi) + (\delta\Phi^+, \mathscr{L}\delta\Phi) \end{aligned} \tag{IV.9c}$$

But, from Equation (IV.8a) we have

$$\mathscr{L}\Phi = \nu'\mathscr{R}\Phi$$

while, from the properties of adjoint operators and Equation (IV.8b) one has†

$$(\Phi^+, \mathscr{L}\delta\Phi) = (\mathscr{L}^+\Phi^+, \delta\Phi) = \nu'(\mathscr{R}^+\Phi^+, \delta\Phi) = \nu'(\Phi^+, \mathscr{R}\delta\Phi).$$

So Equation (IV.9c) becomes, after elimination of equal terms on the left- and right-hand sides:

$$\begin{aligned} \delta\nu'[(\Phi^+, \mathscr{R}\Phi) + (\delta\Phi^+, \mathscr{R}\Phi) + (\Phi^+, \mathscr{R}\delta\Phi) + (\delta\Phi^+, \mathscr{R}\delta\Phi)] \\ = (\delta\Phi^+, \mathscr{L}\delta\Phi) - \nu'(\delta\Phi^+, \mathscr{R}\delta\Phi) \end{aligned} \tag{IV.9d}$$

from which it is clear that the error in the eigenvalue, $\delta\nu'$, is proportional to the products of the errors in Φ and Φ^+, i.e., Equation (IV.8c) is a variational expression for ν'.

† It is important to note that this requires that the trial functions obey correct boundary conditions.

Incidentally, perturbation theory, discussed earlier from the point of view of iterated fission probability, is an application of the variational principle (Equation (IV.8c)) in which the trial functions are taken to be solutions of the "unperturbed" problem (i.e., a simpler problem which one knows how to solve).

The advantages of using Equation (IV.8c) in reactor calculations should now be apparent, and will be discussed in detail later. Again the reader is reminded that expressions of the type (IV.8c) are available in any calculational model, so long as the adjoint equation is properly defined.

4. The Multigroup Equations

Clearly, numerical procedures can be applied to any neutron transport problem, and, indeed, computer solutions to almost every problem of interest in reactor calculations have been obtained. Here, however, we will restrict our interest to criticality problems, that is, to time independent problems. This is not to say that these calculations can be applied only to the problem of calculating which configurations will make a reactor critical. Frequently, the reactor designer wants to know how much *excess reactivity* defined as $(k_{\text{eff}} - 1)/k_{\text{eff}} = \varrho$ is built into a reactor in order that he may be able to insure that the effects of temperature feedback, fission product poisoning, and fuel burn-up can be overcome and that the control rods can safely shut the reactor down. Such problems may be solved without resort to the time-dependent Boltzmann equation in the following manner.

Consider a supercritical reactor ($k_{\text{eff}} > 1$). On the average, each neutron in a given generation gives rise to k_{eff} neutrons in the next generation, according to the definition of the multiplication factor. A steady state neutron flux cannot exist within the reactor unless $k_{\text{eff}} = 1$. If, however, one is not interested in the actual time behavior of the system but wants only to know k_{eff}, possibly for the reasons mentioned above, it is not necessary to solve the time dependent problem. Instead, one frequently allows ν to be an eigenvalue in the manner previously described and

$$k_{\text{eff}} = \frac{\nu}{\nu'} \tag{IV.10}$$

A note of caution should be injected here. As defined in this manner, the effective multiplication factor is the so-called *static reactivity.* It is not, however, the reactivity which arises most naturally in reactor kinetics, which is the *dynamic reactivity.* The difference arises out of the facts that the effective delayed neutron fraction depends on the time behavior of the system and the delayed neutrons have a different energy spectrum from that of the prompt neutrons [5]. The difference is generally quite small.

Because multigroup calculations are expensive, they are rarely carried out for any but the simplest approximations to the Boltzmann equation. Generally, diffusion theory is used for criticality calculations, but other P_L approximations and the S_N approximation are also frequently used. The degree of current knowledge of many cross sections does not warrant use of high order approximations, so that the development will be carried out only for the P_1 approximation.

To begin with, the thermal neutron group will be treated exactly as in Chapter I, i.e., all neutrons with energies below E_T are taken as a single group. It is possible to apply multigroup procedures within the thermal group itself,† but this is outside the range of intent of this book. The problem of up-scattering, that is, of a neutron's gaining energy in a collision, which arises in the treatment of thermal neutrons complicates the problem somewhat and will not be treated here. We shall employ diffusion theory for the thermal group although more rigorous approximations are sometimes used in practice. In a non-uniform reactor one can write

$$\mathbf{J}_T(\mathbf{r}) = -D_T(\mathbf{r})\, \nabla\varphi_T(\mathbf{r}) \tag{IV.11}$$

where the diffusion coefficient is now position dependent. Thus, the thermal group equation becomes:

$$-\nabla \cdot D_T(\mathbf{r})\, \nabla\varphi_T(\mathbf{r}) + \Sigma_{aT}(\mathbf{r})\, \varphi_T(\mathbf{r}) = q_T(\mathbf{r}) \tag{IV.12}$$

The symbols have been previously defined in Chapter I. Some discussion of the methods of obtaining the *thermal group constants* D_T and Σ_{aT} has been given in Chapter I; a more intensive discussion will be given later in this chapter (see Section 10).

For fast neutrons, we begin by writing the P_1 equations, which have been obtained in both Chapters I and III. In general they may

† The techniques employed in setting up the multigroup thermal equations are quite similar to the methods used here. The techniques for solution are, however, slightly modified.

[5] E. E. Gross and J. H. Marable, *Nucl. Sci. Eng.* 7, 281 (1960).

be written:

$$\Sigma_t(\mathbf{r}, u)\,\phi_0(\mathbf{r}, u) + \nabla\cdot\boldsymbol{\phi}_1(\mathbf{r}, u) = \int \Sigma_{s_0}(u' \to u)\,\phi_0(\mathbf{r}, u')\,du' + S(\mathbf{r}, u) \tag{IV.13a}$$

and

$$\Sigma_t(\mathbf{r}, u)\boldsymbol{\phi}_1(\mathbf{r}, u) + \tfrac{1}{3}\nabla\phi_0(\mathbf{r}, u) = \int \Sigma_{s_1}(u' \to u)\,\boldsymbol{\phi}_1(\mathbf{r}, u')\,du' \tag{IV.13b}$$

The multigroup equations are obtained by dividing the lethargy interval from zero to u_T, the thermal cutoff, into several intervals (u_{j-1}, u_j) where $j = 1, 2, \ldots, N-1$, $N-1$ being the number of fast groups desired. Naturally, $u_0 = 0$ and $u_{N-1} = u_T$. The choice of the number of groups for a particular calculation will be discussed later; in general, the larger the number of groups, the more accurate the calculation. On the other hand, the expense of a calculation is approximately proportional to the number of groups so that an optimization is required.

Equations (IV.13) are now integrated in lethargy over group j to obtain:

$$\int_{u_{j-1}}^{u_j} \Sigma_t(\mathbf{r}, u)\,\phi_0(\mathbf{r}, u)\,du + \nabla\cdot\int_{u_{j-1}}^{u_j} \boldsymbol{\phi}_1(\mathbf{r}, u)\,du =$$

$$\int_{u_{j-1}}^{u_j} du \int_0^u du'\,\Sigma_{s0}(u' \to u)\,\phi_0(\mathbf{r}, u') + \int_{u_{j-1}}^{u_j} S(\mathbf{r}, u)\,du \tag{IV.14a}$$

and

$$\int_{u_{j-1}}^{u_j} \Sigma_t(\mathbf{r}, u)\,\boldsymbol{\phi}_1(\mathbf{r}, u)\,du + \tfrac{1}{3}\nabla\int_{u_{j-1}}^{u_j} \phi_0(\mathbf{r}, u)\,du =$$

$$\int_{u_{j-1}}^{u_j} du \int_0^u du'\;\Sigma_{s_1}(u' \to u)\,\boldsymbol{\phi}_1(\mathbf{r}, u') \tag{IV.14b}$$

Now, in analogy to the treatment of the thermal group, group fluxes are defined

$$\phi_{0j}(\mathbf{r}) = \int_{u_{j-1}}^{u_j} \phi_0(\mathbf{r}, u)\,du \tag{IV.14c}$$

$$\boldsymbol{\phi}_{1j}(\mathbf{r}) = \int_{u_{j-1}}^{u_j} \boldsymbol{\phi}_1(\mathbf{r}, u)\,du \tag{IV.14d}$$

Group cross sections are defined by

$$\Sigma_{tj0,1}(\mathbf{r})\,\phi_{j0,1}(\mathbf{r}) = \int_{u_{j-1}}^{u_j} du\,\Sigma_t(\mathbf{r}, u)\,\phi_{0,1}(\mathbf{r}, u) \tag{IV.14e}$$

so that reaction rates are preserved. An approximation is involved here which is frequently overlooked (as indeed it was in Chapter I). In the "1" equation of this set the vectors on the left- and right-hand sides are not necessarily proportional and Σ_{tj1} is actually a second-rank tensor. It usually suffices to ignore this (i.e. to assume that Σ_{tj1} is a scalar); Σ_{tj1} is then calculated as the ratio of the magnitudes of the vectors. Also, transfer cross sections are defined by:

$$\Sigma_{ij_{0,1}}(\mathbf{r})\,\phi_{i_{0,1}}(\mathbf{r}) = \int_{u_{j-1}}^{u_j} \mathrm{d}u \int_{u_{i-1}}^{u_i} \mathrm{d}u'\, \Sigma_{s_{0,1}}(u' \to u)\,\phi_{0,1}(\mathbf{r}, u') \qquad \text{(IV.14f)}$$

When $i = j$ the upper limit on the u' integral is to be taken as u and the approximation noted above is again invoked for the "1" equation. Finally, defining:

$$S_j(\mathbf{r}) = \int_{u_{j-1}}^{u_j} S(\mathbf{r}, u)\,\mathrm{d}u \qquad \text{(IV.14g)}$$

as the group source, one finally arrives at the multigroup P_1 equations:

$$\Sigma_{tj_0}(\mathbf{r})\,\phi_{0_j}(\mathbf{r}) + \nabla\cdot\boldsymbol{\phi}_{1_j}(\mathbf{r}) = \sum_{i=1}^{j} \Sigma_{ij_0}(\mathbf{r})\,\phi_{0i}(\mathbf{r}) + S_j(\mathbf{r}) \qquad \text{(IV.14h)}$$

$$\Sigma_{tj1}(\mathbf{r})\,\boldsymbol{\phi}_{1_j}(\mathbf{r}) + \tfrac{1}{3}\nabla\phi_{0j}(\mathbf{r}) = \sum_{i=1}^{j} \Sigma_{ij_1}\boldsymbol{\phi}_{1i}(\mathbf{r}) \qquad \text{(IV.14j)}$$

Frequently, the terms $\Sigma_{jj}\phi_j(\mathbf{r})$ appearing on the right-hand sides of Equations (IV.14h) and (IV.14j) are subtracted from the $\Sigma_{tj}\phi_j$ terms to yield a somewhat simpler set of equations. The resulting cross section may be interpreted as a cross section for removal from the group:

$$\Sigma_{Rj} = \Sigma_{tj} - \Sigma_{jj} = \Sigma_{aj} + \sum_{k>j} \Sigma_{jk} \qquad \text{(IV.14k)}$$

Aside from the approximation noted above, Equations (IV.14h) and (IV.14j) are rigorously equivalent to the original P_1 equations. The difficulty is that the group cross sections $\Sigma_{tj0,1}$ and the "transfer cross sections" $\Sigma_{ij0,1}$, are defined in terms of the spectra within the groups and cannot rigorously be determined until the fluxes are known, i.e., until the P_1 equations have been solved. The major difficulty in using multigroup theory is thus that of finding suitable fluxes $\phi_0(\mathbf{r}, E)$ and $\boldsymbol{\phi}_1(\mathbf{r}, E)$ over which to average the cross sections. In practice these are obtained in several ways that shall be discussed later.

Note that even in a region of uniform composition, in which the cross sections are not functions of position, the group-averaged cross sections may be functions of position if the flux, ϕ_0, and the current, $\mathbf{\phi}_1$, are not separable in space and energy. Separability was shown in Chapter I to obtain for an asymptotic reactor, i.e., a reactor containing only one region of uniform composition but, of course, for such a reactor there is no need to apply the multigroup scheme at all.

There are several possible solutions to these problems. One which immediately comes to mind is iteration, i.e., assuming flux and current spectra, averaging the cross sections, computing a new spectrum from the multigroup equations, recomputing the averages, using these to compute a new spectrum and continuing until the process converges. The difficulty here is that one obtains only a histogram of a spectrum from the multigroup equations; these must be converted into a continuous spectrum before the averages can be computed; hence this procedure is valid only if a large number of groups are used.

Another possible procedure is to use the methods of the previous chapter to compute the spectrum which would obtain in each region if it were isolated from the other regions. This, the so-called spectral procedure, may also be used iteratively, and is described in greater detail later. The spectrum so calculated can be used to generate group cross sections for the multigroup calculation.

A third procedure, which has been used in fast reactor design, is to assume a single spectrum for all reactors, average the cross sections once and for all, and use the cross sections so obtained in multigroup calculations. This procedure may be valid for fast reactors as cross sections generally do not vary rapidly at high energy. However, it has not been particularly successful in the past and is being discarded as reactor design techniques become more sophisticated.

Ultimately, the test of any of these methods lies in comparison with critical or exponential experiments or with operating reactors. The results lead one to suspect that reactor design is more of an art than a science at the present time; research is still being carried on in this field.

Equation (IV.14j) is frequently simplified by ignoring off-diagonal elements of Σ_1, i.e., $\Sigma_{ij1} = 0$ for $i \neq j$. This corresponds to the neglect of anisotropic slowing down and is thus an inconsistent P_1 approximation. Since inelastic scattering cross sections, at least according to the Hauser–Feshbach [6] theory, are symmetric about 90°, they have no P_1 component—thus the neglect in this approximation is only of the anisotropic component of elastic scattering (see Appendix E). One can then rewrite Equation (IV.14j) as

$$\Sigma_{trj}(\mathbf{r})\,\boldsymbol{\phi}_{1j}(\mathbf{r}) + \tfrac{1}{3}\nabla\,\phi_{0j}(\mathbf{r}) = 0 \tag{IV.14l}$$

where

$$\Sigma_{trj} = \Sigma_{t1j} - \Sigma_{1jj}$$

is seen to be $\Sigma_{tr}(u)$ averaged over $\boldsymbol{\phi}_1(\mathbf{r}, u)$.

If we assume Fick's law to be valid at all energies, Σ_{trj} becomes simply

$$\Sigma_{trj} = \frac{\int\limits_{u_{j-1}}^{u_j} \nabla\,\phi_0(\mathbf{r}, u)\,\mathrm{d}u}{\int\limits_{u_{j-1}}^{u_j} \dfrac{\nabla\,\phi_0(\mathbf{r}, u)\,\mathrm{d}u}{\Sigma_{tr}(u)}} \tag{IV.14m}$$

and if $\phi_0(\mathbf{r}, u)$ is separable in space and energy this reduces further to

$$\Sigma_{trj} = \lambda_{trj}^{-1} \tag{IV.14n}$$

where λ_{trj} is the flux average of the reciprocal of the transport cross section. This definition, while awkward, is frequently used and in another application has been shown to give accurate results in the calculation of the diffusion length for water [7]. (For a more complete discussion of these points see the work of Zweifel and Ball [8].) Under these approximations, the two P_1 multigroup equations can be combined into a single diffusion equation

$$-\nabla \cdot D_j\,\nabla\,\phi_{0j} + \Sigma_{Rj}\phi_{0j} = S_j + \sum_{k<j}\Sigma_{kj}\phi_{0k} \tag{IV.14p}$$

where the subscript "0" on the cross sections have been dropped. This set of N equations for the ϕ_j (assuming N groups) may be written in matrix form (see Section 11). It can be shown that the equations adjoint to the multigroup diffusion equations are simply obtained by transposing this matrix (i.e. interchanging indices).

[6] W. Hauser and F. Feshbach, *Phys. Rev.* **87**, 366 (1952).

[7] C. D. Petrie, M. L. Storm and P. F. Zweifel, *Nucl. Sci. Eng.* **2**, 728 (1957).

[8] G. Ball and P. F. Zweifel, International Conference on the Physics of Fast and Intermediate Reactors, Vienna, 1961.

For variational calculations involving the multigroup equations, this is the correct set of adjoint equations. If, on the other hand, one applies the multigroup method to the P_1 approximation to the adjoint Boltzmann equation, a different set of equations is obtained; for example, the cross sections that appear are averaged over the adjoint spectrum. This set of equations will, in general, yield a better approximation to the true adjoint flux but will not have the variational property with respect to the multigroup equations for the flux. In other words the operations of "multigrouping" and "adjointing" do not commute.

Very often, the multigroup procedure is applied directly to the Age–Diffusion, consistent P_1 Selengut–Goertzel, or Greuling–Goertzel approximations. The major difference between these schemes and the one given above is the presence of derivatives with respect to lethargy in the basic equations from which they are derived.

In these procedures, inelastic scattering must be excluded, or lumped with the source, since the lethargy expansions will certainly not describe it adequately. Ehrlich and Hurwitz [9] treated inelastic scattering by including an additional term Σ_{in} in the absorption cross section and adding a term $\int_0^u \Sigma_{in}(u' \to u)\, \phi(\mathbf{r}, u')\, du'$ to the fission source.

To illustrate the treatment of the derivative terms, let us start with the Age Diffusion equation, rewritten here:

$$\frac{\partial}{\partial u} q(\mathbf{r}, u) + \Sigma_a(\mathbf{r}, u)\, \phi(\mathbf{r}, u) - \nabla \cdot D(\mathbf{r}, u)\, \nabla\, \phi(\mathbf{r}, u) = S(\mathbf{r}, u) \tag{IV.15a}$$

with

$$q(\mathbf{r}, u) = \xi(\mathbf{r}, u)\, \Sigma_s(\mathbf{r}, u)\, \phi(\mathbf{r}, u) \tag{IV.15b}$$

The same procedure is again followed. Equation (IV.15a) is integrated in lethargy from u_{j-1} to u_j and group cross sections are defined as before

$$q(\mathbf{r}, u_j) - q(\mathbf{r}, u_{j-1}) + \Sigma_{aj}(\mathbf{r})\, \phi_j(\mathbf{r}) - \nabla \cdot D_j(\mathbf{r})\, \nabla\, \phi_j(\mathbf{r}) = S_j(\mathbf{r}, u) \tag{IV.15c}$$

where $\phi_j(r, u)$ and $S_j(r, u)$ are defined by Equations (IV.14c) and (IV.14g) respectively. Here the definition of D_j takes the form

[9] R. Ehrlich and H. Hurwitz, Jr., *Nucleonics* **23**, 12 (1952).

(IV.14n) automatically if separability is assumed. Some authors prefer to use

$$\bar{\phi}_j(\mathbf{r}) = \frac{\phi_j(\mathbf{r})}{u_j - u_{j-1}} = \frac{\phi_j(\mathbf{r})}{\Delta u_j} \qquad \text{(IV.15d)}$$

as a variable; the choice is a matter of personal preference.

This set of equations is essentially identical with the set (IV.14p) since, by definition

$$q(u_{j-1}) = \sum_{k<j} \Sigma_{kj}\phi_k + \sum_{\substack{k<j \\ l>j}} \Sigma_{kl}\phi_k \qquad \text{(IV.15e)}$$

and

$$q(u_j) = (\Sigma_{Rj} - \Sigma_{aj})\,\phi_j + \sum_{\substack{k<j \\ l>j}} \Sigma_{kl}\phi_k \qquad \text{(IV.15f)}$$

If these two expressions are inserted in Equation (IV.15c), Equation (IV.14p) results. However, in order to use this formulation of the multigroup equations as given, it is necessary to supply independent relations among $\phi_j(\mathbf{r})$, $q(\mathbf{r}, u_{j-1})$ and $q(\mathbf{r}, u_j)$. Several possible relationships have been suggested; they will merely be listed here as they are described in some detail in Section 9.

One method is to note that $q(\mathbf{r}, u_j)$ is related to $\phi(r, u_j)$ by Equation (IV.15b), that is,

$$q(\mathbf{r}, u_j) = \xi(\mathbf{r}, u_j)\,\Sigma_s(\mathbf{r}, u_j)\,\phi(\mathbf{r}, u_j) \qquad \text{(IV.15g)}$$

Instead of attempting to compute the Σ_{ij} from Equations (IV.14f) using basic cross section data, an alternate procedure is possible and is particularly useful if the cross sections are not well known. This procedure (due to Ehrlich and Hurwitz) is to assume

$$\phi_j(\mathbf{r}) = \omega_1 q(\mathbf{r}, u_{j-1}) + \omega_2 q(\mathbf{r}, u_j) \qquad \text{(IV.15h)}$$

thus yielding a set of equations in the $\phi(\mathbf{r}, u_j)$.

The second approach is to define

$$q_j(\mathbf{r}) = (\xi(\mathbf{r})\,\Sigma_s(\mathbf{r}))_j\,\phi_j \qquad \text{(IV.15i)}$$

an average slowing-down density in group. Then q_j is expressed as

$$q_j(\mathbf{r}) = \omega_1 q(\mathbf{r}, u_{j-1}) + \omega_2 q(\mathbf{r}, u_j) \qquad \text{(IV.15j)}$$

i.e., a linear combination of the end-point values. Of course, this approach is entirely equivalent to that given earlier if ω_1, ω_2 are derived from basic cross section data. A description of how this may be done is given in Section 9.

An alternate derivation of the multigroup equations has been given by Calame and Brooks [10]. They derived variational principles for the Age–Diffusion and Consistent P_1 equations. That is, they found a bilinear functional of $\phi(\mathbf{r}, u)$ and $\phi^+(\mathbf{r}, u)$, which is variationally stationary with respect to deviations of $\phi(\mathbf{r}, u)$ and $\phi^+(\mathbf{r}, u)$ from their correct values. By assuming that $\phi^+(\mathbf{r}, u)$ is constant in each lethargy interval (u_{j-1}, u_j), and by assuming a particular functional form for $\phi(\mathbf{r}, u)$ within that interval, they were able to derive a set of multigroup equations. Usually, this functional form is a linear combination of the fluxes or slowing-down densities at u_{j-1} and u_j. When such a form is employed, the multigroup equations obtained are identical to those obtained by the method used earlier except for the definitions of the average cross sections. This method has not as yet been exploited fully.

Frequently, two group calculations are sufficient for preliminary calculations. However, they are insufficient for many other purposes. The choice of the number of groups rests on a combination of many factors, not the least of which is the cost of computing machine time. In general, the time consumed by a calculation is proportional to the number of groups. There is further advantage, however, in keeping the group coupling to a minimum. If each group is coupled to all others, all of the space dependent group fluxes at lower lethargies must be used in calculation of any particular group flux. In turn, this entails having a machine with a large memory and the calculation of each group flux is slowed considerably by the necessity of using all the other fluxes. Considerable advantage is thus gained in having each group coupled only to the ones immediately adjacent to it and in having fission neutrons born into only one group. All group widths should therefore be at least $\ln 1/\alpha_m$ where α_m is the minimum value of $\alpha = \left(\dfrac{A-1}{A+1}\right)^2$ for any moderator in the reactor.† If hydrogen is present this is clearly an impossibility; hence, water moderated reactor calculations rarely employ

† With the changes in scattering kernel that some of the approximations utilize (see Fig. III.4), there may be coupling between non-adjacent groups even if this criterion is met. This coupling will usually be small, however.

[10] G. P. Calame and H. Brooks, *Trans. Amer. Nucl. Soc.* **2**, 1, Paper 7.4 (June 1959).

more than three or four groups, while calculations for other systems often use more than 20 groups. (These remarks do not necessarily apply to systems in which inelastic scattering is important.)

B. Simple Applications of the Method

5. Analytical Solutions

By their very nature, the multigroup equations are such as to preclude the possibility of obtaining analytical solutions for any but the simplest problems. Such analytical solutions as have been obtained describe systems which do not bear much relation to existing reactors (especially the highly complex power reactors now being designed and built) but they do serve as guides for the reactor designer. Other problems, as we shall see in the next section, are able to give useful information regarding the selection of group constants.

Since the diffusion equation cannot be solved in general unless the coefficients are either independent of position or vary with position in some simple manner, the soluble problems are those in which the reactor consists of a number of geometrically simple uniform regions. Clearly, the simplest such problem is the one-group, one-region problem, i.e., the one-velocity bare reactor problem. From this point one may increase the complexity of the problem by adding more groups and/or more regions. In the M region, N group problem, one is faced with the necessity of solving MN diffusion (or P_1) equations and using boundary conditions to determine the $2MN$ arbitrary constants that arise therefrom. Although symmetries and other conditions (such as finiteness and non-negativity) may serve to reduce the number of constants somewhat, the problem rapidly grows to the point at which hand calculation is not feasible. At this point it becomes necessary to use a computing machine and nearly all reactor design calculations are currently carried out on computers.

The simplest problem which yields all of the qualitative information about reflected thermal reactors is the two group, two region problem. We shall not solve this problem here as its solution is given in other texts [11]. However, we shall solve the multigroup

[11] See, for example, S. Glasstone and M. C. Edlund *The Elements of Nuclear Reactor Theory*, Van Nostrand, 1953, Chapter 8.

bare reactor problem as it yields valuable information about the selection of group constants.

6. Multigroup Age Theory

In this section we consider the problem of a bare reactor using multigroup theory. The rationale for considering this problem is twofold. Firstly, it affords a simple example of the application of the multigroup approach and secondly, and more importantly, it leads to useful approximations for the group constants. The calculation will be designed such that Age–Diffusion theory will result as the number of groups is increased and the size of each group is decreased accordingly. To further simplify the calculation, it is assumed that all fissions take place in the thermal group and that they are born in the first (highest energy) group. (Relaxation of this assumption leads to no additional difficulty, but the increase in mathematical complexity will obscure the desired results.)

Since Age–Diffusion theory implicitly assumes that the lethargy change by a neutron is small in any given collision, neutrons slowing down out of a given group should be scattered only to the next higher lethargy group. It is assumed that each group flux $\phi_j(\mathbf{r})$ is proportional to $\phi(\mathbf{r}, u_j)$, i.e., that ω_1 of Equation (IV.15n) is zero. In the limit of many groups, it does not seriously matter how ω_1 and ω_2 are as chosen so long as they become consistent in the limit of an infinite number of groups, but by choosing them in this fashion, it is possible to write $\phi(r, u_j)$ [or $q(r, u_j)$] in terms of $\phi_j(\mathbf{r})$ alone. Thus, if

$$q(\mathbf{r}, u_j) = \Sigma_{Rj}\phi_j(\mathbf{r})$$

each equation will contain only two $\phi_j(\mathbf{r})$ and for a uniform reactor we can write

$$-D_1\nabla^2\phi_1(\mathbf{r}) + (\Sigma_{a1} + \Sigma_{R1})\,\phi_1(\mathbf{r}) = \nu\Sigma_{fN}\,\phi_N(\mathbf{r}) \tag{IV.16a}$$

for the lowest lethargy group and

$$-D_j\nabla^2\phi_j(\mathbf{r}) + (\Sigma_{aj} + \Sigma_{Rj})\,\phi_j(\mathbf{r}) = \Sigma_{Rj-1}\,\phi_{j-1}(\mathbf{r}) \tag{IV.16b}$$

for the other fast groups with $2 \leqq j \leqq N - 1$. The thermal group equation (the subscript T will be dropped in favor of N), can be written

$$-D_N\nabla^2\phi_N(\mathbf{r}) + \Sigma_{aN}\phi_N(\mathbf{r}) = \Sigma_{R,N-1}\,\phi_{N-1}(\mathbf{r}) \tag{IV.16c}$$

The results obtained earlier in asymptotic reactor theory suggest that we seek solutions of the form

$$\phi_i(\mathbf{r}) = \varphi_i e^{ik \cdot r} \tag{IV.17}$$

Substituting this ansatz into Equations (IV.16a, b, c), we have

$$(D_1 k^2 + \Sigma_{a1} + \Sigma_{R1})\,\varphi_1 = \nu \Sigma_{fN} \varphi_N \tag{IV.18a}$$

$$(D_j k^2 + \Sigma_{aj} + \Sigma_{Rj})\,\varphi_j = \Sigma_{Rj-1} \varphi_{j-1} \quad 1 < j < N \tag{IV.18b}$$

$$(D_N k^2 + \Sigma_{aN})\,\varphi_N = \Sigma_{RN-1} \varphi_{N-1} \tag{IV.18c}$$

which are a set of N homogeneous algebraic equations for the N unknowns. This system of equations will have a solution only if the determinant of the system vanishes. Setting this determinant equal to zero gives the critical condition for the reactor. It may be written

$$\frac{\eta f}{1 + L_N^2 k^2} \prod_{j=1}^{N-1} \frac{\Sigma_{Rj}}{\Sigma_{Rj} + \Sigma_{aj} + D_j k^2} = 1 \tag{IV.19}$$

It is convenient to factor each term of the product:

$$\frac{\Sigma_{Rj}}{\Sigma_{Rj} + \Sigma_{aj} + D_j k^2} = \left(\frac{\Sigma_{Rj}}{\Sigma_{Rj} + \Sigma_{aj}}\right) \left(\frac{1}{1 + \dfrac{D_j}{\Sigma_{aj} + \Sigma_{Rj}} k^2}\right) \tag{IV.20}$$

The first factor is independent of the diffusion coefficient (and hence the leakage) while the second shows only a very weak dependence on the absorption cross section. Thus we may identify the first factor with the resonance escape probability for group

$$p_j = \frac{\Sigma_{Rj}}{\Sigma_{Rj} + \Sigma_{aj}} \approx \frac{1 - \Sigma_{aj}}{\Sigma_{Rj}} \tag{IV.21}$$

and the second factor with the non-leakage probability:

$$\lambda_j = \frac{1}{1 + \dfrac{D_j}{\Sigma_{aj} + \Sigma_{Rj}} k^2} = \frac{1}{1 + L_j k^2} \tag{IV.22a}$$

where we have defined

$$L_j^2 = \frac{D_j}{\Sigma_{Rj} + \Sigma_{aj}} \tag{IV.22b}$$

in analogy to the thermal diffusion length. We can thus expect to identify L_j^2 with the neutron age for group j. Note that in this model the resonance escape and non-leakage probabilities are not entirely independent, a result not obtained in the usual Age–Diffusion approach. Equation (IV.19) can now be written:

$$\frac{\eta f}{1 + L_N^2 k^2} \prod_{j=1}^{N-1} \frac{p_j}{1 + L_j^2 k^2} = 1 \tag{IV.23}$$

Since we are dealing with a uniform bare reactor in the context of diffusion theory, the analysis of Chapter I applies and it is possible to identify:

$$\bar{P}_\infty(k^2) = \prod_{j=1}^{N-1} \frac{p_j}{1 + L_j^2 k^2} = p\lambda(k^2) \tag{IV.24}$$

as the Fourier transform of the slowing-down kernel. Inverting the Fourier transform, the slowing-down kernel itself is the convolution:

$$P_\infty(|\mathbf{r} - \mathbf{r}'|) = \int d^3 r_{N-1} \int \ldots \int d^3 r_1 \left[\prod_{j=1}^{N-2} p_j \frac{\exp\{-|\mathbf{r}_j - \mathbf{r}_{j-1}|/L_j\}}{4\pi L_j^2 |\mathbf{r}_j - \mathbf{r}_{j-1}|} \right] p_{N-1} \frac{\exp\{-|r_{N-1} - r|/L_{N-1}\}}{4\pi L_{N-1}^2 |\mathbf{r}_{N-1} - \mathbf{r}|}$$

As the fast groups are narrowed uniformly one can approximate Equation (IV.24) by

$$\lim_{N \to \infty} \bar{P}_\infty(k^2) = \lim_{N \to \infty} \prod_{j-1}^{N-1} \frac{p_j}{1 + L_j^2 k^2} \approx \lim_{N \to \infty} \prod p_j (1 - L_j^2 k^2)$$

$$\approx \lim_{N \to \infty} \exp\left(- \sum_{j=1}^{N-1} L_j^2 k^2 \right) \prod_{j=1}^{N-1} p_j = \exp\left\{ -k^2 \sum_{j=1}^{\infty} L_j^2 \right\} \prod_{j=1}^{N-1} p_j \tag{IV.25a}$$

which one may readily recognize as the Fourier transform of the Gaussian Age–Diffusion kernel. Therefore, by comparison with the Age–Diffusion result it follows that

$$\tau = \sum_{j=1}^{\infty} L_j^2 = \sum_{j=1}^{\infty} \frac{D_j}{\Sigma_{Rj} + \Sigma_{aj}} \tag{IV.25b}$$

From this we may infer that

$$\tau_j = \tau(u_{j-1} \to u_j) = \frac{D_j}{\Sigma_{Rj} + \Sigma_{aj}} \tag{IV.25c}$$

Equations (IV.21) and (IV.25c) give two relations among the three constants required for group j: D_j, Σ_{Rj}, Σ_{aj}. It is essential that these relationships be obeyed if the multigroup equations are to predict the correct results.

A third equation for the group constants may be obtained as follows. The Fermi age formula:

$$\tau_{th} = \int_0^{u_T} \frac{D(u')}{\xi\Sigma_s(u')} \mathrm{d}u' \tag{IV.25d}$$

can be approximated by the sum

$$\tau_{th} \approx \sum_{j=1}^{\infty} \frac{D_j(u_j)}{\xi\Sigma_s(u_j)} \Delta u_j \tag{IV.25e}$$

since the Δu_j are small. Comparison of Equation (IV.25e) with Equation (IV.25b) leads to

$$\Sigma_{Rj} = \frac{\xi\Sigma_s(u_j)}{\Delta u_j} = \frac{\xi\Sigma_{sj}}{\Delta u_j} \tag{IV.25f}$$

as a consistent definition of the removal cross section. If $\Sigma_s(u)$ is constant, as it frequently is to good aproximation, the group constants are determined.

That Equation (IV.25f) is consistent with the Age–Diffusion formula for the resonance escape probability may be seen as follows. For the case of many groups we have:

$$p = \prod_{j=1}^{\infty} p_j = \prod_{j=1}^{\infty} \frac{\Sigma_{Rj}}{\Sigma_{Rj} + \Sigma_{aj}} = \prod_{j=1}^{\infty} \left[1 - \frac{\Sigma_{aj}}{\Sigma_{Rj} + \Sigma_{aj}}\right] \approx 1 - \sum_{j=1}^{\infty} \frac{\Sigma_{aj}}{\Sigma_{Rj}} \tag{IV.26a}$$

again, in the limit of small absorption. On the other hand, in the Age–Diffusion approximation we had

$$p = \exp\left\{-\int_0^{u_T} \frac{\Sigma_a(u')}{\xi\Sigma_s(u')} \mathrm{d}u'\right\} \tag{IV.26b}$$

By approximating as before, we have

$$p \approx 1 - \int_0^{u_T} \frac{\Sigma_a(u')}{\xi\Sigma_s(u')} \mathrm{d}u' \approx 1 - \sum_{j=1}^{\infty} \frac{\Sigma_a(u_j)}{\xi\Sigma_s(u_j)} \Delta u_j \tag{IV.26c}$$

It is readily seen that the two results [Equations (IV.26a, IV.26c)] are consistent with Equation (IV.25f).

C. Group Constants

7. Fast Group—Analytical Methods

In the previous section, it was shown that, at least in the limit of very small groups, the following formulae give reasonable estimates of the constants required for multigroup calculations:

$$\Sigma_{Rj} = \frac{\xi \Sigma_{sj}}{\Delta u_j} \tag{IV.27a}$$

$$\Sigma_{aj} = \Sigma_{Rj} p_j = \frac{\xi \Sigma_{sj}}{\Delta u_j} p(u_{j-1} \to u_j) \tag{IV.27b}$$

$$D_j = \Sigma_{Rj} \tau_j = \frac{\xi \Sigma_{sj}}{\Delta u_j} \tau(u_{j-1} \to u_j) \tag{IV.27c}$$

where Σ_{sj} can be estimated by a simple lethargy average of the cross section since the scattering cross section is generally nearly constant. An analysis similar to that of the previous section, which allows for fission in groups other than the thermal group would show:

$$\varepsilon - 1 = \nu \sum_j \frac{\Sigma_{fj}}{\Sigma_{Rj} + \Sigma_{aj}} \tag{IV.27d}$$

From which one may infer

$$(\varepsilon - 1)_j = \frac{\nu \Sigma_{fj}}{\Sigma_{Rj} + \Sigma_{aj}} \tag{IV.27e}$$

$$\Sigma_{fj} \approx (\varepsilon - 1)_j \Sigma_{Rj} = \frac{\xi \Sigma_{sj}}{\Delta u_j} (\varepsilon - 1)_j \tag{IV.27f}$$

where $(\varepsilon - 1)_j$ is the ratio of the slowing-down density at u_j to that at u_{j-1} in an infinite medium of the reactor material and in the absence of a capture.

Although the above estimates were obtained for small groups, the analysis should hold for larger groups as well. An alternate, and somewhat intuitive, method of deriving Equation (IV.27a) was given by Wigner [12]. In crossing a lethargy interval of width Δu_j, a neutron can be expected to make $\Delta u_j/\xi$ slowing-down collisions. Now since the scattering collision density in group j is $\Sigma_{sj}\phi_j$,

$$\frac{\Sigma_{sj}\phi_j}{\Delta u_j/\xi} = \frac{\xi \Sigma_{sj}\phi_j}{\Delta u_j} \tag{IV.28}$$

[12] A. M. Weinberg and E. P. Wigner, *op. cit.*, p. 386.

should be a reasonable estimate of the number of neutrons leaving the lethargy interval Δu_j. Hence, by definition, $(\xi\Sigma_{sj}/\Delta u_j)$ should be a reasonable estimate of the removal cross section, in agreement with Equation (IV.27a). The remaining group constants Equations (IV.27b, c, d) can be arrived at in a similar manner.†

It has also been suggested by Wigner [13] that for thermal reactors it is a reasonable procedure to average basic cross section data over a $1/E$ spectrum rather than compute the age and resonance escape probability from the somewhat more complicated formulae suggested above. Thus, one would have

$$\Sigma_{aj} = \int_{\Delta u_j} \Sigma_a(u')\, \mathrm{d}u'/\Delta u_j \tag{IV.29a}$$

$$\Sigma_{fj} = \int_{\Delta u_j} \Sigma_{fj}(u')\, \mathrm{d}u'/\Delta u_j \tag{IV.29b}$$

which is equivalent to replacing the true resonance integral, I, by the infinite dilution resonance integral I_∞ as given in Chapter II. The diffusion coefficient is simply

$$D_j = \int_{\Delta u_j} D(u')\, \mathrm{d}u'/\Delta u_j \tag{IV.30a}$$

in this approximation. In this scheme the age $\tau(u_{j-1} \to u_j)$ is represented by

$$\tau(u_{j-1} \to u_j) = \frac{\int_{\Delta u_j} D(u')\, \mathrm{d}u'}{\xi\Sigma_{sj}} \tag{IV.30b}$$

which is equivalent to the approach used earlier in the Age–Diffusion approximation if $\xi\Sigma_s$ is constant in lethargy. It is clear, however, that errors may be introduced by using these approximations, e.g., if the neutron spectrum is not $1/E$.

8. Fast Group Constants—Numerical Methods

Group constants obtained by the methods just given are sometimes useful for survey type calculations but they are generally unsatisfactory for more detailed calculations. For example, in a

† These formulae are valid for large groups only if Σ_s is nearly constant in a group. Thus, if Σ_s is relatively large near the low energy end of the group, Σ_R should be somewhat larger than $(\xi\Sigma_s/\Delta u_j)$. This was pointed out to the authors by A. B. Reynolds.

[13] *Ibid.*

complex reactor composed of many materials, cross sections vary considerably from point to point as a result of changes in material composition and temperature effects. Hand calculation of position-dependent cross sections by any but the simplest methods is nearly impossible. Furthermore, the cross sections given previously are based on an Age–Diffusion approach; one cannot obtain reasonable results for hydrogen moderated systems by this method.

In much numerical work for obtaining group constants (*spectral* codes), use is made of the Fourier transformed P_1 or B_1 equations, with modifications (*AD*, *SG*, *GG*, etc.) appropriate to the system under consideration. Although this is not strictly a legitimate procedure, it is frequently employed with considerable success. For the Fourier transform variable **k**, one must then use some sort of a buckling. Since the terms containing the transform variable are those that arise out of the spatial dependence of the flux and are therefore those connected with the leakage, the buckling should be chosen so as correctly to represent the leakage.† As used in this manner, the buckling should be lethargy (or group) dependent. If, for example, one is interested in calculating the core group constants for a reflected reactor, the buckling should represent the *net* leakage from the core to the reflector in the group under consideration. It may turn out in some cases that the net leakage is into the core, for example, at thermal energies in a small highly-reflected reactor; a negative value of B^2 may then be indicated. Generally, it is actually difficult to calculate the net leakage so that one frequently uses either a two group calculation and takes as a material buckling the positive root of the criticality condition or uses a bare reactor material buckling. This value is then applied to all fast groups.

Then, a many-group calculation (sometimes more than 200 groups are used) is made and the multigroup Fourier transformed fluxes, currents and slowing-down densities are computed. Cross sections for this problem are usually obtained by numerical averaging of the true cross sections (which is acceptable because of the relatively small size of the groups) except for the case of resonance capture, for which various resonance integral formulae or the results of special calculations are often applied.

† These results may also be obtained by averaging the P_1 equations over a region.

Having obtained the spectrum, it is relatively simple to obtain the required group constants. From Fick's law, written as

$$\phi_1(k, u) = iBD(u)\,\phi_0(k, u) \tag{IV.31a}$$

one obtains, quite simply

$$D_j = -\frac{i}{B}\frac{\int\limits_{\Delta u_j}\phi_1(u')\mathrm{d}u'}{\int\limits_{\Delta u_j}\phi_0(u')\,\mathrm{d}u'} = \frac{1}{3}\frac{\int\limits_{\Delta u_j}\frac{\phi_0(u')\,\mathrm{d}u'}{\Sigma_{tr}(u')}}{\int\limits_{\Delta u_j}\phi_0(u')\,\mathrm{d}u'} \tag{IV.31b}$$

The indicated integrals are approximated by sums over many of the smaller lethargy groups. Removal cross sections are obtained from the definition

$$\Sigma_{Rj}\int\limits_{\Delta u_j}\phi_0(u')\,\mathrm{d}u' = q(u_j) \tag{IV.31c}$$

which insures that $\Sigma_{Rj}\phi_j$ will be the slowing-down density out of each group. Should the various groups be coupled to groups not adjacent to themselves, additional problems are raised: these may usually be avoided by selecting group limits carefully. If this is not possible, hydrogen (and possibly deuterium) cross sections may be treated as inelastic cross sections, thus circumventing the problem. Inelastic multigroup cross sections $\sigma_{m,\,j\to k}$ are evaluated by simply calculating the number of transferred directly to group k from group j by element m and dividing by ϕ_j. Absorption and/or fission cross sections are found by direct averaging.

$$\Sigma_{aj} = \frac{\int\limits_{\Delta u_j}\Sigma_a(u')\,\phi_0(u')\,\mathrm{d}u'}{\int\limits_{\Delta u_j}\phi_0(u')\,\mathrm{d}u'} \tag{IV.31d}$$

The scheme described here is similar to that employed in the MUFT [14] and other codes, which have been used quite successfully in reactor design. Although the physical basis of such codes is far from rigorous, the group constants obtained from them are generally more accurate than those obtained by the simple methods given in the previous section.

Group constants have been complied by Argonne National Laboratory [15] for single elements, in 30 groups. For few group

[14] H. Bohl, Jr., E. M. Gelbard, G. H. Ryan, *MUFT-4, Fast Neutron Spectrum Code*, Report WADP-TM-22 (1957).

[15] *Reactor Physics Constants*, ANL 5800, p. 60 *et seq.*

problems, constants may be obtained by averaging these compiled results over an assumed flux energy spectrum, for example, an "$1/E$" spectrum. Results obtained in this way are not as accurate as those obtained numerically, but they have the obvious advantage of ease of use.

It was mentioned in Section 4 that it has been common practice, especially among fast reactor designers, to average cross sections for each element over an assumed spectrum once and for all and to use the cross sections set developed for all reactor calculations. The inadequacy of this procedure is indicated by the fact that there are as many cross section sets in existence as there are reactor design groups. Clearly, the same spectrum cannot be used for all reactors as this spectrum is itself quite sensitive to the composition of the system, particularly if some of the materials possess many resonances. For example, the average scattering cross section of sodium (which has several hundred scattering resonances below 1 MeV) may vary by a factor of two as the composition of a system changes [16].

While this method is unsatisfactory, it is expensive and inconvenient to use a spectral code calculation before every reactor calculation. Hence a combination of the "spectral code" and "cross section set" methods is often used. That is, one uses a spectral code to compute cross sections for individual elements for a "base reactor" and then uses the cross section set so obtained for calculations involving reactors not very different from the base reactor. In these calculations average cross sections are computed by the methods given above with the exception of the transport cross section. We have seen in Chapter I that the transport cross section must be averaged over the current so that the average transport cross section for element m in group j is given by

$$\sigma^m_{trj} = \frac{\displaystyle\int_{\Delta u_j} \frac{\sigma^m_{tr}(u)}{\Sigma_{tr}(u)}\,\phi(u)\,du}{\displaystyle\int_{\Delta u_j} \frac{1}{\Sigma_{tr}(u)}\,\phi(u)\,du} \qquad \text{(IV.31 e)}$$

which is in agreement with Equation (IV.31 a).

[16] H. Hummel and A. Rago, *Proceedings of the International Conference on Fast and Intermediate Reactors*, Vienna, 1961.

Thus far, three methods for obtaining group constants have been discussed. The first is to average microscopic data over some guess at the reactor spectrum [e.g., Equations (IV.14e, f)]. The second is to make use of integral data such as p_j and τ_j for the jth group, e.g., Equations (IV.21), (IV.25c). The third, just described, is in some sense a combination of the first two methods.

A fourth possibility is to go back to Equations (IV.15h or j) and to choose ω_1 and ω_2 by "guess and by gosh". Ehrlich and Hurwitz use linear relations among slowing-down densities, Equation (IV.15j):

$$q_j(\mathbf{r}) = \omega_1 q(r, u_{j-1}) + \omega_2 q(\mathbf{r}, u_j) \tag{IV.32a}$$

and they mention three possible "schemes" for choosing ω_1 and ω_2

$$\text{(a)} \qquad \omega_1 = \omega_2 = \tfrac{1}{2}$$

$$\text{(b)} \qquad \omega_1 = f, \quad \omega_2 = 0 \tag{IV.32b}$$

$$\text{(c)} \qquad \omega_1 = \frac{(\xi\Sigma_s)_j^{\mathrm{I}}}{(\xi\Sigma_s)_j}\,\omega_1, \quad \omega_2 = \frac{(\xi\Sigma_s)_j^{\mathrm{O}}}{(\xi\Sigma_s)_j}\,\omega_2'$$

Here, the superscript "I" refers to the quantity evaluated at u_{j-1} (that is, "into" the group) and "O" means evaluated at u_j ("out" of the group). The last scheme, the so-called "Oak Ridge Scheme", is equivalent to assuming a linear relation among the fluxes rather than the slowing-down densities [Eqn. (IV.15h)]. If $\xi\Sigma_s$ is essentially constant, there is little difference. Clearly, another possible scheme is to modify (b) above to a flux relation

$$\text{(d)} \qquad \omega_1 = f\frac{(\xi\Sigma_s)_j^{\mathrm{I}}}{(\xi\Sigma_s)_j}, \quad \omega_2 = 0 \tag{IV.32c}$$

Note that in schemes (b), (c), and (d) the ω are, in general, different for each group and for each material region of the reactor. Schemes (b) and (d) have been found by experience to be generally superior although care is required in choosing f. After solving the multigroup problem one should check that the final solution is consistent with the choice of f. This is accomplished by plotting the group fluxes (or slowing-down densities) integrated over a given region, and passing a smooth curve through them. One can then check whether or not the choice was reasonable for the given group and region by comparing the average value in the group with the value at the end of the group. It is easy to see that in groups and regions containing sources, f should be greater than unity, whereas it should be less than unity when there are no sources if there is absorption or leakage.

If one of the above schemes is chosen, it is simple to express $\Sigma_{j\to j+1}$ in terms of the ω. The algebra will not be reproduced here. However, the computation of the $\Sigma_{j\to j+1}$ from basic cross section data as is described in Section I, is seen to be equivalent to still another scheme for the jth group:

$$\omega_1 = 0, \quad \omega_2 = (\xi\Sigma_s)_j/\Sigma_{j\to j+1} \tag{IV.32d}$$

In general, the group constants should be chosen so as to conserve neutrons [17]. Thus

$$(\Sigma_{aj} + D_j B^2)\,\phi_j\,\Delta u_j = \frac{NI_{\mathrm{eff}\,j}\,\mathrm{e}^{-B^2\tau_j}}{(\xi\Sigma_s)_j}\,q_j^I \tag{IV.32e}$$

where $I_{\mathrm{eff}\,j}$ is the effective resonance integral for the jth group, N is the number density of absorber atoms and τ_j is the age for the jth group.

Then from Equations (IV.32a), (IV.32e) and the fact that $q_j^{\mathrm{O}} = q_j^{\mathrm{I}}\,NI_{\mathrm{eff}j}/(\xi\Sigma_s)_j\mathrm{e}^{-B^2\tau_j}$ we find

$$\Sigma_{aj} = \frac{\dfrac{NI_{\mathrm{eff}\,j}}{\Delta u_j}\,\mathrm{e}^{-B^2\tau_j}}{\omega_1 + \omega_2\,\dfrac{NI_{\mathrm{eff}\,j}}{(\xi\Sigma_s)_j}\,\mathrm{e}^{-B^2\tau_j}} - D_j B^2 \tag{IV.32f}$$

This may be taken as a consistency relation between D_j, $(\xi\Sigma_s)_j$, and Σ_{aj} or it may be taken as a definition of Σ_{aj} if the other two group cross sections are assumed known. Note that in Scheme (b) above, in the limit $B^2 \to 0$, Σ_{aj} is defined as

$$\Sigma_{aj} = \frac{NI_{\mathrm{eff}\,j}}{f\Delta u_j} \tag{IV.32g}$$

which may differ appreciably from a frequently used approximation in which Σ_{aj} is taken to be $NI_{\mathrm{eff}\,j}/\Delta u_j$.

The problem of choosing effective cross sections to represent resonance processes in the small groups used to integrate the Boltzmann equation numerically is very similar to the derivation of Σ_{aj} above; $I_{\mathrm{eff}\,j}$ for the small group is related to Σ_{aj} in a manner which depends upon the actual numerical integration scheme of the equation.

[17] P. F. ZWEIFEL and G. BALL. *Proceedings of the International Conference on Fast and Intermediate Reactors*, Vienna, 1961.

9. Thermal Group Constants

As was the case with the fast group, the greatest obstacle to obtaining thermal group constants is usually a lack of knowledge of the exact neutron energy spectrum. Considerable work in this field, called thermalization, has been done in recent years, but we shall only give a brief review here. For a more complete discussion of this field see the work of Corngold [18].

As explained in Chapter I, neutrons in equilibrium with the moderator would have the Maxwell–Boltzmann spectrum (i.e., the canonical distribution) peculiar to the moderator temperature. However, the presence of sources and sinks means that the neutron distribution will not be canonical. In the language of statistical mechanics, the neutron distribution must be described in terms of a Grand Canonical Ensemble [19] which merely means that the statistical distribution function (density in phase space) involves not only the energy of the system but the number of particles (neutrons in our case) as well. This, in general, leads to an energy spectrum which is shifted upwards in energy or "hardened".

Thus, to obtain the correct thermal energy spectrum over which to average cross sections, it is necessary to solve the Boltzmann equation as a function of energy at thermal energies. This is usually done in the diffusion approximation with energy space separability assumed, the spatial part being taken as a fundamental mode of the Helmholtz equation. This merely modifies the infinite medium equation by adding a term DB^2 to the absorption cross section [20].

The solution of this equation is more complicated than is the case of the slowing-down problem for three important reasons:

(1) In chemically bound moderators (e.g., water) the neutron energy may well be below the binding energy of the molecule, which is usually of the order of a few electron volts. Energy transferred from a neutron to a nucleus is then insufficient to break the chemical bond so that the neutron interacts with the entire molecule.

(2) In solids there are crystalline effects which are related to the effect mentioned above. At very low energies the wave properties of the neutron enter and diffraction effects come into play [21].

[18] N. Corngold, *Proc. Neutron Physics Conf.*, at Boyne Mtn., Mich., 1962.
[19] See, e.g., T. Hill, *Statistical Mechanics*, McGraw–Hill, 1956.
[20] See, e.g., D. Parks, *Nucl. Sci. Eng.* **9**, 430 (1961).
[21] M. Nelkin, *Nucl. Sci. Eng.* **2**, 199 (1957).

(3) The presence of up-scattering (scattering with increase in energy) complicates the equation considerably.

The simplest method of obtaining thermal group constants is simply to average the parameters over a Maxwellian spectrum corresponding to the physical temperature. Since tabulations of cross section are usually made for $E = 0.025$ eV (kT at 298°K—corresponding to a velocity of 2200 m/sec) the average absorption or fission cross sections are most often expressed in terms of σ_a (2200 m/sec):

$$\sigma_{aT} = \sigma_a(2200 \text{ m/sec}) \left(\frac{\pi}{4}\right)^{\frac{1}{2}} \left(\frac{T_0}{T}\right)^{\frac{1}{2}} g(T) \qquad \text{(IV.33a)}$$

where g is a factor which corrects for non-$1/v$ behavior and is unity for pure $1/v$ cross sections. Values of g have been tabulated for various isotopes [22]. In Chapter I it was found that the best procedure for determining a thermal diffusion coefficient is to average the transport cross section over the current and set

$$D_T = \frac{1}{3\Sigma_{trT}} \qquad \text{(IV.33b)}$$

At thermal energies scattering is almost entirely elastic and isotropic in the center of mass system (again, ignoring crystalline effects), so we may write:

$$\Sigma_{tr} = \Sigma_a + \Sigma_s\left(1 - \frac{2}{3A}\right) \qquad \text{(IV.33c)}$$

But there is still another point to be considered, namely the chemical binding described earlier which makes both σ_s and A effectively energy dependent. In the first Born approximation one can show that the scattering cross section is given by [23]

$$\sigma_s = \mu^2\sigma_0 = \left(\frac{A}{A+1}\right)^2 \sigma_0 \qquad \text{(IV.34)}$$

where μ is the reduced mass of the neutron–nuclear system and σ_0 is the scattering cross section of the tightly bound atom. For bound

[22] C. H. WESTCOTT, *Effective Cross Section Values for Well Moderated Thermal Reactor Spectra*, Atomic Energy of Canada Ltd., Report CRRP680 (1957).

[23] J. M. BLATT and V. WEISKOPF, *Theoretical Nuclear Physics* (1952). The first application of this formula to the calculation of Σ_{trT} was by A. RADKOWSKY.

hydrogen, this effect causes the scattering cross section to go to 80 barns at low energies whereas it is only 20 barns at high energies at which the hydrogen acts as if it is free. In heavier elements the effect is less dramatic. One may then use Equation (IV.34) and empirical cross section data to calculate an effective mass A_{eff} which may be used in Equation (IV.33c) for purposes of calculating Σ_{tr} as a function of energy at low energies. Calculations using this prescription have been made and are compiled in the literature [24].

The first study of the effect of thermal motions of the moderator on the thermal spectrum was made by Wigner and Wilkins [25], who studied the effect of the thermal motion of the atoms of a moderator gas on the differential scattering cross section. They arrived at a linear integral equation for the spectrum which, unfortunately, is not generally soluble in closed form. We quote only the result. For $E < E'$:

$$\int F(E' \to E)\, \phi(E')\, \mathrm{d}E' = \left[\chi(E) + \frac{\Delta}{2A}\right]\phi(E) \qquad \text{(IV.35)}$$

where

$$F(E' \to E) = \frac{\mathrm{e}^{-\beta(E-E')}}{(2\alpha)^{\frac{1}{2}}}[I(\gamma s - \gamma A t) + I(\gamma A s + \gamma t)] + \frac{\mathrm{e}^{\beta(E-E')}}{(2\alpha)^{\frac{1}{2}}}[I(\gamma s - \gamma A t) - I(\gamma A s - \gamma t)] \qquad \text{(IV.35a)}$$

$I(x)$ is the error function

$$I(x) = \frac{2}{\pi^{\frac{1}{2}}}\int_0^x \mathrm{e}^{-y^2}\,\mathrm{d}y \qquad \text{(IV.35b)}$$

and

$$s = \tfrac{1}{2}(E^{\frac{1}{2}} + E'^{\frac{1}{2}}) \quad t = \tfrac{1}{2}(E'^{\frac{1}{2}} - E^{\frac{1}{2}})$$

$$\beta = \frac{1}{kT} \qquad \gamma = (AkT)^{-\frac{1}{2}} \qquad \text{(IV.35c)}$$

$$\Delta = 2A\frac{\Sigma_a(kT)}{\Sigma_s}$$

$$\chi(E) = \left[(2E)^{\frac{1}{2}} + \frac{kT}{A(2E)^{\frac{1}{2}}}\right] I(\gamma A E^{\frac{1}{2}})$$

[24] C. D. Petrie, M. L. Storm, and P. F. Zweifel, *Nucl. Sci. Eng.* **2**, 728 (1957).

[25] E. P. Wigner and J. E. Wilkins, Report AECD 2275.

For $E > E'$ one can calculate the kernel from the principle of detailed balance:

$$\sigma(E' \to E)\, M(E') = \sigma(E \to E')\, M(E)$$

For the case of hydrogen, or for heavy nuclei, this integral equation can be reduced to a differential equation which may be solved analytically for the case of heavy nuclei. Of course, the advent of high speed digital computers has made possible the numerical solution of Equation (IV.35). This leads to the possibility of obtaining thermal group constants by averaging over a "Wigner–Wilkins spectrum". This was the basis of early codes but more recent work has utilized better models for the calculation of scattering kernels. An extensive literature has grown up in this field but, unfortunately, there is not even room to review it here. For some recent work see the work cited in [26].

For small absorption, the Wigner–Wilkins spectrum must reduce to the Maxwell–Boltzmann distribution. As an aid to hand calculation, it is useful to be able fit this spectrum to a hardened Maxwellian, that is, a Maxwellian–Boltzmann distribution with a "neutron temperature" higher than the physical temperature. K. Cohen [27] suggested that a relation of the form

$$T_{\text{eff}} = T\left[1 + C\,\frac{\Sigma_a(kT)}{\xi\Sigma_s}\right] \tag{IV.36}$$

might be used. Deutsch [28] suggests that a value of $C = 0.75$ be used whereas Coveyou, Bate, and Osborn, [29] using Monte Carlo techniques and a least square fit, find 0.46 as the best value of C.

These simplified calculations are frequently valuable as an aid in hand calculation, especially in cases for which the use of computers is not justified.

[26] *Proc. Brookhaven Conf. on Neutron Thermalization*, BNL-719, 1962.

[27] *Research Reactors, Physics*, New York, McGraw–Hill, p. 375 (1955).

[28] R. W. Deutsch, *Nucleonics* **15**, 47 (1957).

[29] P. R. Coveyou, R. R. Bate, and R. K. Osborn, *J. Nucl. Energ.* **2**, 153 (1956).

D. Solution of the Multigroup Equations

10. A Formal Solution

In principle, the method applied to the solution of the one-region problem could be extended to any multi-region problem in which each region is homogeneous. However, the determinants encountered when this approach is used are too large to be evaluated even on a large computer. Other methods have been found to be faster.

As we shall be dealing with large sets of linear equations, it will be convenient to employ matrix notation in this section. A review of matrix analysis can be found in many standard texts. For applications of matrix theory to neutron problems, see the report by Bilodeau and Hageman [30].

Assuming that the group constants are known, the multigroup equations can be written:

$$- \nabla \cdot D_j \nabla \phi_j(\mathbf{r}) + \Sigma_{tj}(\mathbf{r}) \phi_j(\mathbf{r}) = \sum_{i=1}^{j-1} \Sigma_{ij}(\mathbf{r}) \phi_i(\mathbf{r}) + \chi_j \nu \sum_{i=1}^{N} \Sigma_{fi} \phi_i \tag{IV.37}$$

where $\Sigma_{ij}(\mathbf{r})$ is the cross section for transfer from group i to group j as discussed earlier (inelastic scattering will be included in $\Sigma_{ij}(\mathbf{r})$ for the present, although it may prove convenient in some cases to treat it as indicated earlier). The simplifying assumption of coupling only to adjacent groups is not made, but will obtain if there is no hydrogen present and inelastic scattering is neglected or if these are treated separately. Also, fission is allowed to occur in all groups; χ_j denotes the fraction of the fission spectrum which falls into the jth group:

$$\chi_j = \int_{\Delta u_j} f(u')\, \mathrm{d}u'$$

where $f(u)$ is the fission spectrum discussed in Chapter I. It is assumed, however, that no up-scattering occurs, i.e., that thermal neutrons are treated as one group, denoted by the subscript N.

[30] B. Bilodeau and L. Hageman, *A Survey of Numerical Methods in the Solution of Diffusion Problems*, Report WAPD-TM-64 (1957).

By defining a flux vector† $\mathbf{\Phi}$ as

$$\mathbf{\Phi}(\mathbf{r}) = \begin{pmatrix} \phi_1(\mathbf{r}) \\ \phi_2(\mathbf{r}) \\ \vdots \\ \phi_N(\mathbf{r}) \end{pmatrix} \tag{IV.38a}$$

and the operator matrices

$$\boldsymbol{M} = \begin{pmatrix} -\nabla\cdot D_1\nabla + \Sigma_{t1} & 0 & 0 & & \\ 0 & -\nabla\cdot D_2\nabla + \Sigma_{t2} & & & \\ & 0 & \cdot & & \\ & & & \ddots & \\ & & & & -\nabla\cdot D_N\nabla + \Sigma_{aN} \end{pmatrix} \tag{IV.38b}$$

$$\boldsymbol{F} = \begin{pmatrix} \chi_1\Sigma_{f1} & \chi_1\Sigma_{f2} & \cdots & \\ \chi_2\Sigma_{f1} & \chi_2\Sigma_{f2} & & \\ & & \ddots & \\ & & & \chi_N\Sigma_{fN} \end{pmatrix} \tag{IV.38c}$$

and

$$\boldsymbol{\Sigma} = \begin{pmatrix} 0 & 0 & 0 & & \\ \Sigma_{12} & 0 & & & \\ \Sigma_{13} & \Sigma_{23} & \ddots & & \\ \vdots & & \ddots & 0 & 0 \\ \Sigma_{1N} & & & \Sigma_{N-1N} & 0 \end{pmatrix} \tag{IV.38d}$$

it is possible to write the multigroup equations in matrix form:

$$\boldsymbol{M\Phi} - \boldsymbol{\Sigma\Phi} - \nu_c\boldsymbol{F\Phi} = 0 \tag{IV.38e}$$

Proceeding formally, Equation (IV.38e) can be written as

$$\frac{1}{\nu_c}\boldsymbol{\Phi} = (\boldsymbol{M} - \boldsymbol{\Sigma})^{-1}\boldsymbol{F\Phi} = \boldsymbol{L\Phi} \tag{IV.38f}$$

† Again "vector" in the mathematical sense of an $N \times 1$ matrix.

by operating with the matrix $(\boldsymbol{M} - \boldsymbol{\Sigma})^{-1}$, the inverse matrix to $(\boldsymbol{M} - \boldsymbol{\Sigma})$. That such a matrix exists can be shown and for the moment it is assumed to be a known matrix. Then one needs only to find the eigenvalues $(1/\nu)$ of the matrix $\boldsymbol{L} = (\boldsymbol{M} - \boldsymbol{\Sigma})^{-1} \boldsymbol{F}$. Since all of the matrices involved are of rank N and are non-singular, there will be N eigenvalues, each with a corresponding eigenvector. The largest of these, ν_0, will later be shown to be the significant one and k_{eff} is the ratio of the physical value of ν (e.g., 2.42 for U-235) to this value. The eigenvector belonging to this eigenvalue, i.e., the function satisfying the relation

$$\boldsymbol{L\Phi} = \frac{1}{\nu_0}\boldsymbol{\Phi}$$

is then the steady state flux in the reactor.

All of this is quite formal, but the procedure could be used. In practice, the method is intractable because the problem of inverting the matrix $(\boldsymbol{M} - \boldsymbol{\Sigma})$ is a difficult one. Since $\boldsymbol{M}$ involves the differential operator $\nabla \cdot D\nabla$, the inverse matrix $(\boldsymbol{M} - \boldsymbol{\Sigma})^{-1}$ will involve the inverse of this operator, which must be interpreted in terms of the Green's function. That is

$$(-\nabla \cdot D\nabla + \Sigma_t)^{-1}\,\psi(\mathbf{r}) = \int_R G(\mathbf{r}, \mathbf{r}')\,\psi(\mathbf{r}')\,d^3r' \qquad \text{(IV.39a)}$$

where the Greens' function $G(\mathbf{r}, \mathbf{r}')$ satisfies the equation

$$-\nabla \cdot D\nabla G(\mathbf{r}, \mathbf{r}') + \Sigma_t G(\mathbf{r}, \mathbf{r}') = -\delta(\mathbf{r} - \mathbf{r}') \qquad \text{(IV.39b)}$$

with the boundary condition

$$G(\mathbf{r}, \mathbf{r}') = 0 \qquad \text{(IV.39c)}$$

for $\mathbf{r}$ on the extrapolated boundary of the reactor. The problem of finding the inverse matrix $(\boldsymbol{M} - \boldsymbol{\Sigma})^{-1}$ is thus made extremely difficult and this approach has never been used except for problems involving very few groups and regions. Other methods, which involve iterative procedures, are used more frequently and will be described in succeeding sections.

11. Source Iteration

While the above method is simple and straightforward in theory, it is quite impractical; other schemes must be found. In particular, one would like to have an iterative procedure that converges to the correct effective multiplication factor k_{eff} and the correct space-energy distribution of the flux.

Physically, if a reactor has a known neutron flux distribution in space and energy at time $t = 0$ one expects that this flux will grow, die out, or stay constant according to whether the reactor is super-critical, subcritical, or just critical. Moreover, at much later times the neutron distribution in space and energy is expected to become independent of time; its magnitude may increase or decrease, but the space-energy distribution should remain the same. At this stage each neutron generation will have the same space-energy distribution as the one preceding it but will be multiplied by a factor of k_{eff}. This is, in fact, an operational definition of k_{eff}.

Now one could use the multigroup equations to simulate this behavior. It is, however, somewhat inconvenient to work with the space-energy flux distribution itself as this involves knowledge of a great many quantities. Rather, one notes that the neutron source for the kth generation is

$$S^{(k)}(\mathbf{r}) = \nu_c \sum_{j=1}^{N} \Sigma_{fj}\phi_j^{(k-1)}(\mathbf{r}) \tag{IV.40a}$$

$\phi_j^{(k-1)}(\mathbf{r})$ being the $(k-1)$st generation flux in the jth group, and that $S^{(k)}(\mathbf{r})$ becomes independent of k for large k. Since the spatial shape of the source is energy independent, a considerable savings in effort can be effected by using it rather than the flux itself; only one spatial distribution need be assumed.

We define the source vector:

$$\mathbf{S} = \begin{pmatrix} S_1 \\ S_2 \\ \vdots \\ S_N \end{pmatrix} = \begin{pmatrix} \chi_1 \\ \chi_2 \\ \vdots \\ \chi_n \end{pmatrix} S(\mathbf{r}) = \chi S(\mathbf{r}) \tag{IV.40b}$$

By comparing Equation (IV.40b) with the definition of the matrix $\boldsymbol{F}$, Equation (IV.38c), we have

$$\mathbf{S} = \nu \boldsymbol{F}\boldsymbol{\Phi} \tag{IV.40c}$$

Operating on Equation (IV.38f) with the matrix $\boldsymbol{F}$, we find that the source vector $\mathbf{S}$ satisfies the matrix equation

$$\frac{1}{\nu_c}\mathbf{S} = \boldsymbol{F}(\boldsymbol{M} - \boldsymbol{\Sigma})^{-1}\boldsymbol{F}\boldsymbol{\Phi} = \boldsymbol{F}(\boldsymbol{M} - \boldsymbol{\Sigma})^{-1}\mathbf{S} = \boldsymbol{P}\mathbf{S} \tag{IV.40d}$$

where $\boldsymbol{P}$ is the matrix

$$\boldsymbol{P} = \boldsymbol{F}(\boldsymbol{M} - \boldsymbol{\Sigma})^{-1}$$

In particular, note that if $\boldsymbol{\Phi}$ is the eigenvector of $\boldsymbol{L}$ corresponding to the eigenvalue ν_0 then $\boldsymbol{F\Phi}$ is the eigenvector of $\boldsymbol{P}$ with the same eigenvalue. Thus, either the source or the multigroup fluxes can be used as an iteration quantities in solving the multigroup equations for a reactor. This leads us to the following scheme.

A guess at the neutron source distribution is made, say $S^{(0)}(\mathbf{r})$, and inserted into the multigroup equations. Starting with the first group and working down to the thermal group, one calculates a set of multigroup fluxes, from which a new source can be calculated using Equation (IV.40a). This is the source for the second generation; in carrying out this calculation, we have effectively followed the neutrons from the assumed source through one generation and found the distribution of fission events caused by them. This procedure is continued and is known as *source iteration* and it is physically reasonable that this procedure will converge. Mathematical proofs of the convergence of the method will be cited later. Within each group, iteration procedures are used to calculate the spatial distribution of the flux, i.e., to integrate numerically the group diffusion equation. Such iterations, to be described later, are known as *inner iterations* whereas the entire process of calculating the new source distribution from the previous one is known as an *outer iteration.* Hence, each outer iteration contains N inner iterations. In general, the procedures applied to the various iterations are different owing to the somewhat different nature of the two problems.

A similar procedure can be used for computing adjoint fluxes. Here the quantity guessed is F_H and the calculation proceeds by starting with the thermal group.

12. Outer Iterations

Outer iterations are, as explained above, used to calculate the eigenvalues of an equation of the type

$$\boldsymbol{L\Phi} = \lambda\boldsymbol{\Phi} \tag{IV.41}$$

which is Equation (IV.38f) written with $\lambda = 1/\nu$. Inner iterations are part of a method we shall later use to solve the diffusion equation within each group. Here we are dealing with an equation of the type

$$\mathbf{Q}\phi_j(\mathbf{r}) = T_j(\mathbf{r}) \tag{IV.42}$$

where Q is a space dependent operator and $T_j(\mathbf{r})$ is the source (fission and slowing down) in the jth group; $\phi_j(\mathbf{r})$ and $T_j(\mathbf{r})$ are to vanish on a boundary. Later, it will be shown how an equation of this type can be put into matrix form by means of finite difference techniques. The important point is that the source is an inhomogeneous term in Equation (IV.42). The types of problems solved by inner and outer iterations are thus distinctly different; the former involves an inhomogeneous differential equation (or, in finite differences, a set of inhomogeneous linear algebraic equations) whereas the latter is an eigenvalue problem. It is to be expected, therefore, that the techniques applied to each of these types of iterations will be different. Outer iterations are considered first.

Before proceeding, a brief review of some of the more important properties of matrices will be given and a few theorems will be quoted.

It is known that an $N \times N$ matrix generally possesses N different linearly independent eigenvectors each with its own eigenvalue. That is, there are N different N-dimensional column vectors which satisfy the equations

$$A\psi_n = \lambda_n\psi_n$$

where A is an $N \times N$ matrix, each eigenvector ψ_n corresponds to a definite eigenvalue λ_n (which may be complex and are not necessarily distinct).

Similarly, the adjoint of A, A^+ possesses a set of eigenvectors† defined by:

$$\psi_n^+A^+ = \mu_n\psi_n^+$$

with

$$\mu_n = \lambda_n^*$$

where the (*) denotes complex conjugation. When the vectors are properly normalized they have the property

$$(\psi_n^+, \psi_m) = \delta_{nm}$$

† The adjoint eigenvectors are $1 \times N$, i.e., row vectors.

Finally, the eigenvectors of A form a complete set (any vector with similar structure may be expanded in terms of them).

Further results which apply specifically to the multigroup operators have been derived by Habetler and Martino [31]. They showed that the multigroup operator (the matrix $\boldsymbol{L}$ here) possesses a real eigenvalue λ_0, which is larger in absolute value than any other eigenvalue. Moreover, this eigenvalue is non-degenerate (only one eigenfunction is associated with it); its eigenvector, which we shall call $\boldsymbol{\Psi}_0$, is the only one whose components (elements) are all non-negative throughout the reactor and its adjoint eigenvector, which satisfies the equation

$$\frac{1}{\nu_c}\boldsymbol{\Phi}_0^+ = \boldsymbol{\Phi}_0^+\boldsymbol{L}^+$$

is the only adjoint eigenvector whose elements are everywhere non-negative inside the reactor. These are the only eigenvectors which are acceptable on physical grounds as solutions to the critical reactor problem. Birkhoff and Varga [32] have proven similar results using the finite difference equations which will be given in Section 16.

13. Power Methods

With these preliminaries, we are ready to discuss particular means of carrying out outer iterations. The earlier discussion of iteration methods suggests the following *iteration scheme* known as the power method.

The eigenvalue problem that we wish to solve is given by Equation (IV.40d):

$$\boldsymbol{P}\mathbf{S} = \lambda\mathbf{S} \tag{IV.43a}$$

Now suppose one makes a guess at the function $S(\mathbf{r})$ of Equation (IV.40b); call it $S^{(0)}(\mathbf{r})$. Then

$$\mathbf{S}^{(0)} = \mathbf{S}^{(0)}(\mathbf{r})\,\chi \tag{IV.43b}$$

Now one calculates

$$\mathbf{S}^{(1)} = \boldsymbol{P}\mathbf{S}^{(0)} \tag{IV.43c}$$

[31] G. J. Habetler and M. A. Martino, *The Multigroup Diffusion Equations of Reactor Physics*, Report KAPL-1886 (July 1958).

[32] G. Birkhoff and R. Varga, Reactor Criticality and Non-Negative Matrices, *J. Soc. Ind. App. Math.* **6**, 354 (1959).

In practice this calculation consists of two parts. First one solves Equation (IV.38e) for the flux vector:

$$\mathbf{\Phi}^{(1)} = (M - \Sigma)^{-1}\mathbf{S}^{(0)} = \begin{pmatrix} \phi_0^{(1)}(\mathbf{r}) \\ \vdots \\ \phi_N^{(1)}(\mathbf{r}) \end{pmatrix} \tag{IV.43d}$$

i.e., one inserts the source into the multigroup equations and computes the resulting multigroup fluxes. Then one computes a new (first generation) source from these resulting fluxes:

$$\mathbf{S}^{(1)} = \nu \boldsymbol{F}\mathbf{\Phi}^{(1)} = \chi\nu \sum_j \Sigma_{fj}\phi_j^{(1)}(\mathbf{r}) \tag{IV.43e}$$

Since, by definition

$$\boldsymbol{P} = \boldsymbol{F}(\boldsymbol{M} - \mathbf{\Sigma})^{-1} \tag{IV.43f}$$

the above procedure is entirely equivalent to

$$\mathbf{S}^{(1)} = \boldsymbol{P}\mathbf{S}^{(0)} \tag{IV.43g}$$

i.e., source iteration. By repeating this procedure, i.e., using $\mathbf{S}^{(1)}$ as a guess at the source to compute $\mathbf{S}^{(2)}$ etc., we generate a series of fluxes. Clearly the kth iterate is

$$\mathbf{S}^{(k)} = \boldsymbol{P}\mathbf{S}^{(k-1)} = \boldsymbol{P}^k\mathbf{S}^{(0)} \tag{IV.43h}$$

In light of the physical arguments given above this procedure should converge to the source in the critical reactor. That is, after a sufficient number of iterations $S^{(k)}(\mathbf{r})$ should take the shape of $S(\mathbf{r})$ and

$$\lim_{k\to\infty} \frac{S^{(k)}(\mathbf{r})}{S^{(k-1)}(\mathbf{r})} = \lambda_0 \tag{IV.43i}$$

To demonstrate convergence, we first expand $\mathbf{S}^{(0)}$ in terms of the eigenvectors of $\boldsymbol{P}$.

$$\mathbf{S}^{(0)} = \sum_{n=1}^{N} \sigma_n \boldsymbol{\psi}_n \tag{IV.44a}$$

where the $\boldsymbol{\psi}_n$ satisfy

$$\boldsymbol{P}\boldsymbol{\psi}_n = \lambda_n \boldsymbol{\psi}_n \tag{IV.44b}$$

Substituting this expansion into the iteration formula (IV.43c) we have for the first iterate

$$\mathbf{S}^{(1)} = \sum_{n=1}^{N} \sigma_n \boldsymbol{P}\boldsymbol{\psi}_n = \sum_{n=1}^{N} \sigma_n \lambda_n \boldsymbol{\psi}_n \tag{IV.44c}$$

Repeating this process k times, we have

$$\mathbf{S}^{(k)} = \sum_{n=1}^{N} \sigma_n \lambda_n^k \psi_n = \lambda_0^k \sum_{n=1}^{N} \sigma_n \left(\frac{\lambda_n}{\lambda_0}\right)^k \psi_n \tag{IV.44d}$$

where λ_0 is the real eigenvalue guaranteed by the work cited above. Since λ_0 is larger in absolute value than any of the other λ_n, the eigenfunction corresponding to it will become dominant after a sufficiently large number of iterations, provided that the original guess was such that $\sigma_0 \neq 0$. However, since ψ_0 is the only eigenvector whose elements are all non-negative within the reactor, this can be guaranteed by making $S^{(0)}(\mathbf{r})$ non-negative throughout the reactor volume. Then, after a large number of iterations

$$\mathbf{S}^{(k)} \to \lambda_0^k \sigma_0 \psi_0 \tag{IV.44e}$$

and the eigenvalue λ_0 which is $1/\nu_c$) can be calculated from Equation (IV.43i). The calculation is stopped when the ratio $S^{(k)}(\mathbf{r})/S^{(k-1)}(\mathbf{r})$ is constant throughout the system to within some prespecified limit.

This scheme is a useful one and one sees that the rate of convergence, i.e., the number of iterations required for dominance of the ψ_0 term, is mainly determined by (λ_1/λ_0) where λ_1 is the eigenvalue of $\boldsymbol{P}$ with the second largest absolute value. To a lesser degree it is determined by the ratio of the expansion coefficients (σ_1/σ_0) so that there is an advantage in being able to supply an original source guess which is as accurate as possible. Such a guess might be obtained, for example, from a two group-two region problem.

There are serious drawbacks in the above method. The first is that a scale factor λ_0^k, which may be quite different from unity (e.g. $1/2.42 \cong 0.4$ for U-235), is introduced and may cause the source iterate to grow too small for handling on a computer if many iterations are required. The second drawback is that the convergence criterion employed requires comparison of $S^{(k)}(\mathbf{r})$ with $S^{(k-1)}(\mathbf{r})$ at each point.

The first problem may be overcome by supplying a guess at λ_0 (say λ_0^k) after each iteration. Then instead of using

$$\mathbf{S}^{(k)} = \boldsymbol{P}\mathbf{S}^{(k-1)} \tag{IV.45a}$$

as the source for the $(k + 1)$st iteration, we use $\mathbf{S}^{(k)}/\lambda^{(k)}$. Then instead of Equation (IV.44e) we would have

$$\mathbf{S}^{(k)} \to \frac{\lambda_0^k}{\lambda_0^{(1)}\lambda_0^{(2)} \cdots \lambda_0^{(k)}} \sigma_0 \psi_0. \tag{IV.45b}$$

after a sufficient number of iterations. Since the estimated eigenvalues are not likely to be much different from λ_0, the scale factor introduced is not likely to cause any difficulties. That this procedure converges was proved by Bilodeau and Hageman [33]. We shall not give the proof here as it is essentially the same as the one just given for the power method.

Several methods are available for computing estimates of the eigenvalues. We merely list them here. The simplest method is the Rayleigh quotient method:

$$\lambda_0^{(k)} = \frac{(P\mathbf{S}^{(k-1)}, \mathbf{S}^{(k-1)})}{(\mathbf{S}^{(k-1)}, \mathbf{S}^{(k-1)})} = \frac{(\mathbf{S}^{(k)}, \mathbf{S}^{(k-1)})}{(\mathbf{S}^{(k-1)}, \mathbf{S}^{(k-1)})} \tag{IV.45c}$$

A quite similar result is provided by the Schwartz quotient method

$$\lambda_0^{(k)} = \frac{(P\mathbf{S}^{(k-1)}, P\mathbf{S}^{(k-1)})}{(P\mathbf{S}^{(k-1)}, \mathbf{S}^{(k-1)})} = \frac{(\mathbf{S}^{(k)}, \mathbf{S}^{(k)})}{(\mathbf{S}^{(k)}, \mathbf{S}^{(k-1)})} \tag{IV.45d}$$

A technique which is used less frequently is the component sum method and is essentially similar to Equation (IV.43i):

$$\lambda_0^{(k)} = \frac{\int \mathbf{S}^{(k)}(\mathbf{r})\, d^3r}{\int S^{(k-1)}(\mathbf{r})\, d^3r} \tag{IV.45e}$$

that is, instead of comparing $S^{(k)}(\mathbf{r})$ with $S^{(k-1)}(\mathbf{r})$ pointwise, the integrals of these quantities over the reactor are used.

The second objection to the power method, i.e., the necessity of comparing $S^{(k)}(\mathbf{r})$ and $S^{(k-1)}(\mathbf{r})$ pointwise, is readily overcome in any of the above methods. One merely requires convergence of the eigenvalue rather than the source. That is, one iterates until two successible eigenvalue estimates agree to within some predetermined amount; this, however, does not give as accurate results as source convergence.

14. Adjoint Methods

In addition of the methods given above, there are several schemes which depend on the use of the adjoint flux [34]. One obvious advantage of using the adjoint flux is that one may use variational expressions for the eigenvalue; an accurate estimate of k_{eff} can be obtained with a minimum of computation. Another

[33] BILODEAU and HAGEMAN, *op. cit.*

[34] These were suggested by R. EHRLICH and H. HURWITZ, Jr., *Nucleonics* **12**, 23 (1954).

advantage is that these methods clearly necessitate calculation of the adjoint flux, which is, as we have seen, valuable for perturbation calculations. We have already described the source iteration method for adjoint calculations so that we shall limit ourselves in this section to a consideration of methods of computing the eigenvalue estimates.

Expressions (IV.45c) and (IV.45d) for the eigenvalue were derived by operating on Equation (IV.43h); Equation (IV.45c) results from taking the scalar product with $\mathbf{S}^{(k-1)}$ while Equation (IV.45d) results from taking the scalar product with $\boldsymbol{P}\mathbf{S}^{(k-1)}$. If now we take the scalar product of Equation (IV.43h) with $\mathbf{S}^{+(k)}$, the adjoint source (which is $\boldsymbol{F}_H = \boldsymbol{\Phi}^+\boldsymbol{F}^T$ where $\boldsymbol{F}^T$ is the transpose of the matrix $\boldsymbol{F}$ and $\boldsymbol{\Phi}^+$ is the adjoint flux vector), we arrive at

$$\lambda_0^{(k)} = \frac{(\mathbf{S}^{+(k-1)}, \boldsymbol{P}\mathbf{S}^{(k-1)})}{(\mathbf{S}^{+(k-1)}, \boldsymbol{P}\mathbf{S}^{(k-1)})} = \frac{(\boldsymbol{P}^+\mathbf{S}^{+(k-1)}, \mathbf{S}^{(k-1)})}{(\boldsymbol{P}^+\mathbf{S}^{+(k-1)}, \mathbf{S}^{(k-1)})} \quad \text{(IV.46a)}$$

which is the analog of Equation (IV.45c). Alternatively we could take the scalar product of Equation (IV.43h) with $\boldsymbol{P}^+\mathbf{S}^+$ to obtain

$$\lambda_0^{(k)} = \frac{(\boldsymbol{P}^+\mathbf{S}^{+(k-1)}, \boldsymbol{P}\mathbf{S}^{(k-1)})}{(\boldsymbol{P}^+\mathbf{S}^{+(k-1)}, \mathbf{S}^{(k-1)})} = \frac{(\mathbf{S}^{+(k)}, \mathbf{S}^{(k)})}{(\mathbf{S}^{+(k)}, \mathbf{S}^{(k-1)})} \quad \text{(IV.46b)}$$

Both of these expressions are variationally stationary, i.e., the error in λ_0 is proportional to the product of the errors in $\mathbf{S}$ and $\mathbf{S}^+$.

In using these methods one must guess both $\mathbf{S}$ and $\mathbf{S}^+$. The calculation then proceeds in a "flip–flop" manner; one iterate of the flux is made, the eigenvalue estimated, an iterate of the adjoint flux is made and again the eigenvalue estimated, etc. Suitable convergence criteria can be based on any of the foregoing methods: pointwise convergence of the flux and/or adjoint flux or eigenvalue convergence. In the latter, however, caution is required—the eigenvalue usually converges very rapidly owing to the variational nature of the expressions used to estimate it. Frequently this convergence is so rapid that the fluxes obtained are considerably in error even though the eigenvalue has converged. Of course, if one wants only the eigenvalue this method is ideal.

If the adjoint equations are obtained by "adjointing" the multigroup equations, as they must if Equations (IV.46a, b) are to be stationary, this procedure may be extremely efficient since only one set of group constants is required. The additional computer memory required by this method is not exorbitant so that this scheme is frequently used.

15. Extrapolation Methods

Many attempts have been made to improve the "non-adjoint" schemes [35]. The incentive for improvement lies, of course, in the fact that the power method and its variations require a great many iterations for convergence. Most frequently, convergence of the eigenvalue is the criterion by which multigroup calculations are terminated. In the power method (combined with any of the methods of estimating the eigenvalue) one usually finds that the eigenvalue is always to one side of the correct (converged) value and that the error decreases monotonically as the number of iterations increases.

This result is very helpful in formulating improved convergence schemes. Thus, if the last estimate of the eigenvalue is smaller than the one prior to it, it may be assumed that the current estimate is too high. It is then reasonable to choose an eigenvalue guess which is even lower than the last estimate (and similarly for the flux)—one is practically guaranteed of being closer to the correct value. The danger in such a procedure is, of course, the possibility of "overshoot" i.e. of choosing a value which is lower than the correct result, since this may lead to oscillation in the estimated eigenvalue and an accompanying increase in the number of iterations required.

It may be argued that the rate of convergence (e.g. $\lambda^{(k+1)} - \lambda^{(k)}$ or $\mathbf{S}^{(k+1)} - \mathbf{S}^{(k)}$) should be approximately the same on each iteration. Rationale for this argument for the flux is given by Equation (IV.44a) in which it is seen that the coefficient of the second eigenfunction, ψ_1, which is the major source of variation between $\mathbf{S}^{(n)}$ and the converged source, decreases by a factor of (λ_1/λ_0) on each iteration. Thus, it might be reasonable to set

$$\mathbf{S}^{(n)'} = \mathbf{S}^{(n)} - (\mathbf{S}^{(n-1)} - \mathbf{S}^{(n)}) = 2\mathbf{S}^{(n)} - \mathbf{S}^{(n-1)} \qquad \text{(IV.47a)}$$

where $\mathbf{S}^{(n)'}$ is now the source guess to be used in the $(n + 1)$st iteration. However, this is likely to lead to too large a correction and hence overshoot so that one generally introduces an "extrapolation factor" α:

$$\begin{aligned}\mathbf{S}^{(n)'} &= \mathbf{S}^{(n)} - \alpha(\mathbf{S}^{(n-1)} - \mathbf{S}^{(n)}) = (1 + \alpha)\,\mathbf{S}^{(n)} - \alpha\mathbf{S}^{(n-1)} \\ &= [(1 + \alpha)\,\boldsymbol{P} - \alpha\boldsymbol{I}]\,\mathbf{S}^{(n-1)} \qquad \text{(IV.47b)}\end{aligned}$$

where **I** is the unit matrix

$$I_{ij} = \delta_{ij}$$

[35] Bilodeau and Hageman, *op. cit.*

Generally, in order to avoid overshoot, one chooses α to be less than unity. This is the *extrapolation method.* Clearly, for $\alpha = 0$ the extrapolation method is identical to the power method.

In this procedure the additional question of the choice of the extrapolation factor α arises. Usually α will be different for each iteration and the optimum choice will be that which minimizes the number of iterations required. Means of choosing optimum values of α will be given later in this section.

A variation of this simple extrapolation method is the gradient method. Here one computes the gradient of $\lambda_0^{(k)}$, i.e., finds the "direction" in which it is changing most rapidly, and applies an extrapolation in this direction. This procedure turns out to be equivalent to using a different extrapolation constant at each position and is rather cumbersome.

Having decided that an extrapolation scheme can lead to an improved rate of convergence, one can then generalize the technique. Instead of choosing a $(k+1)$ stiteration source guss $\mathbf{S}^{(k)\prime}$ based only on the results of the two previous iterations, one constructs the guess from the results of all previous iterations:

$$\mathbf{S}^{(k)\prime} = \sum_{j=1}^{k} \beta_j \mathbf{S}^{(j)} \tag{IV.48a}$$

or, since

$$\mathbf{S}^{(k)} = \boldsymbol{P}^k \mathbf{S}^{(0)}$$

$$\mathbf{S}^{(k)\prime} = \sum_{j=1}^{k} \beta_j \boldsymbol{P}^j \mathbf{S}^{(0)} \tag{IV.48b}$$

If the initial source guess $\mathbf{S}^{(0)}$ is again expanded in terms of the eigenfunctions $\mathbf{\Psi}_i$ of $\boldsymbol{P}$ as in Equation (IV.44a)

$$\mathbf{S}^{(k)\prime} = \sum_{j=1}^{k} \sum_{n=1}^{N} \beta_j \sigma_n \lambda_n^j \psi_n = \sum_{n=1}^{N} \sigma_n \psi_n \left(\sum_{j=1}^{k} \beta_j \lambda_n^j \right) \tag{IV.48c}$$

From Equation (IV.48c) it is seen that to obtain the maximum dominance of ψ_0 one should select the β_j such that

$$\sum_{j=1}^{k} \beta_j \lambda_n^j$$

for $n = 0$ is a maximal with respect to this sum for all $n \neq 0$.

Here a theorem of Flanders and Shortley [36] proves extremely useful. Of all polynomials of degree n having the value $+1$ at a

[36] D. A. FLANDERS and G. SHORTLEY, *J. Appl. Phys.* **21**, 132 (1950).

given point, $x = x_0, x_0 > 1$, the polynomial

$$Q_n(x) = \frac{T_n(x)}{T_n(x_0)} \tag{IV.49a}$$

where $T_n(x)$ is the Tschebycheff polynomial of degree n, has the minimum maximum absolute value in the range $|x| < 1$. Any polynomial of degree n will in general have maxima (in its absolute value) in the range $(-1,1)$; the Tschebycheff polynomial has the property that its values at these maxima are lower than the corresponding values of any other polynomial of the same degree. The points $x = \pm 1$ may be included in these maxima if the function is increasing (in absolute value) at ± 1.

This theorem can be used to give a recipe for the β_j of Equation (IV.48a). However, since the eigenvalues may not meet the conditions of the theorem (in general, $\lambda_N \neq -1$, $\lambda_1 \neq +1$, and λ_0 may not be greater than unity) the β_j are not the coefficients of x^j in $Q_n(x)$. Rather, they are the coefficients of x^j in $Q_n(y)$ where:

$$y = \frac{1}{\lambda_1 - \lambda_N}[2x - (\lambda_1 + \lambda_N)] \tag{IV.49b}$$

This is a simple linear transformation that maps λ_N into -1, λ_1 into $+1$ and leaves λ_0 to the right of $+1$.

Unfortunately, use of the Tschebycheff method just described requires the storage of a great many source guesses. Thus the usual procedure is to modify the simple extrapolation procedure (IV.47b) in such a way that the nth order Tschebycheff polynomial is generated after n iterations. If a different extrapolation factor, α_i, is used after each iteration, the result after n iterations will be

$$\mathbf{S}^{(n)'} = \prod_{i=1}^{n} [(1 + \alpha_i)\,\boldsymbol{P} - \alpha_i \boldsymbol{I}]\,\mathbf{S}^{(0)} \tag{IV.49c}$$

On the other hand it is possible to factor $Q_n(y)$:

$$Q_n(y) = \prod_{i=1}^{n} [(1 + \alpha_i')\,x - \alpha_i'] \tag{IV.49d}$$

Thus the operator acting on $\mathbf{S}^{(0)}$ in Equation (IV.49c) will be $Q_n(\boldsymbol{P})$

if the α_i are set equal to the α_i'. Using the explicit expression for $T_n(x)$ it is easily shown that this gives

$$\alpha_i^{(n)} = \frac{\lambda_1 + \lambda_N + (\lambda_1 - \lambda_N)\cos\dfrac{2i+1}{2n}\pi}{2\lambda_0 - \lambda_1 - \lambda_N - (\lambda_1 - \lambda_N)\cos\dfrac{2i+1}{2n}\pi} \qquad \text{(IV.49e)}$$

An obvious difficulty is that the various eigenvalues are not known. Usually, λ_0 is obtained by one of the estimating procedures already given, λ_N is assumed to be zero and λ_1 is either estimated by methods similar to those used for λ_0 or from a guess (supplied by the user) of the ratio λ_1/λ_0 and the estimate of λ_0. This method is used semi-iteratively—calculations are performed in cycles of n iterations and at the end of each cycle the entire procedure is started anew (with new estimates of the eigenvalues). A variation of this method which is based on the recursion relation for Tschebycheff polynomials has also been used [37].

Clearly, only a few of the more commonly used methods have been discussed. A complete discussion of the techniques used is outside the range of this book; the interested reader is referred to Varga's book cited in [37].

16. The Finite Difference Equations—One Dimension

Within each outer iteration, it is necessary to find the spatial dependence of the N group fluxes. Given the fission source, one may begin with the highest energy group and calculate the spatial dependence of the flux in this group. When this flux is known, the source for the next higher energy group is also determined as

$$\Sigma_{R1}\phi_1(\mathbf{r}) + \chi_2 S(\mathbf{r})$$

where the subscript "1" refers to the highest energy group. Using this source, the spatial dependence of the flux in group "2" may be determined. Repetition of the procedure yields all N of the group fluxes and finally a new source

$$\sum_{j-1}^{N} \Sigma_{fj}(\mathbf{r})\, \phi_j(\mathbf{r})$$

which completes one outer iteration. Within each group the problem of calculating the individual group fluxes must be solved.

[37] R. Varga, *Matrix. Iterative Analysis*, Prentice-Hall, Englewood Cliffs, N. J., 1962.

Basically the problem can be stated as one of solving the equation

$$-\nabla \cdot D_j(\mathbf{r}) \nabla \phi_j(\mathbf{r}) + \Sigma_j(\mathbf{r}) \phi_j(\mathbf{r}) = S_j(\mathbf{r}) \qquad \text{(IV.50a)}$$

where we have written

$$\Sigma_j(\mathbf{r}) = \Sigma_{aj}(\mathbf{r}) + \Sigma_{Rj}(\mathbf{r}) \qquad \text{(IV.50b)}$$

and $S_j(\mathbf{r})$ represents the sum of the slowing-down and fission source terms. The boundary conditions that the flux $\phi_j(\mathbf{r})$ and the current

$$-D_j(\mathbf{r}) \nabla \phi_j(\mathbf{r})$$

are continuous at all points in the reactor and that the flux or current obey a given outer boundary condition are to be applied. Generally, for reactor problems the outer boundary condition is that the flux vanish on some surface, but frequently one wishes to apply multigroup calculations to the calculation of group constants for a cell. For example, in a periodic lattice, a cell of the repeating structure may be chosen and it is assumed that the cell is placed in the lattice such that there is no net flow of neutrons into or out of this cell. Such a condition is accurate provided that the cell is in a region in which the spatial variation of the flux (ignoring details within a cell) is nearly flat (such as would obtain, for example, near the center of a large reactor). Then one must solve the (IV.50a) subject to the condition that

$$\nabla \phi_j(\mathbf{r}) = 0 \qquad \text{(IV.50c)}$$

on the outer boundary.

In any case, the important point is that a boundary value problem must be solved. Generally, such problems are considerably more difficult to solve by finite differences methods than are initial value problems; the latter are simplified by the fact that it is generally possible to begin at the initial value and solve for the value of the desired function at each successive point. For boundary value problems the procedure is more likely to be one of guessing the solution and then using the differential equation to correct (*relax*) the guess in such a fashion that the differential equation is satisfied everywhere and the correct boundary values are assumed on the boundary.

Since all of this work will require finite difference approximations, the operator $\nabla \cdot D\nabla$ is first reduced to difference form. First, we note that in Cartesian coordinates

$$\Delta \cdot D\nabla\phi = \frac{\partial}{\partial x}\left(D\frac{\partial \phi}{\partial x}\right) + \frac{\partial}{\partial y}\left(D\frac{\partial \phi}{\partial y}\right) + \frac{\partial}{\partial z}\left(D\frac{\partial \phi}{\partial z}\right) \qquad \text{(IV.51a)}$$

(all of the following may be carried out in any coordinate system with added difficulty). For simplicity we shall work only with the one-dimensional form of this operator

$$\frac{d}{dx}\left(D\frac{d\varphi}{dx}\right)$$

but the generalization to three dimensions is relatively straightforward and will be given later.

Now a one-dimensional *mesh* is laid down, i.e., we pick a set of points $\{x_i\}$ and write a set of difference equations. This mesh need not be uniform — the distance between consequtive points need not be constant. However, the same mesh must be chosen for every group. (This, too, is unnecessary, but is a great convenience.) Generally, the accuracy of the calculation may be improved, for a given number of mesh points, if the points are more closely spaced in regions in which the neutron flux varies rapidly as a function of position and vice versa. For a point midway between x_i and x_{i+1} one can write†

$$D\frac{d\phi}{dx} \cong D_i \frac{\phi_{i+1} - \phi_i}{\Delta x_i} \qquad \text{(IV.51 b)}$$

where ϕ_i is $\phi(x_i)$ (the group labels are suppressed for convenience here) and

$$\Delta x_i = x_{i+1} - x_i \qquad \text{(IV.51 c)}$$

$$D_i = \frac{D(x_i) + D(x_{i+1})}{2}$$

Approximating further we have,

$$\frac{d}{dx} D \frac{d\phi}{dx}\bigg]_{x=x_i} \cong \left[D_i \frac{\phi_{i+1} - \phi_i}{\Delta x_i} - D_{i-1}\frac{\phi_i - \phi_{i-1}}{\Delta x_{i-1}}\right] \times$$

$$\times \frac{2}{\Delta x_i + \Delta x_{i-1}} \qquad \text{(IV.51 d)}$$

which, using the abbreviations

$$a_i^+ = \frac{D_i}{\Delta x_i} \frac{2}{\Delta x_i + \Delta x_{i-1}}$$

$$a_i^- = \frac{D_{i-1}}{\Delta x_{i-1}} \frac{2}{\Delta x_i + \Delta x_{i-1}} \qquad \text{(IV.51 e)}$$

$$a_i = \left(\frac{D_i}{\Delta x_i} + \frac{D_{i-1}}{\Delta x_{i-1}}\right)\frac{2}{\Delta x_i + \Delta x_{i-1}}$$

† An alternate method of replacing Equation (IV.50 a) by a set of algebraic equations is to integrate Equation (IV.50 a) from $x_i - \Delta x_{i-1}/2$ to $x_i + \Delta x_i/2$ and then use Equation (IV.51 b) for the derivative that appears. The coefficients in the resulting equation are slightly different from those given here but the accuracy is similar.

can be rewritten as

$$\frac{\mathrm{d}}{\mathrm{d}x} D \frac{\mathrm{d}\phi}{\mathrm{d}x}\bigg|_{x=x_i} \cong a_i^+\phi_{i+1} + a_i\phi_{i-1} - a_i\phi_i \qquad \text{(IV.51f)}$$

which is the desired difference form of the differential quantity $\mathrm{d}/\mathrm{d}x\, D\, \mathrm{d}\varphi/\mathrm{d}x$ and is known as the three point difference approximation as it gives a relation between the flux at a given mesh point and those adjacent to it. The two- and three-dimensional analogs of Equation (IV.51f) are known as the five and seven point difference approximations. Substituting Equation (IV.51f) into Equation (IV.50a) we have the required difference equation:

$$a_i^+\phi_{i+1} + a_i^-\phi_{i-1} - (a_i + \Sigma_i)\,\phi_i + S_i = 0 \qquad \text{(IV.51g)}$$

where

$$\Sigma_i = \Sigma_j(x_i)$$
$$S_i = S_j(x_i)$$

Equations (IV.51g) are a set of linear inhomogeneous algebraic equations for the unknowns $\boldsymbol{\phi}_i$, the number of equations and unknowns being equal to the number of mesh points.

In the usual calculation, in which the flux is zero at the boundary, the boundary points are left unnumbered and the appropriate ϕ are set equal to zero (i.e. if the boundary points are $n = 1$ and $i = N$, $\phi_0 = \phi_{N+1} = 0$). However, should the boundary condition require zero current at the boundary, the condition can be simulated by adding a fictitious row of additional mesh points outside the boundary and requiring the flux to be the same at these points as at the internal points of which they are images. The symmetry introduced then gives rise to the desired result. However, the boundary points must then be included in the mesh and the equations at the boundary are somewhat different. At the boundary (x_0 and x_n) we have:

$$2a_0^+\phi_1 - (a_0 + \Sigma_0)\,\phi_0 + S_0 = 0 \qquad \text{(IV.52a)}$$

$$2a_N^+\phi_{N-1} - (a_N + \Sigma_N)\,\phi_N + S_N = 0 \qquad \text{(IV.52b)}$$

In order to deal with Equations (IV.51g) in a neat fashion, it is necessary to cast them into matrix form. This is readily done if we define a flux vector:

$$\boldsymbol{\phi} = \begin{pmatrix} \phi_0 \\ \phi_1 \\ \cdot \\ \cdot \\ \cdot \\ \phi_N \end{pmatrix} \tag{IV.53a}$$

a source vector:

$$\mathbf{S} = \begin{pmatrix} S_0 \\ S_1 \\ \cdot \\ \cdot \\ \cdot \\ S_N \end{pmatrix} \tag{IV.53b}$$

and a matrix diffusion operator:

$$A = \begin{pmatrix} (a_0 + \Sigma_0) & -2a_0^+ & 0 & \cdot \\ -a_1^- & (a_1 + \Sigma_1) & -a_1^+ & \cdot \\ 0 & -a_2^- & \cdot & \cdot \\ \cdot & \cdot & \cdot & \cdot \\ \cdot & \cdot & \cdot & \cdot \\ \cdot & \cdot & \cdot & \cdot \end{pmatrix} \tag{IV.53c}$$

Equations (IV.51 g) can then be rewritten as

$$A\boldsymbol{\phi} = \mathbf{S} \tag{IV.53d}$$

the required matrix form.

17. Inner Iterations—Solution of the Finite Difference Equations

Again, the most obvious means of solution involves straightforward matrix methods. Formally, the solution to Equation (IV.53 d) can be written:

$$\boldsymbol{\phi} = A^{-1}\mathbf{S} \tag{IV.54a}$$

where A^{-1} is the inverse of A. A typical problem will use the order of a hundred mesh points so that A will be the order of a hundred by a hundred. In two or three dimensions the size of A may be the square or cube of a hundred, i.e., the order of thousands by thousands. Even with the great number of zero matrix elements, the inversion of a matrix of this magnitude is an overwhelming task. Generally, one-dimensional problems are solved by direct matrix inversion since a good algorithm is available [38] but in general one must resort to an iteration scheme (*inner* iterations).

Equation (IV.53d) is not in a suitable form for iteration as it stands, for on inserting a guess at $\boldsymbol{\phi}$, there is no means for obtaining an iterated flux vector. Hence, some simple transformations are usually applied. The simplest of these is to decompose $\boldsymbol{A}$ into the sum of two matrices, one of which is diagonal (i.e., it has no nonzero elements with different indices). That is, a matrix $\boldsymbol{\Sigma}$ is constructed whose elements are

$$\begin{aligned} \Sigma_{ij} &= A_{ij} \quad i = j \\ &= 0 \qquad i \neq j \end{aligned} \tag{IV.54b}$$

and a matrix $\boldsymbol{B}$ is constructed from the remaining elements of $\boldsymbol{A}$.

$$\begin{aligned} B_{ij} &= A_{ij} \quad i \neq j \\ &= 0 \qquad i = j \end{aligned} \tag{IV.54c}$$

Then clearly

$$\boldsymbol{A} = \boldsymbol{\Sigma} + \boldsymbol{B} \tag{IV.54d}$$

Since $\boldsymbol{\Sigma}$ is diagonal, its inverse is simply:

$$\boldsymbol{\Sigma}^{-1} = \begin{pmatrix} 1/A_{11} & 0 & 0 & \\ 0 & 1/A_{22} & & \\ & & & \\ 0 & & & \\ & & & 1/A_{NN} \end{pmatrix} \tag{IV.54e}$$

Using Equation (IV.54d), we may rewrite Equation (IV.53d) as

$$\boldsymbol{\Sigma\phi} = \mathrm{S} - \boldsymbol{B\phi} \tag{IV.54f}$$

and operating on it with $\boldsymbol{\Sigma}^{-1}$, we obtain

$$\boldsymbol{\phi} = \boldsymbol{\Sigma}^{-1}\mathrm{S} - \boldsymbol{\Sigma}^{-1}\boldsymbol{B\phi} \tag{IV.54g}$$

or denoting

$$\boldsymbol{\Sigma}^{-1}\boldsymbol{B} = -\boldsymbol{M} \tag{IV.54h}$$

$$\boldsymbol{\Sigma}^{-1}\boldsymbol{S} = \boldsymbol{S}'$$

we obtain a set of equations which is amenable to iteration:

$$\boldsymbol{\phi} = \boldsymbol{M\phi} + \mathrm{S}' \tag{IV.54j}$$

[38] J. PEACEMAN and A. RACHFORD, *J. Soc. Ind. Appl. Math.* 3, 28 (1955).

This equation is simply the matrix form of the finite difference equations obtained by dividing Equations (IV.51g) by $(a_i + \Sigma_i)$ and transposing the ϕ_i terms to the right-hand sides of the equations:

$$\phi_i = \frac{a_i^+}{a_i + \Sigma_i}\phi_{i+1} + \frac{a_i^-}{a_i + \Sigma_i}\phi_{i-1} + \frac{S_i}{a_i + \Sigma_i} \tag{IV.54k}$$

Equation (IV.54j) immediately suggests an iteration scheme. Suppose a guess, $\boldsymbol{\phi}^{(0)}$, is made at the flux vector. One then substitutes this guess into the right-hand side of Equation (IV.54j) and calculates

$$\boldsymbol{\phi}^{(1)} = \boldsymbol{M}\boldsymbol{\phi}^{(0)} + \mathbf{S}' \tag{IV.55a}$$

and successive iterates from

$$\boldsymbol{\phi}^{(k+1)} = \boldsymbol{M}\boldsymbol{\phi}^{(k)} + \mathbf{S}' \tag{IV.55b}$$

This, the simultaneous relaxation method (also known as the Gauss or Richardson method) is convergent, but does not converge rapidly compared to methods given later.

This mathematical formalism obscures the nature of the calculation somewhat. To make the nature of the calculation somewhat clearer, we note that we are actually substituting the kth iterate values of ϕ_{i-1} and ϕ_{i+1} into the right-hand side of Equation (IV.54k) and setting the result equal to the $(k + 1)$st iterate flux $\phi_i{}^{k-1}$.

To demonstrate convergence, we define the kth iteration error vector $\mathbf{E}^{(k)}$ as

$$\mathbf{E}^{(k)} = \boldsymbol{\psi} - \boldsymbol{\psi}^{(k)} \tag{IV.55c}$$

i.e., the difference between the correct flux and the kth iterate. Subtracting Equation (IV.55b) from

$$\boldsymbol{\psi} = \boldsymbol{M}\boldsymbol{\psi} + \mathbf{S}' \tag{IV.55d}$$

we find the error vectors $\mathbf{E}^{(k)}$ obey the equation:

$$\mathbf{E}^{(k+1)} = \boldsymbol{M}\mathbf{E}^{(k)} \tag{IV.55e}$$

Now, the error vectors can be expanded in terms of the eigenvectors of the matrix $\boldsymbol{M}$, which satisfy:

$$\boldsymbol{M}\boldsymbol{\chi}_n = \lambda_n\boldsymbol{\chi}_n \tag{IV.55f}$$

$$\mathbf{E}^{(k)} = \sum_{n=1}^{N} a_n^{(k)}\boldsymbol{\chi}_n \tag{IV.55g}$$

Inserting the expansion into Equation (IV.55e) we have:

$$\mathbf{E}^{(k+1)} = \sum_{n=1}^{N} a_n^{(k+1)} \chi_n = \sum_{n=1}^{N} a_n^{(k)} \lambda_n \chi_n \tag{IV.55h}$$

or, in terms of the error in the original guess:

$$\mathbf{E}^{(m+1)} = \sum_{n=1}^{N} a_n^{(0)} \lambda_n^m \chi_n \tag{IV.55j}$$

Thus, the error vector will become zero after many iterations provided that all of the eigenvalues are less than unity in absolute value. In fact, it can be seen that the rate of convergence depends solely on the degree to which the absolute value of the largest eigenvalue (called the *spectral norm*) is less than unity. In general, matrices do not possess this property but Young [39] has shown that the properties of the matrix $\boldsymbol{M}$ are such that it does in fact have the desired properties. Indeed, Young has discovered many useful properties of this type of matrix (said to possess Property A) which will be introduced as required.

As was the case with outer iterations, the most obvious iteration procedure is not the one which converges most rapidly. An easy means of improving the convergence rate is suggested by a rather simple argument. In the simultaneous relaxation method the kth iterate flux is obtained by substituting the $(k - 1)$st iterate flux into the difference equations. An orderly procedure would be to start at one boundary and then step point by point along the mesh to the other boundary. In so doing, we are not taking full advantage of the material at hand.

When computing the flux at x_i by this procedure, we are using the $(k - 1)$st iterate of the flux at x_{i-1} whereas the kth iterate has already been calculated. A better procedure is to use the kth iterate flux whenever it is available.

To establish this method in the matrix formulation we note that the simultaneous relaxation method calculates the elements, ϕ_l, of the flux vector in order of increasing index (i.e., $l = 0, 1, 2, \ldots$) and that the terms of the matrix $\boldsymbol{M}$ (say $M_{ll'}$) which connect an element ϕ_l with those calculated previously (i.e., those with $l' < l$) are those which lie below the diagonal. Hence, it is natural to define two new

[39] D. Young, *Trans. Amer. Math. Soc.* **92**, 76 (1954).

matrices, $\boldsymbol{L}$ and $\boldsymbol{U}$ as the lower and upper halves of $\boldsymbol{M}$:

$$\begin{aligned} L_{ij} &= M_{ij} \qquad i > j \\ &= 0 \qquad\quad i < j \\ U_{ij} &= M_{ij} \qquad i < j \\ &= 0 \qquad\quad i > j \end{aligned} \tag{IV.56a}$$

So that

$$\boldsymbol{M} = \boldsymbol{L} + \boldsymbol{U}$$

In terms of the new matrices, the simultaneous relaxation method can be written

$$\psi^{(k+1)} = \boldsymbol{L}\psi^{(k)} + \boldsymbol{U}\psi^{(k)} + \mathrm{S}$$

The iteration scheme for the new method we are seeking, called the *successive relaxation* (or Seidel–Young or Leibman) method, can then be written

$$\psi^{(m+1)} = \boldsymbol{L}\psi^{(m+1)} + \boldsymbol{U}\psi^{(m)} + \mathrm{S} \tag{IV.56b}$$

or

$$\begin{aligned} \psi^{(m+1)} &= (\boldsymbol{I} - \boldsymbol{L})^{-1}\, \boldsymbol{U}\psi^{(m)} + (\boldsymbol{I} - \boldsymbol{L})^{-1}\, \mathrm{S} \\ &= \boldsymbol{P}\psi^{(m)} + (\boldsymbol{I} - \boldsymbol{L})^{-1}\, \mathrm{S} \end{aligned}$$

The convergence of this method was demonstrated by Geiringer [40] who also showed that the eigenvalues of $(\mathrm{I} - \boldsymbol{L})^{-1}\, \boldsymbol{U} = \boldsymbol{P}$ are the squares of the eigenvalues of $\boldsymbol{M}$. Hence, the spectral norm of $\boldsymbol{P}$ is the square of the spectral norm of $\boldsymbol{M}$ and the successive relaxation method converges twice as fast as the simultaneous method.

It is natural to inquire as to whether the extrapolation method used in the outer iteration problem may be applied to the solution of inhomogeneous equations by inner iterations. Intuitively, it is clear that the procedure is valid and it may be applied to either of the methods already developed. As the simultaneous method does not converge as rapidly as the successive method, it is natural to work only with the latter; the method can, of course, be applied to the simultaneous method if desired. This leads then to the *successive overrelaxation* (*or extrapolated Leibman*) *method.*

As with the extrapolation procedure for the solution of homogeneous equations, we let the kth iterate flux $\psi^{(k)}$ be a linear combi-

[40] H. Geiringer, *Reisner Anniversary Volume*, p. 360 (1950).

nation of the result of the kth mesh sweep:

$$(I - L)^{-1} U\psi^{(k-1)} + (I - L)^{-1} \mathbf{S}$$

and $\psi^{(k-1)}$ itself, i.e., we choose

$$\psi^{(k)} = \{(I - L)^{-1} U\psi^{(k-1)} + (I - L)^{-1} \mathrm{S}\} (1 + \alpha) - \alpha\psi^{(k-1)} \tag{IV.57}$$

In practice, this simply means that one carries out the successive relaxation scheme as outlined earlier and then forms $\psi^{(k)}$ from the result of that calculation and the previous flux iterate $\psi^{(k-1)}$. All of the discussion relative to the choice of α given in Section 15 applies here.

Finally, it is possible to carry over the Tschebycheff polynomial method to inner iterations. That is, after k iterations, one selects as the kth initial flux for the $(k + 1)$st iteration:

$$\psi^{(k)'} = \sum_{j=1}^{m} \beta_j \psi^{(j)} \tag{IV.58a}$$

The error vector in this scheme then becomes:

$$\mathbf{E}^{(k)'} = \sum_{j=1}^{k} \beta_j \mathbf{E}^{(j)} \tag{IV.58b}$$

But, subtracting Equation (IV.56b) from

$$\psi = P\psi - (I - L)^{-1} \mathbf{S} \tag{IV.58c}$$

we have

$$\mathbf{E}^{(k+1)} = PE^{(k)} \tag{IV.58d}$$

So that

$$E^{(k)'} = \sum_{j=1}^{k} B_j P^j \mathbf{E}^{(0)} \tag{IV.58e}$$

Expanding $\mathbf{E}^{(0)}$ in terms of the eigenfunctions χ_n of P where

$$P\chi_n = \lambda_n \chi_n \tag{IV.58f}$$

we have

$$E^{(k)'} = \sum_{n=1}^{N} e_n^{(k)} \chi_n \tag{IV.58g}$$

On substituting this expansion into Equation (IV.58e) we have

$$e_n^{(k)'} = \sum_{j=1}^{N} \beta_j \lambda_n^j e_k^{(0)} \tag{IV.58h}$$

where the linear independence of the χ_n has been applied. Again to minimize $e_n^{(k)}$ for all $\lambda_n < 1$ (note that the λ_n all satisfy this condition as a result of Young's work) we must again choose the β_j as the coefficients of x^j in $Q_j(y)$ of Equation (IV.49 b).†

18. Two- and Three-Dimensional Calculations

The finite differences approximation has been developed only for the one-dimensional case merely as a convenience. In this section the extension to a three-dimensional Cartesian mesh will be given to illustrate the few points of difference that do arise in more than one dimension. Similarly, the extension to curvilinear coordinates is a relatively simple one and can be obtained by the methods used here.

To begin with we lay a mesh of three sets of mutually perpendicular planes on the reactor; call the coordinates of these planes x_i, y_j, z_k where $i = 1, 2 \ldots N_x$; $j = 1, 2 \ldots N_y$; $k = 1, 2 \ldots N_z$. For simplicity we introduce the notation

$$f(x_i, y_j, z_k) = f_{ijk}$$

where $f(x_i, y_j, z_k)$ is any function evaluated at x_i, y_j and z_k. Now all one need do is make the finite difference approximation (IV.51 d) in the y and z-directions as well as the x-direction. The result is the well known seven-point difference formula:

$$\begin{aligned} \nabla \cdot D \nabla \varphi &= a_{ijk}^{+} \phi_{i+1,j,k} + a_{ijk}^{-} \phi_{i-1,j,k} + b_{ijk}^{+} \phi_{i,j+1,k} \\ &\quad + b_{ijk}^{-} \phi_{i,j-1,k} + c_{ijk}^{+} \phi_{i,j,k+1} + c_{ijk}^{-} \phi_{i,j,k-1} \\ &= (a_{ijk} + b_{ijk} + c_{ijk}) \phi_{ijk} \end{aligned} \tag{IV.59 a}$$

where in analogy to the quantities defined by Equations (IV.51 c) and (IV.51 e) we have defined

$$a_{ijk}^{+} = \frac{D_{ijk}^{x}}{\Delta x_i} \frac{2}{\Delta x_i + \Delta x_{i-1}} \qquad a_{ijk}^{-} = \frac{D_{i-1,jk}^{x}}{\Delta x_{-1}} \frac{2}{\Delta x_i + \Delta x_{i-1}} \tag{IV.59 b}$$

$$b_{ijk}^{+} = \frac{D_{ijk}^{y}}{\Delta y_j} \frac{2}{\Delta y_j + \Delta y_{j-1}} \qquad b_{ijk}^{-} = \frac{D_{i,j-1,k}^{y}}{\Delta y_{j-1}} \frac{2}{\Delta y_j + \Delta y_{j-1}} \tag{IV.59 c}$$

$$c_{ijk}^{+} = \frac{D_{ijk}^{z}}{\Delta z_k} \frac{2}{\Delta z_k + \Delta z_{k-1}} \qquad c_{ijk}^{-} = \frac{D_{ijk-1}^{z}}{\Delta z_{k-1}} \frac{2}{\Delta z_k + \Delta z_{k-1}} \tag{IV.59 d}$$

† Again only a few methods have been discussed. For a more complete discussion see the work of Varga cited in footnote [37], p. 260.

$$\left.\begin{aligned} a_{ijk} &= \left(\frac{D^x_{ijk}}{\Delta x_i} + \frac{D^x_{i-1jk}}{\Delta x_{i-1}}\right)\frac{2}{\Delta x_i + \Delta x_{i-1}} \\ b_{ijk} &= \left(\frac{D^y_{ijk}}{\Delta y_j} + \frac{D^y_{ij-1k}}{\Delta y_{j-1}}\right)\frac{2}{\Delta y_j + \Delta y_{j-1}} \\ c_{ijk} &= \left(\frac{D^z_{ijk}}{\Delta z_k} + \frac{D^z_{ijk-1}}{\Delta z_{k-1}}\right)\frac{2}{\Delta z_k + \Delta z_{k-1}} \end{aligned}\right\} \qquad \text{(IV.59e)}$$

$$\left.\begin{aligned} D^x_{ijk} &= \frac{D(x_i, y_j, z_k) + D(x_{i+1}, y_j, z_k)}{2} \\ D^y_{ijk} &= \frac{D(x_i, y_j, z_k) + D(x_i, y_{j+1}, z_k)}{2} \\ D^z_{ijk} &= \frac{D(x_i, y_j, z_k) + D(x_i, y_j, z_{k+1})}{2} \end{aligned}\right\} \qquad \text{(IV.59f)}$$

Substituting Equation (IV.59c) into the diffusion Equation (IV.50a) we finally arrive at:

$$\begin{aligned} &a^+_{ijk}\phi_{i+1jk} + a^-_{ijk}\phi_{i-1jk} + b^+_{ijk}\phi_{ij+1k} + b^-_{ijk}\phi_{ij-1k} + \\ &c^+_{ijk}\phi_{ijk+1} + c^-_{ijk}\phi_{ijk-1} - (a_{ijk} + b_{ijk} + c_{ijk} + \Sigma_{ijk})\,\phi_{ijk} = S_{ijk} \end{aligned} \qquad \text{(IV.59g)}$$

Again a large set of linear algebraic equations results. Here the number of equations and unknowns is $N_xN_yN_z$, but the principles involved in solving them are essentially the same as those already discussed, except that here direct matrix inversion methods are out of the question and iterative procedures are always used.

To cast Equations (IV.59g) into matrix form it is necessary to relabel the ϕ_{ijk} so that they only have a single index, say ϕ_m. The best manner in which to do this is suggested by the mesh sweep technique. Since it is necessary to use iterative procedures, an orderly scheme is to start at one corner of the mesh, step along the edge, say in the x-direction (i.e., the line $y = y_1$, $z = z_1$), then come back and sweep along a second line adjacent to the first (the line $y = y_2$, $z = z_1$) and continue until one boundary plane (here $z = z_1$) has been completely covered. Then the procedure is repeated for a second plane (e.g., $z = z_2$), a third ($z = z_3$), etc., until the entire reactor volume is covered. In this manner we are calculating the ϕ_{ijk} in the order

$$\phi_{111}, \phi_{211} \cdots \phi_{N_x 11}, \phi_{121}, \phi_{221} \cdots \phi_{N_x 21}, \phi_{131} \cdots \phi_{1N_y1} \cdots$$
$$\cdots \phi_{N_xN_y1}, \phi_{112}, \phi_{212} \cdots \phi_{N_xN_yN_z}$$

Thus it is natural to define the flux vector as

$$\boldsymbol{\phi} = \begin{bmatrix} \phi_{111} \\ \phi_{211} \\ \vdots \\ \phi_{N_x11} \\ \phi_{121} \\ \vdots \\ \phi_{N_x21} \\ \vdots \\ \phi_{1N_y1} \\ \vdots \\ \phi_{N_xN_yN_z} \end{bmatrix} \tag{IV.60a}$$

so that ϕ_{ijk} now finds itself as the mth element of the vector where

$$m = i + (j - 1)\, N_x + (k - 1)\, N_xN_y \tag{IV.60b}$$

In a similar fashion one constructs a matrix A with elements A_{ij} defined

$$\left.\begin{aligned} A_{m,m} &= -[a_{ijk} + b_{ijk} + c_{ijk} + \Sigma_{ijk}] \\ A_{m,m+1} &= a^{+}_{ijk} \\ A_{m,m-1} &= a^{-}_{ijk} \\ A_{m,m+N_x} &= b^{+}_{ijk} \\ A_{m,m-N_x} &= b^{-}_{ijk} \\ A_{m,m+N_xN_y} &= c^{+}_{ijk} \\ A_{m,m-N_xN_y} &= c^{-}_{ijk} \end{aligned}\right\} \tag{IV.60c}$$

The source vector is defined in analogy to the flux vector

$$\mathbf{S} = \begin{pmatrix} S_{111} \\ S_{211} \\ \vdots \\ S_{N_x11} \\ \vdots \\ S_{N_xN_yN_z} \end{pmatrix} \tag{IV.60d}$$

i.e., S_{ijk} is placed in the position given by Equation (IV.60b).

By defining the vectors and coefficient matrix in this manner it is possible to cast Equations (IV.59g) into the matrix form:

$$A\boldsymbol{\phi} = S \tag{IV.60e}$$

Formally Equation (IV.60e) is identical to Equation (IV.53d) except for the immense increase in the size of the vectors and the matrix A. Thus all of the procedures discussed for the one-dimensional case may be carried over to the three- (or two-) dimensional case *mutatis mutandis*.

Synthesis Methods

Although many two- and three-dimensional multigroup diffusion theory computer programs have been written [41, 42], they generally require extremely large expenditures of computer time even for problems with only a few groups. While this expenditure might be justified for final design calculations, it is hard to justify for routine survey type calculations. On the other hand, one-dimensional calculations are clearly unable to yield the detailed flux and power distributions required for the engineering design of a reactor. Hence, there has been considerable interest in methods of *synthesizing* two- and three-dimensional flux distributions from the results of one-dimensional calculations.

The simplest approach, and hence the first to be developed, is quite similar to the method for computing group constants described in Section 8. One assumes that the leakage in the y and z-directions may be represented by "equivalent bucklings" B_y^2 and B_z^2, so that the multigroup equations may be written

$$- D_j \frac{d^2\phi_j(x)}{dx^2} + [D_j(B_{yj}^2 + B_{zj}^2) + \Sigma_{tj}]\,\phi_j(x) = \sum_{i=1}^{j-1} \Sigma_{ij}\phi_i(x) + \nu\chi_j \sum_{i=1}^{N} \Sigma_{fi}\phi_i(x) \qquad \text{(IV.61 a)}$$

These equations are identical to the original one-dimensional multigroup equations provided one replaces Σ_{aj} by $(\Sigma_{aj} + D_jB_y^2 + D_jB_z^2)$, i.e., $D(B_y^2 + B_z^2)$ is a "leakage cross section" for the y and z directions. This is similar to the approach used in Section 8 in which the leakage in all three dimensions was treated by means of an equivalent geometric buckling.

Now, one can solve Equation (IV.61 a) by the methods already given to obtain flux distributions in the x-direction. Accurate values of the multiplication constant will be obtained only if the perpendicular bucklings assumed correctly represent the leakage in the other two directions. In the simplest procedure one merely cal-

[41] T. B. FOWLER and M. TOBIAS, *EQUIPOISE*-3, *A Two Dimensional, Two Group Neutron Diffusion Code*, ORNL 3199 (1962).

[42] T. B. FOWLER and M. TOBIAS, *WHIRLAWAY, A Three Dimesional Neutron Diffusion Code*, ORNL 3150 (1961).

culates these bucklings using elementary arguments. Most commonly, one assumes that the flux is given by

$$\phi(x, y, z) = \phi(x) \cos B_y y \cos B_z z \qquad \text{(IV.61 b)}$$

with B_x and B_y given by Table 1 of Chapter I and uses the same bucklings for all groups.

A more accurate procedure is a simple generalization of the above. After completing the calculations outlined above, one makes a similar calculation in the y-direction. Here one need not guess a buckling for the x-dimension, as the leakage in the x-direction may be evaluated from the results of the previous calculation. Thus one can be assured that B_x^2 does accurately represent the leakage in the x-direction. Repetition of this procedure in the z-direction (evaluating both B_{xj}^2 and B_{yj}^2 from the results already obtained) one lobtains three-dimensional representation of the flux and, presumably, an accurate value of k_{eff}.

A further improvement is obtained by using the method described in the previous paragraph iteratively. After completing one set of three calculations, one repeats the calculation in the x-direction using the more accurate bucklings obtained from the y and z calculations. Then the y and z calculations are repeated and the entire procedure is carried on until the bucklings obtained on successive iterations converge, i.e., agree to within some predetermined amount. This is the *buckling iteration method.*

Variations of the above using both one- and two-dimensional calculations have also been developed. Here, after the one-dimensional calculation in the x-direction, one does a two-dimensional calculation in the y and z-directions and iteratively does one-dimensional calculations in the x-direction and two-dimensional y–z calculations.

While considerable success has been obtained with the buckling iteration method, it is not too satisfactory for multi-region reactors, e.g., reflected or blanketed reactors. For example, consider a totally reflected cylindrical reactor. The flux shapes in the z-direction in the core and radial reflector are not likely to be very similar; neither are the radial flux shapes in the core and axial reflector. To account for this, one usually assumes different bucklings for the core and reflector regions. In turn, this means that each region must be treated separately. In practice this method is very much like ordinary buckling iteration. One does a one-dimensional calculation in the radial direction for each zone—in the example given above, the core and the axial reflector. From the results of these calculations, bucklings are computed for the axial calculations; these bucklings are functions of the axial (z) coordinate. Axial one-dimensional calculations are then carried out for each region—here, the core and the radial reflector. Again the iteration is carried out until the bucklings have converged [43].

Although it represents a considerable improvement over simple synthesis schemes, this last method is still not entirely satisfactory. The major difficulty is that the fluxes calculated by the zonal buckling iteration method need not be continuous across the zone boundaries. Another scheme which does not suffer from this difficulty was suggested by Kaplan [44]. This method starts out very much like the zonal buckling iteration method. Radial (or axial) flux shapes are computed for each zone using approximate axial (or radial) bucklings; let us denote these flux shapes in group j by $H^j_k(r)$, $k = 1, 2 \ldots M$, where M is the number of zones. Then the flux is assumed to have the shape

$$\phi(r, z) = \sum_{k=1}^{M} G^k_j(z)\, H^k_j(r) \qquad \text{(IV.62a)}$$

where the $G_j(z)$ are "mixing coefficients", i.e., they determine the amount of each flux component as a function of z. This ansatz is then substituted into the multigroup diffusion equations and the equations are multipled by $H^k_j(r)$, each in turn, and integrated over the radial coordinate. This results in a set of coupled equations for the mixing coefficients, $G^k_j(z)$, which can then be solved by the methods already given.

This method appears to give excellent results. In fact, test problems indicate that the fluxes synthesized by this method are almost as accurate as full two-dimensional calculations.

There are many variations on this method which are currently undergoing extensive development. Many of these are fairly obvious generalizations of the above method but they are too numerous to be described here in any detail. An important recent application of the method is to time-dependent problems (burn-up and kinetics).†

[43] E. L. WACHSPRESS, R. D. BURGESS and S. BACON, *Nucl. Sci. Eng.* **12**, 381 (1962).

[44] S. KAPLAN, *Nucl. Sci. Eng.* **13**, 22 (1962).

† S. KAPLAN, private communication.

APPENDIX A

TIME DEPENDENT ASYMPTOTIC REACTOR THEORY†

In this appendix we shall treat the time dependent problems alluded to in Chapter I. For the sake of brevity the asymptotic (bare homogeneous thermal) reactor model is adhered to. The rationale for so doing lies in the fact that all the basic principles may be discovered within the framework of the asymptotic theory; the development for more complex models yields essentially the same information, albeit in a more rigorous form.‡ We shall also make the further, and more serious, approximation of ignoring the effects of delayed neutrons.

Within the framework of the diffusion approximation to asymptotic reactor theory, we may infer from Chapter I that the time-dependent thermal flux obeys the following equation:

$$\frac{1}{v}\frac{\partial\varphi}{\partial t}(\mathbf{r}, t) - D\nabla^2\varphi(\mathbf{r}, t) + \Sigma_a\varphi(\mathbf{r}, t)$$
$$= S(\mathbf{r}, t) + \nu\Sigma_f \int_0^t \mathrm{d}t' \int_{\text{all space}} \mathrm{d}^3r'\varphi(\mathbf{r}', t')\, P(|\mathbf{r} - \mathbf{r}'|, t - t') \qquad (A.1)$$

where $P(|\mathbf{r}|, t)$ is the time-dependent slowing-down kernel of Chapter I and $S(\mathbf{r}, t)$ is to represent only non-fission sources.

As in Chapter I, the flux and source are expanded in terms of the eigenfunctions of the wave equation, (I.26):

$$\varphi(\mathbf{r}, t) = \sum_{n=0}^{\infty} a_n(t)\,\chi_n(\mathbf{r}) \qquad (A.2a)$$

$$S(\mathbf{r}, t) = \sum_{n=0}^{\infty} \sigma_n(t)\,\chi_n(\mathbf{r}) \qquad (A.2b)$$

† This appendix closely follows T. J. Krieger and P. F. Zweifel, *Nucl. Sci. Eng.* **5**, 21 (1959).

‡ For more complex discussions of time dependent reactor problems, the reader is referred to A. Henry, *Nucl. Sci. Eng.* **3**, 52 (1958); A. Ussachoff, *PICG* **5**, 503 (1955).

On substituting the expansions (A.2) into Equation (A.1), applying lemma (I.34), and equating the coefficients of each of the $\chi_n(\mathbf{r})$, we have the following equation for the $a_n(t)$:

$$\frac{1}{v}\frac{\partial a_n(t)}{dt} + (\Sigma_a + DB_n^2)\,a_n(t) = \sigma_n(t) + \nu\Sigma_f\int_0^t a_n(t')\,\overline{P}(B_n^2, t-t')\,dt' \tag{A.3}$$

In general, the source $S(\mathbf{r}, t)$ may be one of either fast of thermal neutrons. In practice all sources emit fast neutrons so that it is sufficient to consider only that case. Furthermore, one need only consider a pulsed source

$$\sigma_n(t) = \sigma_n\,\delta(t)$$

since the solution for this case is the Green's function for the general case. However, this type of source gives rise to a source

$$\sum_{n=0}^{\infty}\sigma_n\int_{\text{all space}}\chi_n(\mathbf{r})\int_0^t\delta(t')\,P(|\mathbf{r}-\mathbf{r}'|, t-t')\,dt'\,d^3r' = \sum_{n=1}^{\infty}\sigma_n\overline{P}(B_n^2, t)\,\chi_n(\mathbf{r})$$

of thermal neutrons. Finally, since the solution for this source can be written in terms of the solution for a pulsed thermal source, one need consider a source of this type. Hence, there will be no loss of generality if we simplify Equation (A.3) to:

$$\frac{1}{v}\frac{da_n(t)}{dt} + (\Sigma_a + DB_n^2)\,a_n(t) = c_n\,\delta(t) + \nu\Sigma_f\int_0^t a_n(t')\,\overline{P}(B_n^2, t-t')\,dt' \tag{A.4}$$

Equation (A.4) may be solved by any of the standard techniques for solving Volterra integral equations, but we shall choose the Laplace transform method. Since $a_n(t) = 0$, $t < 0$, we have, on taking the Laplace transform of (A.4):

$$\left(\frac{s}{v} + \Sigma_a + DB_n^2\right)\overline{a}_n(s) = \nu\Sigma_f\overline{\overline{P}}(B_n^2, s)\,\overline{a}_n(s) + c_n \tag{A.5}$$

where

$$\overline{a}_n(s) = \int_0^\infty dt\,e^{-st}\,a_n(t) \tag{A.6a}$$

$$\overline{\overline{P}}(B_n^2, s) = \int_0^\infty dt\,e^{-st}\,\overline{P}(B_n^2, t) \tag{A.6b}$$

It is noteworthy that the calculation of $\bar{\bar{P}}(B^2, s)$ is simply that of calculating $\bar{P}(B^2)$ (as in Chapter I) with Σ_a replaced by $\Sigma_a + s/v$. On solving for $\bar{a}_n(s)$ we have

$$\bar{a}_n(s) = \frac{vc_n}{s + v(\Sigma_a + DB_n^2) - v\nu\Sigma\,\bar{\bar{P}}(B_n^2, s)} \tag{A.7}$$

which may be inverted formally to yield

$$a_n(t) = \frac{c_n v}{2\pi i}\int_{c-i\infty}^{c+i\infty} \frac{e^{st}\,ds}{s + v(\Sigma_a + DB_n^2) - v\nu\Sigma_f\bar{\bar{P}}(B_n^2, s)} \tag{A.8}$$

The only singularities of the integrand are the roots of

$$s + v(\Sigma_a + DB_n^2) - v\nu\Sigma\,\bar{\bar{P}}(B_n^2, s) = 0 \tag{A.9}$$

In a critical reactor it was shown in Chapter I that the neutron flux is all in the first mode and is constant in time. This solution arises from the fact that Equation (A.9) with $n = 0$ then has a root $s = 0$, which then makes Equation (A.9) the criticality condition (I.25b). For a reactor not far from critical, this root can be expected to move only slightly from zero and since it is a single root it is reasonable to expect that the asymptotic solution will be due to the pole arising from this zero. For small s we may expand†

$$\bar{\bar{P}}(B_n^2, s) = \bar{\bar{P}}(B_n^2, 0) + s\left(\frac{\partial\bar{\bar{P}}(B_n^2, s)}{\partial s}\right)_{s=0} \tag{A.10}$$

so that the solution of Equation (A.9) becomes

$$s_{no} = \frac{\nu\Sigma\,\bar{\bar{P}}(B_n^2, 0) - (\Sigma_a + DB_n^2)}{\dfrac{1}{v} - \nu\Sigma_f\left(\dfrac{\partial\bar{\bar{P}}(B_n^2, s)}{\partial s}\right)_{s=0}} = \frac{k_n - 1}{l_n - \langle t\rangle_n k_n} \tag{A.11}$$

Here we have defined

$$\langle t\rangle_n = \frac{\left(\dfrac{\partial\bar{\bar{P}}(B_n^2, s)}{\partial s}\right)_{s=0}}{\bar{\bar{P}}(B_n^2, 0)} = \frac{\int_0^\infty t\bar{P}(B_n^2, t)\,dt}{\int_0^\infty \bar{P}(B_n^2, t)\,dt} \tag{A.12}$$

† Although the above argument justifies this expansion only for $n = 0$, we shall use it for all n. The expansion is clearly invalid for large n [cf. Equation (A. 14)].

the mean slowing-down time for a neutron in the nth mode. In accord with the results of Chapter I [cf. Equation (I.41 g)]

$$k_n = \frac{\nu\Sigma_f \bar{\bar{P}}(B_n^2, 0)}{\Sigma_a + DB_n^2} \tag{A.13a}$$

$$l_n = \frac{1}{v(\Sigma_a + DB_n^2)} \tag{A.13b}$$

are the multiplication constant and lifetime for the nth mode respectively. The condition for the expansion (A.10) to be valid is that the second term be much smaller in magnitude that the first, i.e., that

$$|k_{n-1}| \langle t \rangle_n \ll l_n + \langle t \rangle_n k_n \tag{A.14}$$

Since k_n falls off rapidly with increasing n, this approximation is clearly valid only for small n. This is to be expected as the higher modes yield large values of s.

The asymptotic behavior of $a_n(t)$ is then

$$a_n(t) \cong \alpha_n \, e^{s_{n_0} t} \tag{A.15}$$

where α_n is the residue to the integrand of Equation (A.5) at $s = s_{n_0}$:

$$\alpha_n = c_n v \left(1 - v\nu\Sigma_f \left(\frac{\partial \bar{\bar{P}}(B_n^2, s)}{\partial s}\right)_{s=0}\right)^{-1} \tag{A.16}$$

APPENDIX B

VECTOR IDENTITIES

IN CHAPTER I six vector identities were used which are proved here.

(1) $$\int d\mathbf{\Omega} = 4\pi$$

Proof: Note that $d\mathbf{\Omega}$ is just differential area on a unit sphere (see Fig. B.1):

$$d\mathbf{\Omega} = \sin\theta \, d\theta \, d\varphi = d(\cos\theta) \, d\varphi$$

or with

$$\mu = \cos\theta$$

$$d\mathbf{\Omega} = d\mu \, d\varphi$$

The azimuthal angle runs from 0 to 2π while θ runs from $-\pi$ to 0 (or -1 to 1 in μ) in covering the sphere. Hence

$$\int d\mathbf{\Omega} = \int_{-1}^{1} d\mu \int_{0}^{2\pi} d\varphi = 4\pi$$

(2) $$\int \mathbf{A} \cdot \mathbf{\Omega} \, d\Omega = 0$$

Choose as the polar axis in Fig. B.1 a vector in the $\mathbf{A}$ direction. Then

$$\mathbf{A} \cdot \mathbf{\Omega} = |A| \cos\theta = |A| \, \mu$$

$$\int \mathbf{A} \cdot \mathbf{\Omega} \, d\mathbf{\Omega} = |A| \int_{-1}^{1} \mu \, d\mu \int_{0}^{2\pi} d\varphi = 0$$

(3) $$\int \mathbf{\Omega} \, d\mathbf{\Omega} = 0$$

This result follows from the fact that Identity (2) must hold for arbitrary $\mathbf{A}$. It is also readily proven by means of a symmetry argument.

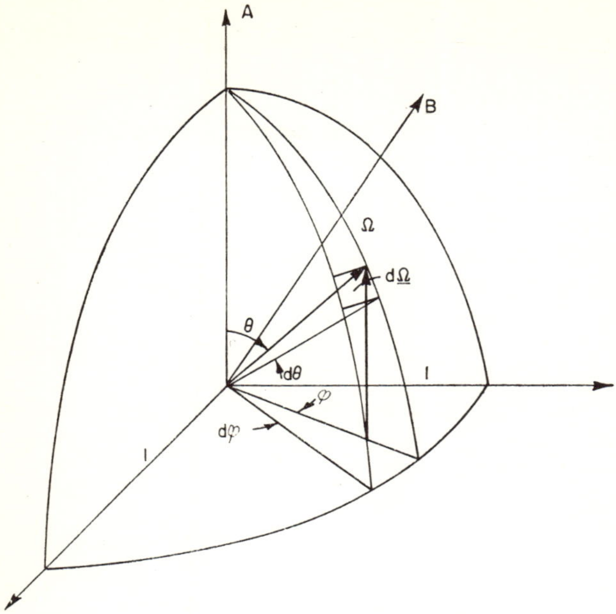

FIG. B.1.

(4) $$\int \mathbf{A} \cdot \mathbf{\Omega\Omega}\, d\mathbf{\Omega} = \frac{4\pi}{3}\mathbf{A}$$

This integral is readily evaluated by noting that the only vector not integrated over is **A**. Hence, the result must be proportional to **A**:

$$\int \mathbf{A} \cdot \mathbf{\Omega\Omega}\, d\mathbf{\Omega} = C\mathbf{A}$$

To evaluate C one takes the scalar product of this equation with **A** to get

$$|A|^2\, C = \int (\mathbf{A} \cdot \mathbf{\Omega})^2\, d\mathbf{\Omega}$$

and again choosing **A** as the polar axis

$$|A|_n\, C = |A^2| \int_{-1}^{1} \mu^2\, d\mu \int_{0}^{2\pi} d\varphi = \frac{4\pi}{3}|A^2|$$

or

$$C = \frac{4\pi}{3}$$

(5) $$\int (\mathbf{A} \cdot \mathbf{\Omega})\, (\mathbf{B} \cdot \mathbf{\Omega})\, d\mathbf{\Omega} = \frac{4\pi}{3}\mathbf{A} \cdot \mathbf{B}$$

This is easily proved by taking the scalar product of (4) with $\mathbf{B}$.

$$\int (\mathbf{A} \cdot \mathbf{\Omega})\,(\mathbf{B} \cdot \mathbf{\Omega})\,\mathbf{\Omega}\,d\mathbf{\Omega} = 0 \tag{6}$$

This proof follows from a symmetry argument. Since $\mathbf{\Omega} \cdot \mathbf{A}$ and $\mathbf{B} \cdot \mathbf{\Omega}$ both change sign when $\mathbf{\Omega}$ is changed to $-\mathbf{\Omega}$, the entire integrand changes sign under this transformation. However, integration over $-\mathbf{\Omega}$ is equivalent to integration over $\mathbf{\Omega}$. Thus the integral is its own negative and is therefore zero.

APPENDIX C

THE VALIDITY OF ASYMPTOTIC REACTOR THEORY†

IN CHAPTER I it was shown how the major results of bare reactor theory could be derived quite simply by means of asymptotic reactor theory. However, the development of asymptotic reactor theory presented was, at best, highly intuitive and the limitations inherent in the theory were not at all obvious. In this appendix a set of sufficient conditions for the validity of asymptotic reactor theory is derived.

In essence, asymptotic reactor theory is the following. The criticality problem is that of solving the thermal diffusion equation

$$D\nabla^2\varphi(\mathbf{r}) + \Sigma_a\varphi(\mathbf{r}) = \nu\Sigma_f \int_R \varphi(\mathbf{r}')\, P(\mathbf{r}' \to \mathbf{r})\, \mathrm{d}^3r' \qquad \text{(C.1)}$$

together with the boundary condition

$$\varphi(\mathbf{r}) = 0 \quad \mathbf{r}\varepsilon S \qquad \text{(C.2)}$$

where R is the reactor volume bounded by S. Asymptotic reactor theory claims that it is possible to find a displacement kernel K such that the solution to the infinite medium equation:

$$-D\nabla^2\varphi'(\mathbf{r}) + \Sigma_a\varphi'(\mathbf{r}') = \nu\,\Sigma_f \int_{\text{all space}} \varphi'(\mathbf{r}')\, K(|\mathbf{r} - \mathbf{r}'|)\, \mathrm{d}^3r' \qquad \text{(C.3)}$$

with the boundary condition (C.2) is identical to the solution of Equation (C.1) (with the same boundary condition) for $\mathbf{r}\varepsilon R$, i.e.,

$$\varphi'(\mathbf{r}) = \varphi(\mathbf{r}) \quad \mathbf{r}\varepsilon R \qquad \text{(C.3a)}$$

If K is indeed the infinite medium kernel, $P_\infty(|\mathbf{r} - \mathbf{r}'|)$, then the results derived in Chapter I follow. That this is the case will now be demonstrated under certain conditions.

† The development closely follows K. M. CASE, J. H. FERZIGER and P. F. ZWEIFEL, *Nucl. Sci. Eng.* **10**, 352 (1961).

The proof is as follows. If Equation (C.3a) holds then the left-hand sides of Equations (C.1) and (C.3) are equal for $\mathbf{r}\varepsilon R$. Then

$$\int_R \varphi(\mathbf{r}')\, P(\mathbf{r}' \to \mathbf{r})\, \mathrm{d}^3 r' = \int_{\text{all space}} \varphi'(\mathbf{r}')\, K(|\mathbf{r} - \mathbf{r}'|)\, \mathrm{d}^3 r' \quad \mathbf{r}\varepsilon R \tag{C.4}$$

Now $\varphi(\mathbf{r})$ may be expanded in terms of the eigenfunctions $\chi_n(\mathbf{r})$ which satisfy the Helmholtz equation

$$\nabla^2 \chi_n(\mathbf{r}) + B_n^2 \chi_n(\mathbf{r}) = 0 \tag{C.5}$$

and vanish on S; the set so defined is complete and orthogonal in R†. Thus formally

$$\varphi(\mathbf{r}) = \sum_n \left[\int_R \varphi(\mathbf{r}')\, \chi_n(\mathbf{r}')\, \mathrm{d}^3 r'\right] \chi_n(\mathbf{r}) \quad \mathbf{r}\varepsilon R \tag{C.6}$$

Since $\varphi' = \varphi$ for $\mathbf{r}\varepsilon R$ and φ' may be assumed to have no singularities other than simple poles, $\varphi'(\mathbf{r})$ must be the unique analytic continuation of $\varphi(\mathbf{r})$ to all space. The $\chi_n(\mathbf{r})$ are readily continued to all space and these continuations are, in fact, just the solutions of Equation (C.5) in all space with the boundary condition that they vanish on S. Hence

$$\varphi'(\mathbf{r}) = \sum_n \left[\int_R \chi_n(\mathbf{r}')\, \varphi(\mathbf{r}')\, \mathrm{d}^3 r'\right] \chi_n(\mathbf{r}) \quad \text{all } \mathbf{r} \tag{C.7}$$

Substituting Equation (C.7) into Equation (C.4) we have, after interchanging the order of the operations:

$$\int_R \varphi(\mathbf{r}')\, P(\mathbf{r}' \to \mathbf{r})\, \mathrm{d}^3 r'$$
$$= \int_R \sum_n \chi_n(\mathbf{r}'')\, \varphi(\mathbf{r}'')\, \mathrm{d}^3 r'' \int_{\text{all space}} \chi_n(\mathbf{r}')\, K(|\mathbf{r} - \mathbf{r}'|)\, \mathrm{d}^3 r' \tag{C.8}$$

An application of the lemma (I.34) yields

$$\int_R \varphi(\mathbf{r}')\, P(\mathbf{r}' \to \mathbf{r})\, \mathrm{d}^3 r' = \int_R \varphi(\mathbf{r}'') \sum_n \chi_n(\mathbf{r}'')\, \chi_n(\mathbf{r})\, \overline{K}(B_n^2)\, \mathrm{d}^3 r'' \tag{C.9}$$

Hence

$$P(\mathbf{r}' \to \mathbf{r}) = \sum \chi_n(\mathbf{r})\, \chi_n(\mathbf{r}')\, \overline{K}(B_n^2) \tag{C.10}$$

Equation (C.10) yields much important information. First, it allows calculation of $P(\mathbf{r}' \to \mathbf{r})$ once $K(|\mathbf{r} - \mathbf{r}'|)$ is known; it also

† R. Courant and D. Hilbert, *Methods of Mathematical Physics*, Interscience (1953).

shows that the kernel $P(\mathbf{r}' \to \mathbf{r})$ is symmetric and that if the kernel P can be written in this form, asymptotic reactor theory is valid.

Now $P(\mathbf{r}' \to \mathbf{r})$ is the slowing-down density of neutrons at $\mathbf{r}$ at the thermal cutoff energy E_T due to a unit source of fission neutrons at $\mathbf{r}'$. Thus, it is related to the flux $\varphi(\mathbf{r}' \to \mathbf{r}, E)$ by Equation (I.32b):

$$P(\mathbf{r}' \to \mathbf{r}) = \int_0^{E_T} \mathrm{d}E \int_{E_T}^{\infty} \mathrm{d}E' \Sigma_s (E' \to E)\, \varphi(\mathbf{r}' \to \mathbf{r}, E) \quad \text{(C.11)}$$

Since the relationship between $\varphi(\mathbf{r}' \to \mathbf{r}, E)$ and $P(\mathbf{r}' \to \mathbf{r})$ does not involve the spatial variable. Equation (C.10) will be satisfied if

$$\varphi(\mathbf{r}' \to \mathbf{r}, E) = \sum_n \bar{h}(B_n^2, E)\, \chi_n(\mathbf{r})\, \chi_n(r') \quad \text{(C.12)}$$

where $\bar{h}(B_n^2, E)$ is a function such that

$$\bar{K}(B^2) = \int_0^{E_T} \mathrm{d}E \int_{E_T}^{\infty} \mathrm{d}E' \Sigma_{s_0}(E' \to E)\, \bar{h}(B_n^2, E) \quad \text{(C.13)}$$

The flux $\varphi(\mathbf{r}' \to \mathbf{r}, E)$ is the total flux at point $\mathbf{r}$ at energy E due to a unit fission source at $\mathbf{r}'$ and may therefore be obtained from the solution of the Boltzmann equation with a source of this type. But, Davison† has shown that for the one speed case in the P_L approximation, with L odd, the total flux obeys an equation of the form

$$F_{(L+1)/2}(\nabla^2)\, \varphi(\mathbf{r}, E) = G_{(L-1)/2}(\nabla^2)\, S(\mathbf{r}, E) \quad \text{(C.14)}$$

where F_N and G_M are polynomials of degree N and M respectively. For the multi-speed case, the "coefficients" become operators with respect to energy, i.e., integrals containing the kernels $\Sigma_{sj}(E' \to E)$. For $\varphi(\mathbf{r}' \to \mathbf{r}, E)$ Equation (C.14) becomes

$$F_{(L+1)/2}(\nabla^2)\, \varphi(\mathbf{r}' \to \mathbf{r}, E) = G_{(L-1)/2}(\nabla^2)\, \delta(\mathbf{r} - \mathbf{r}') f(E) \quad \text{(C.15)}$$

Now if we attempt to find a solution of the form

$$\varphi(\mathbf{r}' \to \mathbf{r}, E) = \sum \alpha_n(\mathbf{r}', E)\, \chi_n(\mathbf{r}) \quad \text{(C.16)}$$

and note that

$$\delta(\mathbf{r}' - \mathbf{r}) = \sum \chi_n(\mathbf{r})\, \chi_n(\mathbf{r}') \quad \text{(C.17)}$$

† B. DAVISON, *Neutron Transport Theory*, Oxford, 1955.

we have

$$F_{(L+1)/2}(-B_n^2)\,\alpha_n(\mathbf{r}', E) = G_{(L-1)/2}(-B_n^2)\,f(E)\,\chi_n(\mathbf{r}') \tag{C.18}$$

From this equation it is seen that

$$\alpha_n(\mathbf{r}', E) = \bar{\varphi}_\infty(B_n^2, E)\,\chi_n(\mathbf{r}') \tag{C.19}$$

and where $\bar{\varphi}_\infty(B_n^2, E)$ is seen to be the solution the Fourier transform of Equation (C.15) evaluated at $k^2 = B_n^2$. Hence,

$$\varphi(\mathbf{r}' \to \mathbf{r}, E) = \sum_n \bar{\varphi}_\infty(B_n^2, E)\,\chi_n(\mathbf{r})\,\chi_n(\mathbf{r}') \tag{C.20}$$

which when inserted into Equation (C.11) shows that $P(\mathbf{r}' \to \mathbf{r})$ does indeed have the form (C.10). Furthermore, the fact that $\bar{\varphi}_\infty(B_n^2, E)$ is the Fourier transform of the flux in an infinite medium shows that $K(|\mathbf{r} - \mathbf{r}'|)$ is the infinite medium slowing-down kernel as asserted earlier.

To recapitulate, it has been shown that if the expansion (C.6) is a legitimate representation of the neutron flux in a reactor, then the slowing-down kernel does have the form predicted for it by asymptotic reactor theory. Hence the proof of the validity of asymptotic reactor theory is complete provided an interpretation of the expansion (C.6) [or (C.16)] can be given. This interpretation is most readily obtained by an investigation of the rigorous solution of the Milne problem, i.e., the problem of a half space bounded by a plane that contains a source in the deep interior. In the Milne problem it is found that the flux far from the boundary has an asymptotic hyperbolic sine shape and that this flux extrapolates to zero a short distance from the physical boundary of the system. Near the surface, however, there is a large non-asymptotic component of the flux which tends toward zero in the interior. Since the asymptotic flux obeys the boundary condition implied by an expansion of the type (C.6), namely Equation (C.2), it must be this flux which the asymptotic theory calculates; this, in fact, must be the origin of the term "asymptotic reactor theory".

Asymptotic reactor theory is therefore a sensible approach so long as the asymptotic flux from a point source is a reasonable approximation to the actual flux. In general, this condition obtains for systems larger than a mean free path and asymptotic reactor theory can therefore be applied even to relatively small systems.†

Finally, it may be noted that the asymptotic theory has meaning not only in diffusion theory but in higher order approximations as well. For, while diffusion theory does not correctly predict the asymptotic behavior of the flux from a point source (i.e., it gives the wrong exponential), asymptotic reactor theory does predict this exponent correctly.

† For an application of asymptotic reactor theory, see S. Yip and P. F. Zweifel, *Nucl. Sci. Eng.* **10**, 361 (1961).

APPENDIX D

EXTENSIONS OF ASYMPTOTIC REACTOR THEORY

ALTHOUGH the development of asymptotic reactor theory has heretofore generally been restricted to completely thermal reactors and to diffusion theory (Appendix C shows that this restriction is not necessary), the same methods may be applied to any model for a homogeneous reactor. Thus we may take as a starting point the Boltzmann Equation (I.11):

$$\mathbf{\Omega} \cdot \nabla \Phi(\mathbf{r}, E, \mathbf{\Omega}) + \Sigma_t(E)\,\Phi(\mathbf{r}, E, \mathbf{\Omega}) = \frac{\nu}{4\pi}\int dE'\,f(E')\int d^3r' \times$$

$$\times P(\mathbf{r}' \to \mathbf{r}, E' \to E, \mathbf{\Omega}) \cdot \int dE''\,\Sigma_f(E'')\int \Phi(\mathbf{r}', E'', \mathbf{\Omega}')\,d\mathbf{\Omega}' \quad (D.1)$$

$$+\frac{\nu f(E)}{4\pi}\int \Sigma_f\,(E')\,dE'\int \Phi(r, E', \mathbf{\Omega}')\,d\mathbf{\Omega}'$$

For convenience, thermal neutrons will not be treated separately, but their behavior is to be included in the details of calculation of the kernel P appearing on the right-hand side of Equation (D.1). As in Chapter I, the asymptotic reactor model is now invoked: all cross sections are independent of position, the kernel $P(\mathbf{r}' - \mathbf{r}, E' \to E; \mathbf{\Omega})$ is replaced by the infinite medium kernel $P_\infty(\mathbf{r}' \to \mathbf{r}, E' \to E, \mathbf{\Omega})$ and the integral over $\mathbf{r}'$ is extended to all space.† Then, with the definition:

$$P_\infty(\mathbf{r}' - \mathbf{r}, E, \mathbf{\Omega}) = \int f(E')\,P(\mathbf{r}' - \mathbf{r}, E' \to E, \mathbf{\Omega})\,dE' \qquad (D.2)$$

† The following may be shortened somewhat by use of the properties of the infinite medium kernel. See J. H. FERZIGER and P. F. ZWEIFEL, *Trans. Amer. Nucl. Soc.* **6**, 274 (1963).

Equation (D.1) becomes

$$\mathbf{\Omega}\cdot\nabla\Phi(\mathbf{r},E,\mathbf{\Omega}) + \Sigma_t(E)\,\Phi(\mathbf{r},E,\mathbf{\Omega}) = \frac{\nu}{4\pi}\int_{\text{all space}} P_\infty(\mathbf{r}'-\mathbf{r},E,\mathbf{\Omega})\,\mathrm{d}^3\mathbf{r}'\times \tag{D.3}$$

$$\times\int\Sigma_f(E')\,\mathrm{d}E'\int\Phi(\mathbf{r}',E'\,\mathbf{\Omega}')\,\mathrm{d}\Omega' + \frac{\nu f(E)}{4\pi}\int\Sigma_f(E')\int\Phi(\mathbf{r},E,\mathbf{\Omega}')\,\mathrm{d}\mathbf{\Omega}'$$

The symmetry properties of P_∞ need not be applied at this point but will be useful later: P_∞ is a function only of the source to field point distance $|\mathbf{r}-\mathbf{r}'|$ and the cosine of the angle between $(\mathbf{r}-\mathbf{r}')$ and $\mathbf{\Omega}$, i.e., of $(\mathbf{r}-\mathbf{r}')\cdot\mathbf{\Omega}$.

As before, the use of the asymptotic model renders the Boltzmann equation amenable to Fourier transformation. On taking the Fourier transform of Equation (D.3) we have

$$(\Sigma_t(E) + i\,\mathbf{\Omega}\cdot\mathbf{k})\,\bar{\Phi}(\mathbf{k},E,\mathbf{\Omega}) = \frac{\nu\bar{P}_\infty(\mathbf{k},E,\mathbf{\Omega})}{4\pi}\int\Sigma_f(E')\,\mathrm{d}E'$$

$$\int\bar{\Phi}(\mathbf{k},E',\mathbf{\Omega}')\,\mathrm{d}\mathbf{\Omega}' + \frac{\nu f(E)}{4\pi}\int\Sigma_f(E')\mathrm{d}E'\int\bar{\Phi}(\mathbf{k},E',\mathbf{\Omega}')\,\mathrm{d}\mathbf{\Omega}' \tag{D.4}$$

The critical condition is now easily derived by dividing Equation (D 4) by $\Sigma_t + i\mathbf{\Omega}\cdot\mathbf{k}$, multiplying by $\Sigma_f(E)$ and integrating over energy and angle:

$$\int\Sigma_f(E)\,\mathrm{d}E\int\bar{\Phi}(\mathbf{k},E,\mathbf{\Omega})\,\mathrm{d}\mathbf{\Omega} = \left[\frac{\nu}{4\pi}\int\Sigma_f(E)\frac{\bar{P}_\infty(\mathbf{k},E,\mathbf{\Omega})}{\Sigma_t(E)+i\,\mathbf{\Omega}\cdot\mathbf{k}}\,\mathrm{d}\mathbf{\Omega}\,\mathrm{d}E\right.$$

$$\left.+\frac{\nu}{4\pi}\int\frac{f(E)\,\Sigma_f(E)\,\mathrm{d}E}{\Sigma_t(E)+i\mathbf{\Omega}\cdot\mathbf{k}}\,\mathrm{d}\mathbf{\Omega}\right]\int\Sigma_f(E')\,\mathrm{d}E'\int\Phi(\mathbf{k},E,\mathbf{\Omega}')\,\mathrm{d}\mathbf{\Omega}' \tag{D.5}$$

so that $\bar{\Phi}(k,E,\mathbf{\Omega}) = 0$ unless

$$1 = \frac{\nu}{4\pi}\int\Sigma_f(E)\,\mathrm{d}E\int\frac{\bar{P}_\infty(\mathbf{k},E,\mathbf{\Omega})}{\Sigma_t(E)+i\mathbf{\Omega}\cdot\mathbf{k}}\,\mathrm{d}\mathbf{\Omega} + \frac{\nu}{4\pi}\int\frac{f(E)\,\Sigma_f(E)}{\Sigma_t(E)+i\mathbf{\Omega}\cdot\mathbf{k}}\,\mathrm{d}E\,\mathrm{d}\mathbf{\Omega} \tag{D.6}$$

which is the desired critical condition. It still remains to show that the critical condition depends only on k^2 and has only one positive real root for k^2. Now since $P(\mathbf{r},E,\mathbf{\Omega}) = P(r,E,\mathbf{r}\cdot\mathbf{\Omega})$, its Fourier transform can be written $P(\mathbf{k}, E, \mathbf{\Omega}) = P(k, E, \mathbf{k} \cdot \mathbf{\Omega})$ so that by choosing the axis for the integration over $\mathbf{\Omega}$ in the direction of $\mathbf{k}$, the integration over azimuthal angle can be performed immediately:

$$1 = \frac{\nu}{2} \int \Sigma_f(E) \, \mathrm{d}E \int \frac{\overline{P}_\infty(k, E, \mu)}{\Sigma_t + ik\mu} \, \mathrm{d}\mu + \frac{\nu}{2} \int f(E) \, \Sigma_f(E) \, \mathrm{d}E \times$$

$$\times \int \frac{\mathrm{d}\mu}{\Sigma_t(E) + ik\mu} \qquad \text{(D.7)}$$

where

$$\mu = \frac{\mathbf{\Omega} \cdot \mathbf{k}}{k}$$

Now the slowing-down kernel may be expanded in Legendre polynomials:

$$\overline{P}_\infty(\mathbf{k}, E, \mu) = \sum \frac{(2l+1)}{2} P_l(\mathbf{k}, E) \, P_l(\mu) \qquad \text{(D.8)}$$

Upon inserting this expansion into Equation (D.7) we see immediately that the μ integrals are just the definitions of the $A_{0l}(\eta)$ (see Equation (III.69a)), so that Equation (D.7) becomes

$$1 = \nu \sum_{l=0}^{\infty} (2l+1) \int \Sigma_f(E) \, P_l(k, E) \, A_{0l}(\eta) \, \mathrm{d}E$$

$$+ \nu \int \Sigma_f(E) f(E) \, A_{00}(\eta) \, \mathrm{d}E \qquad \text{(D.9)}$$

Now the slowing-down kernel P has been defined in Chapter I, Section 3, as the number of neutrons *appearing* per unit lethargy at per unit volume at $\mathbf{r}$ due to a point source of neutrons. In other words, it is the degradation integral for the point source problem. Thus $P_l(k, E)$ is identical to $J_l(k, E)$ for this problem. Hence, using the basic equation of the B_L method (III.68a), Equation (D.9) becomes

$$1 = \nu \int \Sigma_f(E) \, \varphi_0(\mathbf{k}, E) \, \mathrm{d}E \qquad \text{(D.10)}$$

where $\varphi_0(\mathbf{k}, E)$ is the Fourier Transform of the total flux generated by a point source of fission neutrons in an infinite medium. However, in an infinite medium the flux $\varphi_0(\mathbf{r}, E)$ depends only on the distance r so that $\varphi(\mathbf{k}, E)$ depends only on the magnitude of $\mathbf{k}$, i.e., only on k^2. Finally, since $\varphi_0(k, E)$ is related to the slowing-down kernel $\bar{P}(k^2, E)$ by an operator that acts only on the energy variable, it will have the same behavior as a function of k^2 as $\bar{P}(k^2, E)$.

Now if $\bar{P}(k^2, E)$ is a monotonically decreasing function of k^2, Equation (D.10) will have only one root and all of the results of asymptotic reactor theory will follow. In particular, we will have the "Second Fundamental Theorem" discussed in Chapter I — $\bar{P}_0(B_g^2, E)$ will be the non-leakage probability to energy E for a reactor with geometric buckling B_g^2. Since this quantity must, on physical grounds, be a decreasing function of B_g^2, the argument is consistent, i.e., $\bar{P}_0(k^2)$ is monotonically decreasing, Equation (D.10) has only one root and asymptotic reactor theory is valid. For the particular case of one speed theory, one can calculate $\bar{\varphi}_0(k)$:

$$\bar{\varphi}_0(k) = \frac{1}{\dfrac{k}{\tan^{-1}k} - c} \tag{D.11}$$

which is indeed a monotonically decreasing function of k^2. This result has not been proven for the energy dependent case except by means of the plausibility argument given above.

The question of whether the replacement of the finite medium kernel by the infinite medium kernel is valid has been treated in Appendix C. The results given there, i.e., that the flux must go to zero as the reactor boundary has been proven only for the special case in which the flux depends on a single coordinate but they appear to hold in the more general case.

APPENDIX E

IMPROVEMENTS IN CRITICALITY CALCULATIONS

MOST criticality calculations performed by reactor designers are done within the framework of diffusion theory, generally in the multigroup approximation. The reasons for the choice of diffusion theory as a basis for calculations are, as pointed out in Chapter IV, the simplicity of the equations involved, the short computation time required, and, to a lesser extent, a lack of fundamental data. Frequently the justification lies no deeper than the reasons given above, and several ill-evaluated approximations inherent in the theory are often applied. In this appendix we shall derive simple formulae by means of which one can test whether a consistent P_1 calculation will yield a significant improvement over diffusion theory insofar as the multiplication constant is concerned.

In the previous appendix a critical condition was derived by applying the asymptotic reactor model to a general bare homogeneous system. Although the use of an angular dependent kernel, i.e., $P(\mathbf{r}' - \mathbf{r}, E, \mathbf{\Omega})$, in asymptotic reactor theory has not been justified explicitly, we shall assume that the critical condition (D.9) is a reasonable approximation to the true critical condition.

We start by rewriting Equation (D.9):

$$\int \Sigma_f(E)\,\mathrm{d}E \sum_{l=0}^{\infty} (2l+1)\,A_{0l}\left(\frac{\Sigma_t}{k}\right)[P_l(\mathbf{k}, E) + f(E)\,\delta_{l0}] = 1 \qquad \text{(E.1)}$$

For simplicity, first collision fission shall be neglected, i.e., we shall drop the second term in the bracket. Within the framework of diffusion theory, slowing-down collisions are assumed isotropic in the laboratory system, so that diffusion theory is equivalent to retaining only the first term of the series in Equation (E.1). Diffusion theory also requires that $A_{00}(\eta)$ be approximated by an expansion in powers of η up to η^2. Thus, a measure of the change in the multiplication constant obtained by considering anisotropic slowing down is:

$$\left|\frac{A_{10}(\eta)\,P_1(\mathbf{k}, E)}{A_{00}(\eta)\,P_0(\mathbf{k}, E)}\right| \tag{E.2}$$

In order to evaluate (E.2) we note that to order η^2

$$A_{00}(\eta) = 1 + \frac{\eta^2}{3} \tag{E.3}$$

and

$$A_{01}(\eta) = \frac{A_{00}(\eta) - 1}{\eta} = \frac{\eta}{3} \tag{E.4}$$

Now, from the definitions of the expansion coefficients $P_l(\mathbf{r}, E)$ we have at energies above thermal:

$$P_0(\mathbf{r}, E) = \int \Sigma_{s0}(E' \to E)\,\varphi_0(\mathbf{r}, E')\,\mathrm{d}E' \cong \Sigma_s(E)\,\varphi_0(\mathbf{r}, E) \tag{E.5}$$

and

$$P_1(\mathbf{r}, E) = \int \Sigma_{s1}(E' \to E)\,\varphi_1(\mathbf{r}, E')\,\mathrm{d}E' \cong \Sigma_s(E)\,\bar{\mu}\,(E)\,\varphi_1(\mathbf{r}, E)$$

$$\cong -\,\Sigma_s\bar{\mu}\,D\nabla\varphi_0$$

Hence,

$$|P_1(\mathbf{k}, E)| \cong \Sigma_s\bar{\mu}kD\varphi_0 \cong D\bar{\mu}k|P_0(k, E)| \tag{E.6}$$

Substituting the values into the expression (E.2) we have

$$\left|\frac{A_{01}P_1}{A_{00}P_0}\right| \cong \left|\frac{ik}{3\Sigma_t}\,D\bar{\mu}k\right| = \frac{D\bar{\mu}k^2}{3\Sigma_t} \cong \left(\frac{k}{\Sigma_t}\right)^2\frac{\bar{\mu}}{a} \cong D^2k^2\bar{\mu} \tag{E.7}$$

which is the desired result to the lowest order of approximation. In a criticality expression, k^2 would have the significance of a geometric buckling B_g^2 and the criterion (E.7) then states that $D^2B_g^2\bar{\mu}$ $\left(\text{or } \frac{\bar{\mu}}{9}\frac{B_g^2}{\Sigma_t^2}\right)$ must be small. This in turn implies that for the P_1 approximation to have any advantage over diffusion theory, the system must be small *or* scattering must be quite anisotropic in the laboratory system.

The above argument (i.e., the approximations (E.5) and (E.6)) is useful only if all fissions take place at energies above thermal. The opposite case, in which all fissions take place at thermal, is best treated by means of the thermal Boltzmann equation:

$$\mathbf{\Omega}\cdot\nabla\phi_T(\mathbf{r}, \mathbf{\Omega}) + \Sigma_{tT}\phi_T(\mathbf{r}, \mathbf{\Omega}) = \int \Sigma_s(\mathbf{\Omega}\cdot\mathbf{\Omega}')\,\phi_T(\mathbf{r}, \mathbf{\Omega}')\,\mathrm{d}\mathbf{\Omega}'$$

$$+\,\nu\Sigma_{fT}\int \mathrm{d}^3r'\,\phi_T(\mathbf{r}_0')\,P(\mathbf{r}' - \mathbf{r}, \mathbf{\Omega}) \tag{E.8}$$

where $P(\mathbf{r}' - \mathbf{r}, \mathbf{\Omega})$ is the angular dependent thermalization, or slowing-down density, kernel at the thermal cutoff energy. Now, proceeding as in Appendix D, the thermal flux $\phi_T(\mathbf{r}, \mathbf{\Omega})$ and the thermalization kernel can be expanded in Legendre polynomials, the Fourier transform taken and a critical condition extracted.† Since we are interested only in the first term beyond diffusion theory we quote only the P_1 result:

$$\frac{\nu\Sigma_{fT}}{\Sigma_{aT}} \frac{[\bar{P}_0(k^2) - 3D\mathbf{k}\cdot\nabla_k\bar{P}_1(k^2)]}{1 + L^2k^2} = 1 \tag{E.9}$$

where D and L^2 retain the meanings ascribed to them in Chapter III and $\bar{P}_0(k^2)$ and $\bar{P}_1(k^2)$ are the Fourier transforms of the first two Legendre moments of $P(\mathbf{r} - \mathbf{r}', \mathbf{\Omega})$, i.e., the first two moments of the slowing-down density at the thermal cutoff energy. To simplify Equation (E.9) somewhat we will use the expansions (III. 14a) and, additionally, Fick's rule:

$$\begin{aligned} P_0(\mathbf{r}' \to \mathbf{r}) &\cong \xi\Sigma_s\varphi_0(\mathbf{r}' \to \mathbf{r}) \\ P_1(\mathbf{r}' \to \mathbf{r}) &\cong -D\lambda\Sigma_s\,|\nabla\varphi_0(\mathbf{r}' \to \mathbf{r})| \end{aligned} \tag{E.10}$$

Finally a Gaussian kernel is assumed for φ_0:

$$\xi\Sigma_s\varphi_0(\mathbf{r}' \to \mathbf{r}) = \frac{e^{-|\mathbf{r}-\mathbf{r}'|^2/4\tau}}{(4\pi\tau)^{3/2}} \tag{E.11}$$

Under these approximations Equation (E.9) becomes (with $k^2 = B^2$)

$$k_{\text{eff}} = \frac{\nu\Sigma_{fT}}{\Sigma_{aT}} \frac{e^{-B^2\tau}}{1 + L^2B^2}\left[1 - \frac{3\lambda}{\xi}D^2B^2\right] \tag{E.12}$$

The corrections for some common moderators are given below:

Element	$\frac{3\lambda}{\xi}D^2$
H	+0.0435
D	−0.140
Be	−0.310
C	−0.754

Since typical bucklings for reactors are less than 10^{-3} cm^{-2} the correction is at most 1 per cent.

† The details of this calculation are given in P. F. Zweifel and J. H. Ferziger, *Nucl. Sci. Eng.* **10**, 357 (1961).

APPENDIX F

EQUIVALENCE OF THE B_L AND P_L APPROXIMATIONS

IN SECTION 19 of Chapter III it was stated without proof that the P_L approximation is equivalent to the B_L approximation if the $A_{jl}(\eta)$ of the latter are evaluated by the $L + 1$ point Gauss quadrature scheme. That is, if the statement is true, the P_L equations can be written in the form:

$$\Sigma_t \phi_j = \sum_{l=0}^{L} (2l + 1) A'_{jl} [J_l + S\delta_{0l}] \tag{F.1}$$

where

$$A'_{jl} = \frac{1}{2} \int_{G_{L+1}} \frac{d\mu}{1 - \eta\mu} P_l(\mu) P_j(\mu) \qquad A'_{L+1,l} = A'_{l,L+1} = 0$$

However, the P_L equations are already of the form

$$J_l = \sum_{m=0}^{L} B_{lm}\phi_m - S\delta_{0l} \tag{F.2}$$

where the B_{lm} may be read from Equations (III.10a)

$$B_{lm} = \Sigma_t \delta_{lm} - \frac{ikl}{2l + 1} \delta_{m,l-1} - \frac{ik(l + 1)}{2l + 1} \delta_{m,l+1} (1 - \delta_{l,L}) \tag{F.3}$$

Substituting Equation (F.2) into Equation (F.1) we have

$$\begin{aligned} \Sigma_t \phi_j &= \sum_{l=0}^{L} (2l + 1) A'_{jl} \left[\sum_{m=0}^{L} B_{lm}\phi_m - S\delta_{0l} + S\delta_{0l} \right] \\ &= \sum_{m=0}^{L} \phi_m \sum_{l=0}^{L} (2l + 1) A'_{jl} B_{lm} \end{aligned} \tag{F.4}$$

However, this equation holds only if

$$\sum_{l=0}^{L} (2l + 1) A'_{jl} B_{lm} = \delta_{jm}\Sigma_t \tag{F.5}$$

and, on inserting the expression (F.3) for B_{lm}, we have

$$(2m+1)\frac{A'_{jm}}{\eta} - mA'_{j,m-1} - (m+1)A'_{j,m+1}/\eta = \delta_{jm} \qquad \text{(F.6)}$$

It is simple to show that the A'_{jl} obey the recursion relation (F.6). The proof is identical with the proof that the

$$A_{jl} = \int_{-1}^{1} \frac{P_j(\mu)\,P_l(\mu)}{1-\eta\mu}\,\mathrm{d}\mu$$

satisfy a similar recursion relation, Equation (III.69 a). We write the recursion relation for Legendre polynomials

$$(2m+1)\,\mu P_m(\mu) - (m+1)\,P_{m+1}(\mu) - mP_{m-1}(\mu) = 0 \qquad \text{(F.7)}$$

and multiply through by

$$\frac{1}{2}\,\frac{P_j(\mu)}{1-\eta\mu}$$

to obtain

$$(2m+1)\,P_mP_j\,\frac{\mu}{1-\eta\mu} - (m+1)\frac{P_{m+1}P_j}{1-\eta\mu} - m\frac{P_{m-1}P_j}{1-\eta\mu} = 0 \qquad \text{(F.8)}$$

Now

$$\frac{\mu}{1-\eta\mu} = -\frac{1}{\eta}\left[1 - \frac{1}{1-\eta\mu}\right]$$

Thus, Equation (F.8) becomes

$$\frac{2m+1}{2}\,\frac{P_mP_j}{\eta(1-\eta\mu)} - \frac{(m+1)}{2}\,\frac{P_{m+1}P_j}{1-\eta\mu} - \frac{m}{2}\,\frac{P_{m-1}P_j}{1-\eta\mu} = \frac{2m+1}{2}\,\frac{P_mP_j}{\eta} \qquad \text{(F.9)}$$

If we now integrate over μ, we obtain Equation (III.69 a), the recursion relation for the A_{jm}. However, if we integrate over μ by the $(L+1)$ point Gauss quadrature formula we obtain

$$(2m+1)\frac{A'_{mj}}{\eta} - (m+1)\,A'_{m+1,j} - mA'_{m-1,j} = \int_{G_{L+1}} \frac{2m+1}{2}\,\frac{P_mP_j}{\eta}\,\mathrm{d}\mu = \frac{\delta_{mj}}{\eta} \qquad \text{(F.10)}$$

for $m < L$ and $j \leqq L$. The last equality in Equation (F.10) follows from the fact that $P_m P_j$ is a polynomial of degree $(m + j)$ and is thus integrated rigorously for $m + j \leqq 2L + 1$. For $m = L$, this equality still holds for $j < L$ (actually $j < L + 1$), but the term $A_{m+1,j} = 0$ because the integrand is evaluated at the zeroes of $P_{L+1}(\mu)$ so that the factor $P_{m+1}(\mu)$ is identically zero at each discrete ordinate. Since no values of A'_{jm} enter the P_L approximation if either index is greater than L, this proves the recursion relation. This also completes the proof because the recursion relation (F.3) defines a unique set of functions A'_{jm} for any given value of L since two of the A'_{jm} have been defined (i.e., $A'_{j,L-1} = A'_{j,L+1} = 0$).

INDEX